309 CIRCUITS

309 CIRCUITS

Uitgeverij Segment B.V.
Postbus 75
6190 AB Beek (L)

British Library Cataloguing in Publication Data
A catalogue record for this book is available from the British Library

ISBN 978-0-905705-69-9

Prepress production: Autronic, Blaricum
First published in the United Kingdom 2007
Printed in the Netherlands by Wilco, Amersfoort
© Segment BV 2007

059012/UK

Contents

Miscellaneous

Generators and Oscillators

Hobby and Model Making

High-Frequency

Home and Garden

Measuring and Testing

Microprocessor

Power Supplies and Battery Chargers

001
Car-Stereo LED Power (VU) Meter

This circuit senses AC audio voltage supplied to the car-radio loudspeakers and displays it as power using a LED bar graph, achieving at the same time an attractive visual effect. It is designed to cover common car-radio output power ranges, but can easily be modified to suit different needs. It is supplied from the 12 V car electrical system and is suitable for classical CC (Capacitor-coupled) as well as BTL (Bridge Tied Load) types of amplifier with no changes to the circuitry or connections at all. In fact, only the

meaning of the LEDs changes – with BTL, the LED increments equal four times the CC value on the same load. CC-type amplifiers have the loudspeaker connected via a DC-decoupling capacitor at the output and ground (negative). BTL-type amplifiers, on the other hand, have the loudspeaker DC-coupled and 'stretched' between two equal, parallel, but phase-reversed outputs. The result compared to 'CC' is twice the voltage swing, hence quadrupling the power being fed to the same loudspeaker load. It is necessary to know to which of the two types this circuit is connected to only in order to correctly assign power levels (W) to the LEDs. CC-type have no DC voltage to ground at the outputs and return wires. The return wires are actually connected to the common ground (negative). BTL-type have approximately Vcc/2 at outputs and on the return wires too, explaining at the same time why no DC-decoupling capacitors are needed. The

LM3915N integrated circuit used in this circuit has been the subject of numerous publications in this magazine so will not will not be discussed again. In this application, the two LM3915Ns are configured as a LED bar graph drivers (pin 9 connected to pin 3), The ICs share the same power supply section. The audio input signal is fed via network C1/C2, R1, R2 (C3/C4, R5, R6) to pin 5 of IC1 (IC2). Only positive half-waves are processed by the ICs. Internally, the buffered input voltage is compared using comparators to the voltages along a resistor ladder network. The nominal +1.25 V reference source voltage (between pins 7 and 8) is applied across R3 (R7) to program the LED current. The programming current flows through R4 (R8) to achieve the desired reference voltage between pin 7 and ground. Here, only 2.0 V is developed, allowing this circuit to be used with low power amplifiers too. This voltage is applied to the 'top' of the on-chip resistor array (pin 6) and so determines the threshold at which the LED connected to the L10 output comes on. The other (low) side of the array (pin 4) is connected to ground. So, for an input voltage equal to or greater than the voltage at pin 6, all LEDs are on. At input voltages below the threshold set up for the lowest LED (89.3 mV or –27 dB below the top LED) all LEDs are off. In order to limit power dissipation of IC1 and IC2, the LED voltage is stepped down to +5 V using IC3, C6 and C7. Diode D1 protects the circuit against reversed polarity. If a 'dot' mode graph is preferred pins 9 of IC1 and IC2 should be left open circuit. Using the listed value for R1 (R5), the indicator range covers audio power levels of 10 W into 4 Ω (CC) or 40 W into 4 Ω (BTL). Each 'lower' LED indicates half the power of the previous 'higher' LED Only R1 (R5) needs to be redimensioned for different power levels. The value can be calculated from

$$R1 = \left(R2 \cdot \sqrt{\frac{P_O - Z_L}{k \cdot V_{RefOut}}} \right) - R2$$

P_O = maximum output power to be indicated (LED D2 or D12)
Z_L = loudspeaker impedance
$R2$ = R6
V_{RefOut} = 2.0 V
k = constant; 2 for BTL, 1 for CC

The condition $\sqrt{(P_O - Z_L)/(k \cdot V_{RefOut})} \geq 1$ must be met. A small printed circuit board

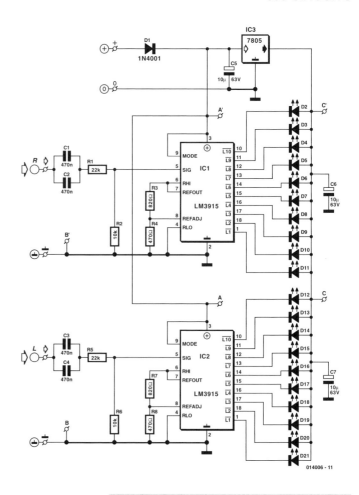

COMPONENTS LIST

Resistors::
R1,R5 = 22kΩ
R2,R6 = 10kΩ
R3,R7 = 820Ω
R4,R8 = 470Ω

Capacitors::
C1-C4 = 470nF, lead pitch 5mm
C5,C6,C7 = 10µF 63V radial

Semiconductors::
D1 = 1N4001
D2-D21 = LED, 3mm dia. or rectangular-face
IC1,IC2 = LM3915N-1 (National Semiconductor)
IC3 = 7805

Miscellaneous::
Heatsink for IC3 (10 K/W)

has been designed to allow a stereo version of the power indicator to be built. The board is cut in two to separate the channels. The boards may be assembled in a sandwich construction with three inter-board connections A-A', B-B' and C-C' made in stiff wire. IC3 should be secured to a small heatsink (10 K/W). Rectangular-face LEDs are recommended for this circuit.

If on the other and 3 mm dia. LEDs are used, these may have to be filed down a bit to be able to fit them in a straight row. The connection to the car radio should not present any problems.

The audio signal is taken from the (+) loudspeaker connector for each channel and ground. The power supply leads to the indicator circuit are connected in parallel

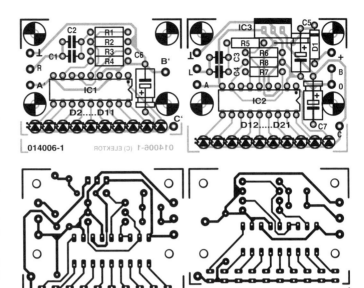

with car radio power supply. At a supply voltage of 14.4 V, the maximum and quiescent current consumption of the circuit was measured at 171 and 22 mA respectively.

R. Lalic   (014006-1)

002
EEDTS Pro Budget Booster

For many people, the existing EEDTS booster, which has an output current capacity of 10 A, is perhaps a bit too much of a good thing. Although the existing booster is significantly superior to the 'Budget Booster' described here in many respects, the latter booster is more suitable for making the transition from analogue to digital control for your model railway. You can even use your old reliable train transformer (AC) from the analogue era to power the new booster! As can be seen from the schematic diagram, the output is driven by a

complementary pair of Darlington transistors (T3 & T4). These Darlingtons are wired somewhat differently than in the 10 A booster, with the load resistors R10 and R11 being placed in the emitter leads. This has the advantage that the two collectors are connected together. Since this junction is also connected to the mounting tabs, the two Darlingtons can be fitted to a heat sink without using insulating washers. A small aluminium plate (3 × 8 cm, 3 mm thick) provides a more than adequate heat sink. The combination of R10/R11 and di-

here (0.22 Ω). The Darlingtons are driven by a complementary pair of normal transistors (T1 & T2). In the absence of an input signal, both of these transistors will be cut off, and the Darlingtons will also not conduct, causing the output to be in a high-impedance state. Besides current limiting, the booster also features a disable circuit built around relay RE1. As long as a negative voltage is present at the output of the booster, this circuit will allow the booster to be driven (the Motorola format is negative most

odes D7–D12 provides current limiting for the booster. Current limiting will take effect at a voltage drop of approximately 0.6 V, which means at a current of approximately 2.5 A with the resistor value used

the time, so the input signal will cause the output to be almost continuously negative). If the output is shorted, the relay drops out, disconnecting the control signal and causing the output to assume a high-impedance

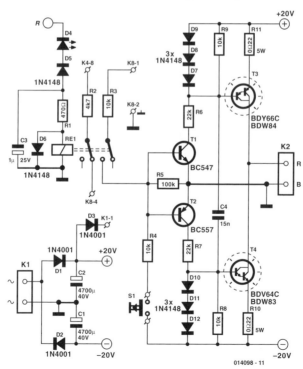

014098 - 11

state. Once the short has been removed, the relay can be re-energised by briefly applying a negative input voltage (by pressing S1, the 'Go' button). After this, the input signal can again drive the booster. Capacitor C3 provides a certain amount of delay to prevent the relay from dropping out too quickly. An LED in series with diode D5 is provided as a short-circuit indicator. This LED will be illuminated when the booster is outputting a signal and extinguished when there is a short. If desired, a pushbutton switch can be wired in series with R1 to serve as an emergency stop switch. This addition can be easily made by fitting R1 vertically on the circuit board. In accordance with an unwritten convention, you should use a green pushbutton for the Go switch and a red pushbutton for the Stop switch.

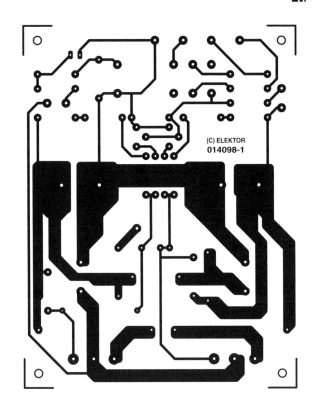

(C) ELEKTOR
014098-1

The circuit has a remote short-circuit indicator in the form of resistor R2 connected to relay RE1. This allows the new controller to sense whether there is a signal on the track. A 16 V / 32 VA train transformer connected to K1 provides an excellent source of power. Symmetric DC supply voltages are derived from the transformer secondary in a simple manner using D1, D2, C1 and C2. With a small modification, the EEDTS Pro controller can also be powered from the train transformer by fitting a small heat sink (2 × 3 cm, 2 mm

Note: although the Budget Booster can also be used with other controllers, since a symmetric signal can be connected to K1 and K2, this circuit is intended to be connected to the EEDTS Pro controller. The connectors are labelled to match the connectors on the EEDTS Pro controller board.

Components list

Resistors:
R1 = 470 Ω
R2 = 4k 7
R3,R4,R8,R9 = 10k
R5 = 100k
R6,R7 = 22k
R10,R11 = 0 22Ω 5W

Capacitors:
C1,C2 = 4700µF 40V radial
C3 = 1µF 25V radial
C4 = 15nF

Semiconductors:
D1,D2,D3 = 1N4001
D4 = LED   D5-D12 = 1N4148
T1 = BC547   T2 = BC557
T3 = BDY66C   T4 = BDV64C

Miscellaneous:
K1,K2 = 2-way PCB terminal block, lead pitch 7.5mm
S1 = on/off switch
RE1 = 12V, changeover contacts, e.g., Schrack V23037
PCB, order code 014098-1

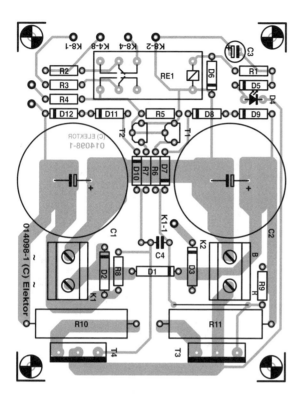

thick) to the 5 V regulator IC on the controller board and tapping off the supply voltage via D3. Most of the space on the printed circuit board for the booster is taken up by C1, C2 and RE1. Since the actual circuit consists of only a few components, the board can be built relatively quickly.

S. van de Vries   (014098)

003
Control Box with Status for Märklin Railway Points

The model rail manufacturer Märklin produces the turnout mechanism (or points) type 74490 for C track, the 7549 for K track (H0 gauge) and also the 5625 (1 gauge) all fitted with limit switches. The circuit here shows a control box to inter--face to these mechanisms and also display the status of the turnout using LEDs. All these turnouts are fitted with two actuating coils, when energy is supplied to coil L1 the turnout moves to the 'straight ahead' position while energising L2 moves them so that the train 'turns out' onto a branch line. The standard Märklin control box for controlling these devices is fitted with a green push button to activate L1 and a red push button for L2. This colour convention is also used here for connections. Yellow indicates the power supply connection and brown the earth connection. The limit switches interrupt power to the coils when the turnout reaches the end of its travel. These ensure that the coils do not become overloaded and also divert the supply to the other coil, enabling the points to be moved back later. When the pushbutton is pressed, current flows through the coil and also the status LED. The only other additional com-ponents are the LED current limiting resistors, protection diodes and a rectifier.

The low-current LEDs require an additional 2 mA (red) or 4 mA (green) from the transformer. Series resistors R1 and R2 limit current through the LEDs while D1 and D2 protect the LEDs from the back-emf generated when current through the coil is interrupted. Diode D3 rectifies the supply to both LEDs and can also sup-ply additional LEDs if more than one turnout control is built. A feature of the circuit is that the status LED will illuminate when its corresponding pushbutton is pressed even if power to its coil is not connected through the limit switch. This does not affect operation but can be remedied if necessary by fitting separate rectifiers for each coil.

Additional low-current or standard LEDs or even optocouplers can be added in series with the LEDs if more status indicators are needed but it will be necessary to change the value of the series resistors to suit. A

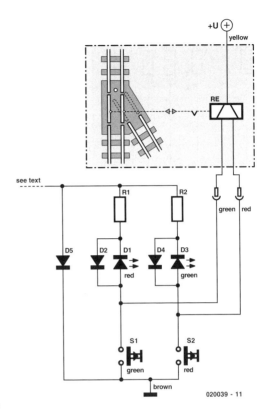

conventional control box like the Märklin 7272 can be substituted for the pushbuttons shown in the circuit and the LEDs can be fitted to provide visual positional feedback.

N. Körber   (020039-1)

004
Light Gate with Counter

The circuit described here counts the number of times that an infrared beam is interrupted. It could be used to count the number of people entering a room, for instance, or how often a ball or another object passes through an opening (handy for playing shuffleboard). The heart of the cir-

cuit consists of – you guessed it – a light gate! Diode D1 is an IR diode that normally illuminates IR transistor T1. The light falling on T1 causes it to conduct to a cer-tain extent. The resulting voltage on the collector of T1 should be just low enough to prevent the following transistor

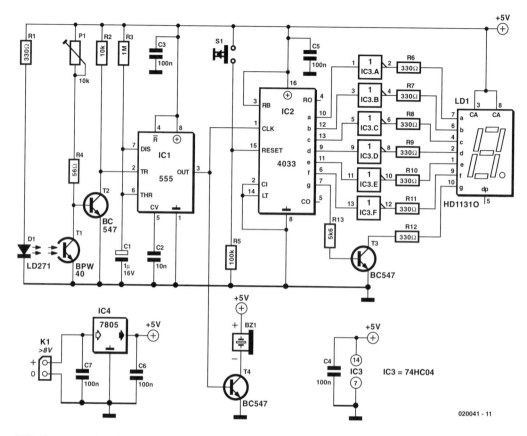

020041 - 11

(T2) from conducting. This voltage can be adjusted within certain limits using P1. As soon as an object comes between D1 and T1, the light shin-ing on T1 will be partially or fully blocked, causing the IR tran-sistor to conduct less current. As a result, the voltage on its col-lector will increase, producing a brief rise in the voltage on the base of T2. This will cause T2 to conduct and generate a negative edge at IC1. This negative edge will trigger the monostable multivibrator, which will then hold the output signal on pin 3 'high' for a certain length of time (in this case, one second). At this point, two things will occur. First, a buzzer will be ener-gised by the output of IC1 and produce a tone for approximately one second. When the buzzer stops, a negative edge will be applied to the clock input of IC2, causing the counter in IC2 to be incremented by 1. IC2 is conveniently equipped with an internal bi-

nary-to-BCD decoder, so its outputs only have to be buffered by IC3 and T3 to allow the state of the counter to be shown on the 7-segment display. Switch S1 can be used to reset the counter to zero. If a one-second interval does not suit your wishes, you can modify the values of R3 or C1 to adjust the time. Increasing the value of R3 lengthens the interval, and decreasing it naturally shortens the interval.

The same is true of C1. When building the circuit, make sure that T1 is well illuminated by the light from D1, while at the same time ensuring that T1 'sees' as little ambient light as possible. This can best be done by fitting T1 in a small tube that is precisely aimed toward D1. The longer the tube, the less ambient light will reach T1. The sensitivity of the circuit can be adjusted using P1.

T. Hareendran   (020041-1)

005
Processor Fan Control

Fans in PCs can be objectionably loud. In many cases, the amount of noise produced by the fan can be considerably reduced by lowering its speed. Although this will decrease the amount of cooling, this need not be a problem as long as you don't go overboard with slowing down the fan. Particularly with older-model processors, which consume quite a bit less power than the latest models, this trick can be used without any problems. This circuit is anyhow intended to be used with relatively old PCs, since more recent models generally have a fan control circuit already integrated into the motherboard. These controllers ensure that the amount of cooling is increased if the processor becomes too warm and decreased if the processor tempera-ture is relatively low. The circuit described here consists of only a handful of components, which you will probably already have in a drawer somewhere. Transistors T1 and T2 are driven into conduction by the base current flowing to the fan via P1 and D1. There will always be a current flowing through R1, and it will be approximately 120 times as large as the current through R2. R3 has been added to prevent the base

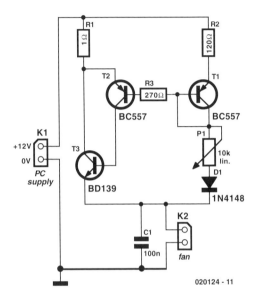

current of T2 from becoming too large when P2 is set to its minimum resistance. D1 ensures that even at this extreme setting, the voltage on the base-emitter junction of T3 will still be large enough to allow it to conduct.

B. v. Gijzen   (020124-1)

006
RC5 Repeater

The designer of this circuit had fitted two (waterproof) loud-speakers in his bathroom and connected them to the stereo system in the living room via a long cable.

Naturally, this promptly led to the desire to be able to use the remote control unit from the bathroom. Commercially available extension sets for this purpose were judged to be unsatisfactory, primarily because they require an additional IR transmitter diode to be fitted in front of the amplifier. Although the repeater shown here requires a length of coaxial cable, it provides a simpler, and above all more reliable, solution to the problem. The signal transmitted by the remote con-trol unit is received by IR receiver IC2, and the (nearly) open-collector output of T1 is connected to the RC5 bus of the stereo system. This proved to work excellently with Philips equipment, and it will probably also work with equipment from other manufacturers with a few small modifications. Voltage regulator IC1 is used here to allow the supply voltage to range from 8 V to 30 V, and diode D1 provides protection against a reverse-polarity supply voltage connection. A nice side benefit arose from the fact that the loudspeakers in question (Conrad models) have transparent cones and protection grilles with rather large openings. This made it possible to fit the tiny circuit, which was built on a piece of prototyping board, to the frame of one of the speakers, behind the cone. The whole arrangement is thus hidden, but the remote control still works

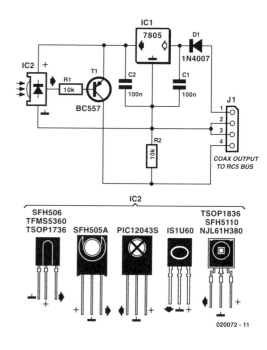

perfectly if it is aimed towards the speakers.

H. Tempelman (020072-1)

007
Simple Alarm System

The circuit presented here is a very simple and yet highly effective alarm system for protecting an object. The circuit requires no special devices and can be built using components that you will no doubt be able to find in the junk box. The alarm-triggering element is a simple reed switch. To generate the alarm signal itself any optical or acoustic device that operates on 12 V can be used: for example a revolving light, a siren, or even both. In the quiescent state the reed switch is closed. As soon as the reed switch opens, the input to IC1.B will

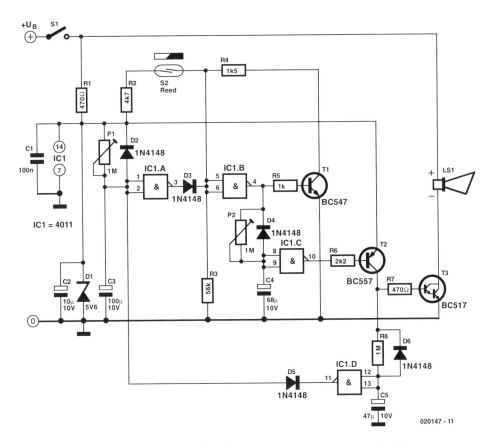

020147 - 11

go low (previously the potential divider formed by R2 and R3 held the input at 5.17 V, a logic high level). A turn-on delay of between 0 and approximately 90 s can be set using P1, and a turn-off delay of between 0 and approximately 20 s can be set using P2. When the system is turned on (using S1), the turn-on delay is activated, giving the user of the system at most 90 s to leave the object alone before the system goes into the armed state, and the object is then protected. Once the reed switch opens the turn-off delay of at most 20 s starts: this allows the rightful owner of the object to turn the system off before the alarm is triggered. If some unauthorised person causes the reed switch to open, the alarm will be trig-

gered after the turn-off delay. Also, even if the reed switch is briefly opened and then closed again, the alarm will still be triggered.

Once the alarm is triggered, T3 will conduct for about 45 s (because of R8 and C5). The turning off of the alarm is necessary to avoid the nuisance caused by a perma-

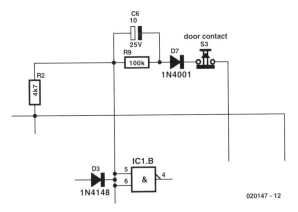

020147 - 12

nently sounding alarm system. The system then returns to the armed state, which means that the next time the reed switch is opened the alarm will trigger again. If it is not desired that the duration of the alarm be limited, for example if a visual indication is used, D5 should not be fitted. The system can be extended by fitting multiple reed switches in series. As soon as any one is opened, the alarm is triggered. When S1 is closed C3 charges via P1. Depending on the potentiometer setting, it takes between 0 and 90 s to reach the input threshold voltage of IC1.A. The output of IC1.A then goes low and D3 stops conducting. Assuming the reed switch is closed, the inputs of IC1.B stay high and the output therefore low. If the reed switch is opened after the turn-on delay expires the output of the gate will change state and turn on T1. This ensures that the output of the gate remains high even after the reed switch is closed again. C4 now starts charging via P2, reaching the input threshold voltage of IC1.C after between 0 and 20 s, again according to the potentiometer setting. The

output of IC1.C goes low, and T2 and T3 are turned on – and the siren sounds. Any Darlington transistor can be used for T3. At the same time, C5 charges via R8, reaching the input threshold of IC1.D in about 45 s. When the output of IC1.D swings low, it pulls the inputs of IC1.A low via diode D5: the siren stops and the system returns to the armed state. If the potentiometers P1 and P2 are replaced by fixed resistors it is possible to build the circuit small enough to fit in a match-box, without the need to resort to SMD components. This is ideal if the circuit is to be used to protect a motorbike. If the alarm system is to be used in a car, an existing door switch contact can be used instead of the reed switch. In this case an RC combination needs to be added to prevent false triggering. Use a 10 µF/25 V electrolytic for C6, a 100 kΩ resistor for R9 and a 1N4001 for D7. It is again possible to wire multiple door switch contacts in parallel: as soon as one contact closes, IC1.B will be triggered.

L. Libertin   (020147)

008
Universal Keypad

Keypads are needed for a variety of applications and DIY projects. Particularly for selecting the measurement range of an instrument or setting the output voltage of a laboratory power supply, as well as for many other uses, potential-free relay outputs are desirable. In addition, break-before-make switching should be used to avoid short circuits. This cannot be guaranteed by simple relay circuits, since the pull-in speed of most relays is greater than the dropout speed! In many cases, an expensive or hard-to-obtain special-purpose IC is used for this purpose, but most such ICs limit the number of keys that can be

used, and break-before-make switching is often an unheard-of luxury! The circuit shown here, which is about as simple as possible and can be built using only a handful of parts from your 'junk box', shows that that there are other ways to solve this problem. This design has the following advantages:
• break-before-make operation;
• potential-free relay outputs;
• any desired number of keys;
• any desired operating voltage;
• well-defined initial state (all relays disengaged) – only one stage switches if multiple keys are pressed.

Each of the relays must have a spare working contact for the changeover function. The pushbutton switches must have changeover (double-throw) contacts that can switch the coil currents of the relays. If the current rating of the pushbutton switch contacts is insufficient (particularly with high-power circuitry), a small relay can be connected in series with the switch to supply power to the load relay. Operating power is supplied to the switching stages via a series connection of the normally-closed (nc or 'break') contacts of all of the pushbutton switches. When any one of the switches is actuated, the supply voltage to all stages is interrupted, causing all of the relays to drop out, and the base capacitor associated the actuated switch is charged. When the switch is released, power is again applied to all of the relay stages, but only the capacitor of the selected switching stage provides sufficient energy to drive the connected transistor into conduction and engage the relay.

The working contact of the relay latches the relay. LEDs with series resistors connected in parallel with the relay coils indicate their switching states. When a different key is pressed, the above process is repeated. First, the supply voltage is disconnected from all of the relays, which ensures break-before-make operation. When the key is released, the relay whose capacitor has been charged is actuated. If several keys are pressed at the same time, only one capacitor receives a charging voltage. This is the capacitor for the only pushbutton switch in the series circuit that is still connected to the positive terminal, which means the first switch in the series. This ensures that only one stage can switch.

The values of the components marked with a * must be chosen to match to the supply voltage used or the required switching current. If power relays are used, Darlington transistors with suitable power ratings should be used. The number of switching stages can be extended to almost any de-

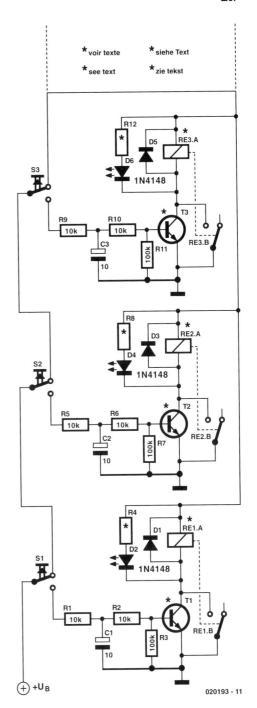

sired limit to allow you to fit your equipment with a convenient, reliable keypad.

H. Kraus   (020193-1)

009
Lights On!

This circuit ensures that you will never again forget to switch on the lights of your car. As soon as the engine is running, the dipped beams and the sidelights are automatically switched on. The circuit also causes the dipped beams to be extinguished as soon as the main beams are switched on. As you can see from the schematic diagram, no special components are needed. When the engine is running, the alternator will generate a voltage of more than 14 V. Diode D1 reduces this voltage by 5.6 V and passes it to the base of T1 via R1. Due to the resulting current, T1 conducts. The amplified current flows via R3, the base of T3 and D3 to ground. This causes

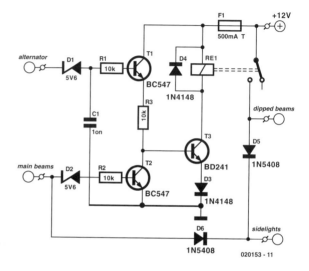

T3 to also conduct and energise relay Re1. If the driver now switches on the main beams, a current flows through D2 and R2 into the base of T2, causing this transistor to conduct.

As a result, the voltage on the base of T3 drops, causing T3 to cut off and the relay to drop out. When the main beams are switched off, the previous situation is restored, and the relay again engages. The dipped beams and the sidelights are switched by the contacts of relay Re1. Diodes D5 and D6 ensure that the sidelights are illuminated if either the dimmed beams or the main beams are switched on. In practice, this means that the sidelights will be on whenever the engine is running, regardless of whether the main beams are switched on.

R. 't Hooft

(020153-1)

010
In-Vehicle Voltage Regulator

In vehicles it is often required to have a powerful yet stabilised voltage that's not affected in any way by fluctuations of the battery voltage. The circuit shown here does the job using discrete and inexpensive parts only. While its low cost is a def-

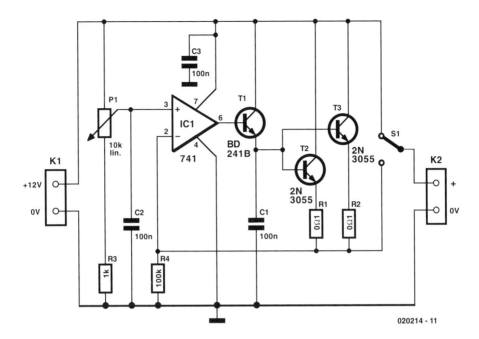

020214 - 11

COMPONENTS LIST

Resistors:
R1,R2 = 0Ω1, 5W
R3 = 1kΩ
R4 = 100kΩ
P1 = 10kΩ linear potentio-
meter

Capacitors:
C1,C2,C3 = 100nF

Semiconductors:
IC1 = 741CN
T1 = BD241B
T2,T3 = 2N3055 (TO-3 case)

Miscellaneous:
K1,K2 = 2-way PCB terminal
block, lead pitch 5mm
S1 = switch, heavy-duty, 1
change-over contact (see
text)

inite advantage over just about any kind of regulator IC, on the downside we have a minimum voltage drop of 2 volts – in fact the output voltage can be set to any value between 1.8 V and about 10 V. Continuous loads up to 100 watts can be handled, while peak values of 140 W should not present problems.

The power stage consists of two parallel-connected 2N3055 transistors in TO-3 cases. Because of their high base current requirement, a driver transistor type BD241B is incorporated.

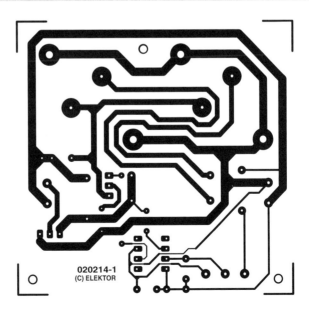

020214-1
(C) ELEKTOR

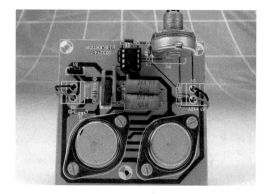

The feedback voltage arrives at the inverting input of the regulator IC, a type 741 opamp. The level of the reference voltage at the invert-ing input is adjusted with potentiometer (or preset) P1.

The circuit board, of which the layout is given here, accommodates all parts including the two 2N3055 power transistors. As a matter of course, they should be properly cooled. Remember, the case of a 2N3055 is connected to the collector which is at battery-positive potential. If necessary the voltage regulator may be bypassed by an external switch connecting the battery + terminal with the output terminal. The switch, if used, should be capable of passing considerable currents – at relatively low output voltages (up to about 6 V) currents of up to 15 A (continuous) is reduced to 10 A when the 10 V level is approached, it is betor 20 A (peak) may be expected. Although the output current ter to be safe than sorry.

L. Libertin   (020214-1)

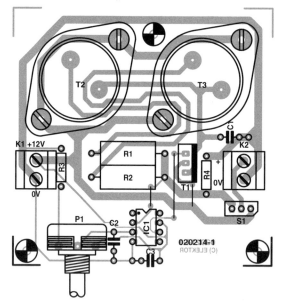

011
Three Hour Timer

Manufacturers of cordless drills generally recommend a battery charging time of three hours. Once the charging time is up the battery must be disconnected from the charger: if you forget to do this there is a danger of overcharging the battery. This circuit, which sits between the charger circuit and its battery socket, prevents that possibility:

the contact of relay Re1 interrupts the charging current when the three hours are up. Ten LEDs show the remaining charging time in steps of 20 minutes. The timer is reset each time power is applied and it is then ready for a new cycle. When power is applied IC3 is reset via C4 and R5. When the charging time has elapsed, Q9 (pin 11) goes

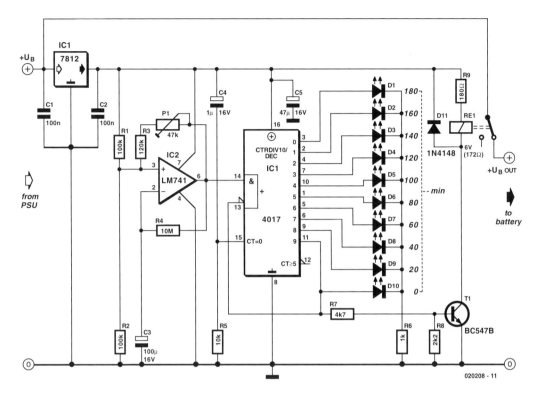

020208 - 11

high, which turns the relay on and interrupts the charging current. Since Q9 is connected to the active-low EN (enable) input, the counter will now remain in this state. The charging time can be adjusted from about 2 hours 15 minutes to 4 hours 30 minutes using P1. The author set P1 to 30 kΩ, giving a charging time of 3 hours 7minutes. The greater the resistance of P1, the shorter the charging time. The timing of the circuit is

not particularly pre-cise, but its accuracy is entirely adequate for the job. When adjusting the charging time it is worth noting that the first clock cycle after the circuit is turned on (from Q0 to Q1) is longer than the subsequent ones. This is because initially capacitor C3 has to be charged to around half the supply voltage.

L. Libertin   (020208-1)

012
Automatic Mains Disconnect

Downloading and CD-burning programs usually provide the option of automatically shutting down the PC on completion of their tasks. However, this energy-saving feature is of little benefit if even after the PC has

been switched off, all of the peripheral equipment remains connected to the mains and happily consumes watt-hours. The circuit shown here provides a solution to this dilemma. It is connected ahead of the power

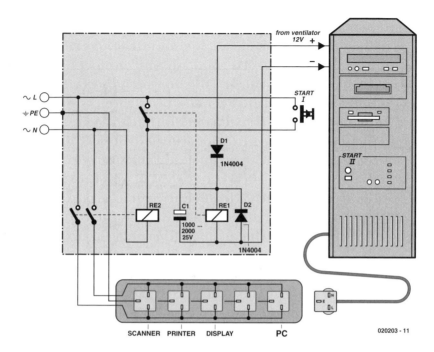

020203 - 11

strip and connects or disconnects mains power for all of the equipment via a power relay. A connection to a 12 V PC fan (which may be the processor fan or the fan for the chipset, if the latter is present) indicates whether the PC is switched on. If you are certain that the 12 V power supply voltage is switched off when the PC is in the sleep mode, you can use this connection instead. To switch everything on, press the Start button to cause the power relay to be energised and provide mains voltage to all of the equipment. If the PC has an ATX board, its Power switch must be pressed at the same time to cause the PC to start up. When the PC fan starts to run, low-power relay Re1 engages and takes over the function of the Start switch, which can then be released. This state is stable. If the PC switches to the sleep state, the 12 V voltage drops out. The electrolytic capacitor ensures that Re1 remains engaged for a short time, after which it drops out, followed by the power relay. D1 prevents the electrolytic capacitor from discharging through the connected fan, and D2 is the usual freewheeling diode. The system is disconnected from both mains leads and is thus completely de-energised.

Be sure to select components that are suitable for their tasks. Naturally, the contacts of Re2 should be rated to handle the total current drawn by all of the peripheral equipment and the PC, and the relay coil must be suitable for use with mains voltage (6 mm minimum separation between coil and contacts). A low-power 12 V relay that can switch mains voltage is adequate for Re2. The Start pushbutton switch is connected to the mains voltage, so a 230 V type must be used. The circuit board layout and enclosure must also be designed in accordance with safety regulations. A separation of at least 6 mm must be maintained between all components carrying mains voltage and the low-voltage components, and the enclosure must be completely free of risk of electrical shock. With a bit of skill, the circuit can be fitted into a power bar with a built-in switch, if the switch is replaced by a pushbutton switch having the same mounting dimensions. Note that the circuit is not suitable for use with deskjet printers that can only be switched on and off by a front panel button.

A. Brandon

(020203-1)

013
Add a Sparkle

This circuit was developed to provide novel coloured lighting effects for a crystal glass figurine. The circuit shown in Figure 1 consists of an 8-pin (SMD package) PIC microcontroller together with two blue, one red and one green LED. By using two blue LEDs and different valued series resistors it is possible to produce all the basic colours of equal brightness. The microprocessor slowly dims each LED individually in a pseudo-random pattern to produce interesting lighting effects. Power comes from a standard mains unit supplying 9 V to 15 V while D1 protects the circuit against inadvertent reverse polarity. IC1 provides a regulated 5 V for the circuit. The circuit was fitted into the base of the figurine so it was necessary to keep the PCB

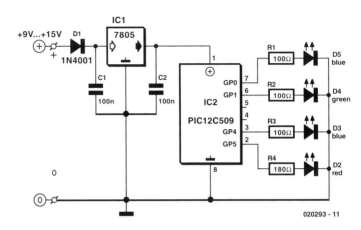

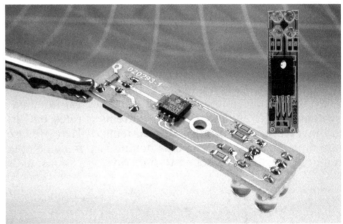

COMPONENTS LIST

Resistors:
R1,R2,R3 = 100Ω (SMD size 1206)
R4 = 180Ω (SMD size IC1 = 7805 1206)

Capacitors:
C1,C2 = 100nF (SMD size 1206)

Semiconductors:
D1 = 1N4001
D2 = LED, red, (Sharp GL-code 020293-11

or Free 5HD23)
D3,D5 = LED, blue (Kingbright L53BC)
D4 = LED, green (Sharp GL-5EG23) (all LEDs superbright, in colourless transparent enclosure)
IC1 = 7805
IC2 = PIC12C509A-04/SM, programmed, order code 020293-41
Disk, PIC software, order Download 1

(Figure 2) as small as possible and use SMD components where possible. The microcontroller is clocked internally and does not require an external crystal. Mounting the components on the PCB should be straightforward and begins by

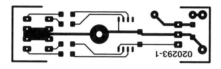

fitting all the SMD components onto the copperside and then fitting the voltage regulator to the other side, the mounting bolt through the regulator heat sink will be at earth potential and can also be used as a convenient fixing point for the whole PCB. The source and hex files (along with the PCB layout) can be freely downloaded from the Elektor Electronics website. The software can also be ordered as disk 020293-11 (see Readers Services). If your PIC programmer is not fitted with an SMD socket it may be worth ordering the pre-programmed controller (order code 020293-41), also from our Readers Services.

C. Wendt   (020293)

014
DC Control for Triacs

If a circuit is to switch a mains voltage, a relay is a simple solution in cases where switching times are long and high currents are involved. However, at lower currents, and in particular where rapid switching is required, such as in sound-to-light sys-

tems, a relay no longer fills the bill. Electrical isolation is often a requirement, which rules out driving a triac via a transistor. Here we use the MOC3041 optocoupler, which is specially designed for such applications, to drive a power triac.

The control circuit therefore remains galvanically isolated from the mains. The internals of the optocoupler are somewhat more complex than appears from the circuit diagram. A special zero-crossing detector circuit in the optocoupler ensures that the connected triac is only triggered when the alternating mains voltage goes through zero. This has the advantage of generating less interference compared to switching the triac at arbitrary phase in a cycle. Indeed, it means that we can dispense

with the sup-pressor choke at the output that would otherwise be necessary. If very brief pulses are likely to be present at the input to the optocoupler, a 220 nF capacitor should be connected between the input of the circuit and the emitter of T1 to lengthen the drive pulses.

This ensures that the triac will be triggered even with very short input pulses, which might otherwise miss the zero-crossing point of the mains waveform. The triac should be an AW-suffix type. These types are less sensitive, but have higher dv/dt and di/dt specifications. The gate resistance must be constructed from two resistors connected in series, since normal resistors are not suitable for direct use with mains voltages. It is also necessary to exercise care around the optocoupler.

In order to guarantee Class II isolation the solder pads on the input and output sides must be separated by at least 6 mm. The leads may therefore need to be bent outwards when soldering.

L. Libertin   (020279-1)

015
Microscope 2.5-D Lighting

Interesting, surprising as well as scientifically useful visual effects can be obtained by employing coloured LEDs instead of the normal lamp fitted in the lighting assembly in a microscope base. Of course, the lamp can be simply replaced by a white LED with appropriate changes in the supply current/voltage etc., but it is more interesting to have three colours available – here, red, green and orange (amber) which are individually adjustable for intensity. The dashed line around LEDs D2, D3 and D4 indicates that they are fitted in a 10 mm long section of 1 inch plastic (PVC) pipe.

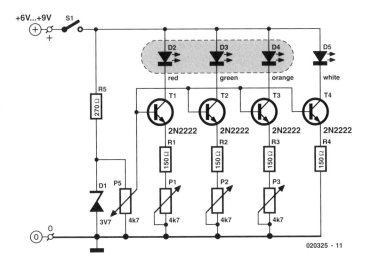

020325 - 11

The 5 mm holes for the LEDs are drilled at an angle so that the LED beams are pointed downwards at the centre of the holder. As shown in the photograph, as small piece of veroboard can be fitted under the ring, with a 1-inch hole in it so that the centre of the ring remains open, allowing the assembly to be placed on top of the glass specimen carrier. The PCB, which is held secured to the ring by the LED wires, allows wires to be run between the lighting assembly and the driver electronics fitted in the microscope base. The white LED, D5, takes the place of the (6 V) lamp in the lamp holder,

and is connected to the driver circuit by wires. The potentiometers P1, P2 and P3 allow the intensity of each colour to be individually adjusted, while P5 acts as the master intensity control. Space allowing, the pots may be fitted in the microscope base.

If there are problems with reflections in the ring, paint the inside black. Finally, in most microscopes there is a need for strong lighting which pleads for high-intensity LEDs to be applied in this circuit.

A. Bauer   (020325-1)

016
Tube Box Upgrade

The Tube Box published in the June 2002 issue of Elektor Electronics opens up many new expressive sonic possibilities for the electric guitarist. The Tube Box can replace Hall, Echo, Pitch Shift, Phaser and the like: it offers so many possibilities that you may not need any other effects units.
The author has adapted and extended the Tube Box circuit, building it into the existing electronics compartment of a guitar. What was missing was the ability automatically to switch the unit on when the jack is plugged in. The usual method used in effects units is to connect one battery terminal to the middle pin of a stereo jack socket and so connect that termi-nal to ground when the jack is inserted, thus switching on the power to the circuit. This method cannot be used here, however, because ground in the Tube Box circuit is not connected directly to the battery. The small additional circuit shown here solves that problem by switching the supply voltage as the jack is

inserted using a transistor, connected as a switch. The battery, a 9 V PP3 type, is fitted inside the guitar and powers the circuit. The negative terminal is connected to K3 while the positive terminal is taken to transistor T1 via an optional bypass switch. If no cable is plugged in the transistor is turned off. Only a tiny current (in the region of a few microamps) flows. If a cable is now plugged in, a small current flows through R1, R2 and the internal resistance of the Tube Box circuitry, turning on T1. T1 goes into saturation. C1 prevents interference from mechanical contact noise on the jack socket as well as when the circuit switches on and off.
The bypass switch can be connected in a different way from the original circuit, namely directly between the pickup and the jack socket. This is a particularly good idea for stage use, since it means that in case of 'emergency' (severe interference or a flat battery) one can still play without the

Tube Box circuit. In the 'bypass' position the battery is disconnected and the signal from the pickup appears directly on the output socket. With the switch in the 'on' position, the output signal comes from K2 of the Tube Box circuit. The input to the Tube Box circuit is in any case connected directly to the pickup. In the author's case the 'original' signal that appeared on the output jack of the instrument before it was modified was used.

The tone control capacitors (in parallel with the pickups via a potentiometer) are removed. Instead, the three new potentiometers are fitted to the guitar.

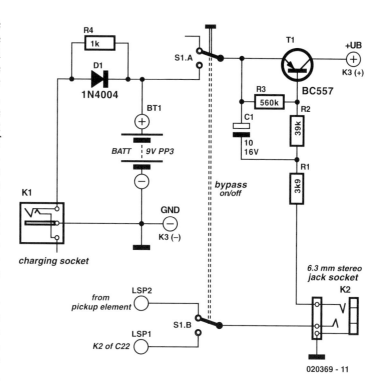

020369 - 11

The following controls:
• the potentiometers for each of the two pickups (to control the mixing of the signals);
• the switch to select between the two pickups, or to switch them in parallel remain connected as before, whereas the following are new:
  – bypass/on switch;
  – Tube Box tone switch;
  – Tube Box 'Mid Cut/Boost' switch (existing 'tone' switch can be used);
  – Tube Box drive control potentiometer;
  – Tube Box volume control potentiometer;
  – Tube Box tone control potentiometer.

A file containing photographs of the modifications made to the author's guitar can be downloaded from www.elektor-electronics.co.uk/dl/dl.htm, reference number 020369-11, under month of publication of this article. Great care

was taken in construction to ensure adequate shielding. Self-adhesive copper foil (available, for example, from Conrad Electronics) was used and the enclosure for the electronics was completely screened, the individual pieces of copper foil being soldered together. This is very important, as the Tube Box has a high gain and is very sensitive to external interference. Conductive spray can also be used around the switches, potentiometers and pickups: the effort is definitely worthwhile. D1 and R4 are provided to connect the battery to the outside world.

The battery can be charged via connector X1 and D1. R4 allows the battery voltage to be measured without having to dismantle the instrument. D1 and R4 prevent a short on X1 from damaging the battery. An old mains adaptor can be used as a charger: a 150 mAh battery can be charged at around 20 mA.

M. Graeber   (020369)

017
Railway Points Sequencer

Dedicated model rail enthusiasts using sophisticated train and points controllers often have the problem that as their layouts get bigger and more complex, the transformer supplying power to the points does not have enough current to switch several points at the same time. The actuators in the points are designed for ac operation so it doesn't help by rectifying the supply and adding reservoir capacitors, the coils can overheat and burn out if they get jammed during their travel (ac operation actually helps to overcome friction in the mechanism).

The circuit shown here solves this problem by using a sequencer to ensure than only one points actuator can be active at any point in time. During operation the controller will switch all the points on one line at the same time as usual, but the other connection to each coil is connected to the sequencer unit. This circuit will only allow current to flow through one coil at a time. The sequencer circuit consists of a 555 timer configured as an astable multivibrator clocking a 4017 Johnson counter where the ten outputs are used to switch ten triacs in

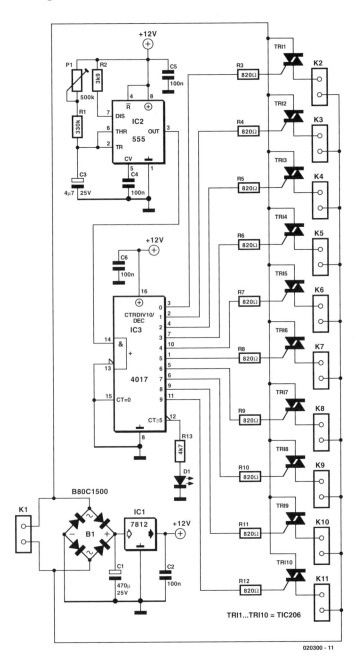

020300 - 11

sequence, enough for ten sets of points. P1 alters the oscillator frequency of the 555 timer and can be adjusted so that each time interval of the sequencer is long enough to allow the points to switch. The switching time varies depending on the type of points but is typically between 1 s and 1.5 s. Any points that jam during switching give out a characteristic humming noise in time to the switching frequency so it makes them easier to find. The eleventh output of the 4017 can

be connected to an LED (together with a series resistor). This will flash to give a visual indication of the sequencers operation. Power for the circuit is provided by 15 V ac from the points transformer.

The B80C1500 bridge rectifier (80 Vpiv, 1.5 A) and regulator IC1 produce a stabilised 12 V for the circuit. Current consumption is only a few milliamps.

F. Pascher   (020300-1)

018
Simple Amplifier with Surround System

The AN7147 Dual 5.3 watt Audio Power Amplifier from Panasonic is listed as a 'replacement type' so hopefully will be around for some time to come. Together with some extra components, it can represent a simple surround-sound system requiring no opamps or a negative voltage supply. As shown by the circuit diagram the basic stereo amplifier is changed into a surround-sound system by a trick called 'adding feedback from the opposite channel'. When surround sound is required, the negative feedback signals supplied by C13-R3 and C12-R4 are fed to the inputs of the 'other' amplifier. The resulting phase difference causes the surround effect. If surround sound is not required, the effect can be disabled by pressing

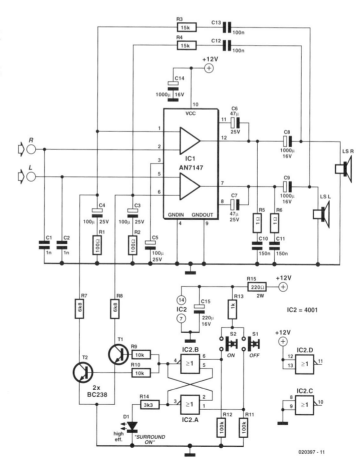

pushbutton S1. This causes the bistable built around IC2.A and IC2.B to toggle and drive transistors T1 and T2 such that the above mentioned negative feedback signals are effectively shunted to ground.

A high-efficiency LED and a 3.3 kΩ series resistor (R14) should be used to make sure the maximum output current of the CMOS

4001 device is not exceeded. The amplifier should not be loaded with impedances smaller than 3 Ω. The AN7147 will typically supply up to 4.3 watts into 4 Ω. The SIL-12 case needs to be cooled wit a small heatsink of about 6 K/W or better. The quiescent current is modest at just 19 mA.

H. Sadeghi   (020397-1)

019
EL Lamp Driver using HV832MG

An EL lamp is a solid state, low power, uniform area light source. Because of its thin profile (as thin as 0.3 mm) and the fact that it can be built into almost any size and shape, EL lamps are an ideal way to provide backlighting for LCD dis-plays, membrane keypads and a variety of other applications. Electroluminescent (EL) lamps offer significant advantages over point light sources such as LEDs, incandescent, and fluorescent lighting systems. As a result of these advantages, EL lamps are seeing growing use. Many wireless phone and pager manufacturers are converting to EL lighting systems in keypads and displays. The typical lamp consists of light emitting phosphor sandwiched between two conductive electrodes with one of the electrodes transparent so allowing light to escape. As an AC voltage is applied to the electrodes, the electrical field causes the phosphor to rapidly charge and discharge, resulting in the emission of light during each cycle. Since the number of light pulses depends on the magnitude of the applied voltage, the brightness of EL lamps can generally be controlled by varying the operating voltage. Because EL lamps are a laminate, they exhibit a capacitance of the order of 2.5 nF to 3.5 nF per square inch. When high voltage is applied across the electrodes, the resulting electric field ex-

cites the phosphor atoms to a higher energy state. When the electric field is removed, the atoms fall back to a lower energy state, emitting photons in the process. The wavelength of the emitted light is determined by the type of phosphor used and the frequency of the excitation voltage. With most phosphors, the spectrum of emitted light will tend to shift towards blue with an increase in excitation frequency. Colour is usually controlled, however, by selecting the phosphor type, by adding fluorescent dyes in the phosphor layer, through the use of a colour filter over the lamp, or a combination of these. EL lamp brightness increases approximately with the square of applied voltage. Increasing frequency, in addition to affecting hue, will also increase lamp brightness, but with a nearly linear relationship.

Most lamp manufacturers publish graphs depicting these relationships for various types of lamps. Excitation voltages usually range from 60 VPP to 200 VPP at 60 Hz to 1 kHz. Increased voltage and/or frequency, however, adversely affects lamp life, with higher frequencies generally decreasing lamp life more than increased voltage. EL lamps, unlike other types of light sources, do not fail abruptly. Instead, their brightness gradually decreases through use. For intermittent use, lamp life is seldom a con-

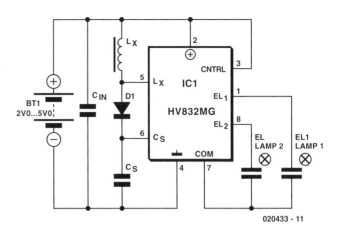

020433 - 11

cern. For example, if a lamp is used 20 minutes per day, over the course of 10 years the lamp will be activated for a total of 1,216 hours, well within the useful life of almost any EL lamp available. When designing a drive circuit, a balance needs to be struck between lamp brightness, hue, useful life, and supply current consumption. To generate the high voltages needed for driving EL lamps, dedicated ICs like the Supertex HV832MG employ switch-mode converters using inductive flyback. By integrating high voltage transistors on-chip, this driver avoids the need for expensive, bulky, and noisy transformers to generate high out-put voltages. The HV832MG employs open-loop conversion. This EL driver incorporates a lamp drive oscillator that is separate from the power conversion oscillator. This allows setting lamp drive frequency independently from the power conver-sion frequency and so optimise overall performance. The power conversion cycle begins when a MOSFET switch in the HV832MG is turned on and current begins to rise in inductor Lx.

When the switch is turned off, inductive flyback causes the voltage across the inductor to reverse polarity and rise until it reaches the level of the storage capacitor CS, (plus diode drop) at which point the rectifier conducts and the energy contained in the magnetic field of the inductor is transferred to CS. When all the inductor energy is transferred and inductor current drops to zero, the rectifier stops conducting and inductor voltage drops to zero, ready for the next cycle. Output power is simply the amount of energy transferred per cycle multiplied by the number of cycles per second. It is important to select the inductor and conversion frequency to provide the required output voltage while assuring that the inductor current does not approach saturation levels. If the inductor saturates, excessive current will flow, potentially lead-

ing to device failure. Ideally, the inductor current should be allowed to return to zero between cycles. If inductor current is not allowed to return to zero, a higher average current will be needed to meet output power requirements, increasing I^2R losses, and decreasing conversion efficiency. On he other hand, if too much time is allowed between zero inductor current and the start of the next cycle, more energy will need to be transferred each cycle to maintain out-put power, thus risking inductor saturation and increasing I^2R and core losses. This circuit provides an output of 130 volts at 300-450 Hz, draws just 30 mA current, yet is capable of driving EL lamps with a surface area of up to 9 cm^2. This design has excellent drive capability and provides a symmetrical bipolar drive, resulting in a zero-bias signal. Many lamp manufacturers recommend a zero-bias drive signal to avoid potential migration problems and increase lamp life. The supply voltage should be bypassed with a capacitor located close to the lamp driver. Values can range from 0.1 µF to 1 µF depending on supply impedance. The HV832MG may be obtained from Supertex Semiconductors, USA, http://www.supertex.com. For very large lamps representing much larger capacitances, a FET follower circuit may be employed to boost the output drive capability of the lamp driver.

D. Prabakaran   (020433)

020
Mini Running Text Display

This charming little circuit is a genuine four-digit running-text display, complete with a Christmas / New Year's greeting. Naturally, any competent programmer can easily arrange to have a different text scroll across the display. The associated software, including the source code, can be downloaded from the Free Downloads section of our website or obtained from Readers Services on diskette (order number 020365-11). As can be seen from the schematic, the hardware con-

sists of little more than an AT90S1200 microcontroller, a 4-digit LED display and a 5 V voltage regulator. The only external circuitry needed by the microcontroller consists of a reset circuit and a 4 MHz crystal, and the remainder of the components are limited to a few decoupling capacitors. In the prototype model, an Osram SLO2016 display module was used. Although this four-digit module measures only 10 × 20 mm, it provides an especially clear and bright display. In order to give the 7805 voltage regulator sufficient 'breathing room', the

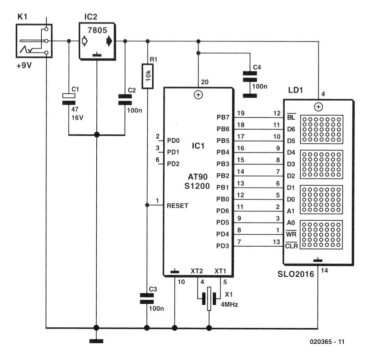

020365 - 11

supply voltage should be at least 8 V. A standard 9 V mains adapter should thus be perfectly adequate. The supply voltage does not have to be stabilised, and the adapter does not have to provide an espe-

cially large amount of current, since the running-text display draws scarcely more than 50 mA.

R. van Arem   (020365-1)

021
Seconds and Minutes Clocks from DCF77

Nowadays, the availability of inexpensive receiver components makes it attractive to use the DCF77 signal at 77.5 kHz (transmitted from Mainflingen, Germany) for trivial applications. The circuit shown in Figure 1 can relatively easily extract seconds and minutes clocks from the DCF77

signal. As is well known, the 59th second pulse is missing every minute in the DCF77 signal, with the missing pulse marking the minute. To obtain an uninterrupted seconds clock, we must first generate minutes pulses from these gaps and then derive the missing seconds pulses

1

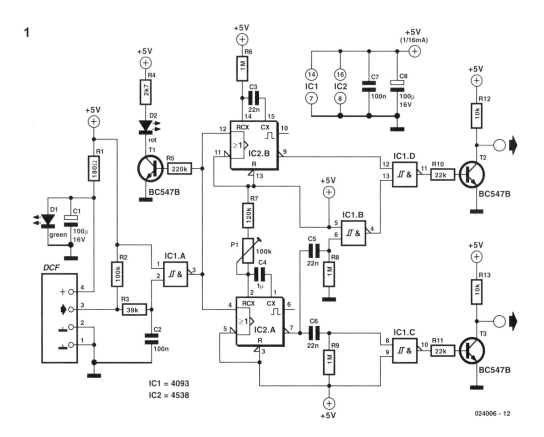

IC1 = 4093
IC2 = 4538

024006 - 12

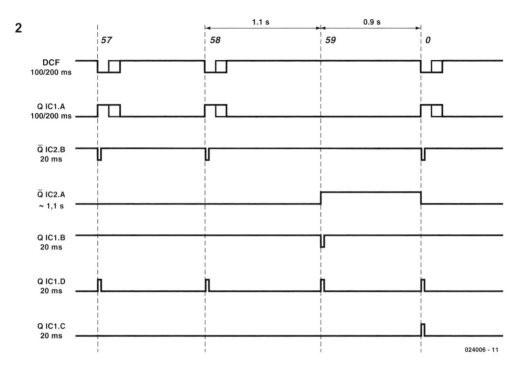

024006 - 11

from the minutes clock. The circuit input is suitable for use with DCF77 receivers provid-ing negative output pulses. The pin assignment of the receiver corresponds to the Conrad Electronics HK 009 module; other modules may use different arrangements. After passing through a low-pass network that attenuates any residual high-frequency components, the signal arrives at gate IC1a, which is wired as an inverter and serves as an input buffer. Following this gate, the signal branches to the two monostables of IC2. IC2b is triggered by the positive edges of the 100 ms or 200 ms pulses of the DCF77 signal and generates 20 ms pulses at a one-second rate. However, the 59th pulse is still missing! IC2a acts as a missing-pulse detector. This monostable 2 is also triggered by the positive edges of the DCF77 signal, but it is retriggerable. The period of this mono-DCF stable is approx-imately 1100 ms. During the string of seconds pulses, its pin-7 output remains low, but at the time of the missing 59th pulse, it jumps high for around 1100 ms. IC1b de-

tects the leading edge of this pulse and converts it into a 20 ms pulse that is added to the seconds clock signal via logical NOR gate IC1d, in order to provide the 59th pulse. IC1c detects the trailing edge of the monostable pulse and generates a 20 ms pulse, which is applied to output inverter T2 to provide a minutes clock. If you find this all a bit difficult to follow, have a look at the timing diagram in Figure 2 (not drawn to scale). In practice, the monostable period is adjusted so that with 'normal' seconds pulses, the circuit just avoids adding an extra seconds pulse via the logical NOR gate. When the circuit is operating, the low-current LED driven by T3 shows the 'ticking' of the DCF signal. The receiver is powered from the supply voltage via a series resistor. A (normal) green LED (the colour is important) limits the voltage to approximately 2.2 V. The current consumption of the circuit, ex-cluding the DCF module, is very low at around 1 mA.

W. M. Köhler   (024006-1)

022
HEX Display

When experimenting with digital and microcontroller-based circuits it is often more convenient to display a value as a number and not as a bit pattern on a row of LEDs. This circuit is designed for exactly that purpose. Thanks to its low current consumption (a little under 10 mA) and its compact construction it can be plugged directly

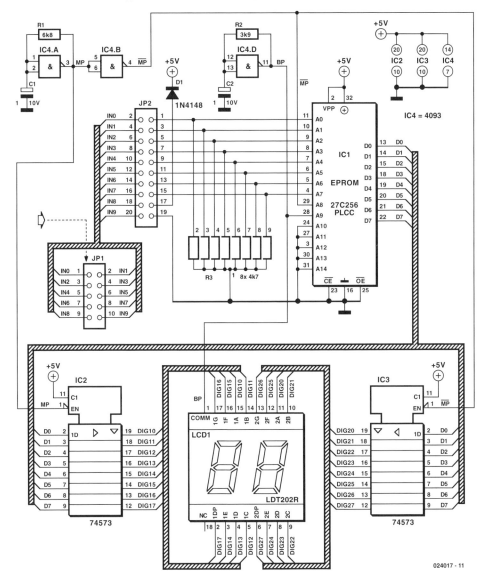

024017 - 11

into a header on a microcontroller board, taking its supply from that board. An EPROM, which stores all the required values for display, is used as the decoder. The 256 kbit EPROM is considerably larger than strictly necessary for this job – 1 kbit would be enough – but the type shown is much easier to obtain. The binary data (and power, if available) are taken from JP1 to one side of JP2. The binary data and power signals can be permuted using a suitably-wired plug fitted into JP2. Also, power can be provided at this point, if it is not available from JP1. The data are now taken directly to address signals A0 to A7. The incoming data byte is viewed as two 4-bit values. The clock generator consisting of IC4.A and IC4.B switches between banks in the memory using address line A8. At the same time this signal provides a multiplex clock for IC2 and IC3. At any given moment the outputs of exactly one of these latches are enabled while those of the other

are put into a high-impedance state. In this way the high and low halves of the input byte are separated and switched through to the correct digit. Oscillator IC4.D produces the backplane signal for the LCD. An LCD segment only 'lights' when its front and rear electrodes are driven with signals of opposite phase. The back-plane signal is therefore also taken to A9 on the EPROM as a further bank-switching signal (we can afford to be generous with memory space!). This memory bank stores mask values for each possible input value.

A. Barthel   (024017-1)

The author has designed a printed circuit board for this circuit. The layout, in Eagle format, is available for download from the Elektor Electronics website, along with the HEX file for the EPROM. The file numbers are 024017-1 and 024017-11 respectively.

023
Adjustable 3 A Regulator

By combining a common 78L05 with an integrated audio amplifier of the type TDA2030, an adjustable voltage regulator can be constructed in a very simple manner that works very well. The output voltage is adjustable up to 20 V, with a maximum current of 3 A. Since the TDA2030 comes complete with a good thermal and short-circuit protection circuit, this adjustable regulator is also very robust. As illustrated by the schematic, the design of this circuit is characterised

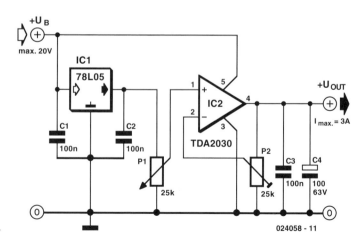

by simplicity that is hard to beat. In addition to the two ICs, the regulator contains actually only two potentiometers and a few capacitors.

The adjustment is done by first turning potentiometer P1 to maximum (wiper to the side of the 78L05) and subsequently turning trimpot P2 until the desired maximum output voltage is reached. P1 is then used to provide a continuously adjustable volt-age between this maximum and nearly zero volts. At relatively small output currents there are no specific requirements regarding the cooling. However, when the out-put current exceeds 1 A, or if the input to output voltage difference is quite large, the amplifier IC has to dissipate too much power and a small heatsink is certainly appropriate.

G. Baars   (024058-1)

024
TX Check for GSM/DECT Phones

A fieldstrength meter is all you need for a quick go/non-go check on the transmitter section (TX) in a GSM phone (900 MHz or 1800 MHz), a DECT phone (1900 MHz), or a base station. The antenna used here is a half-wave dipole designed for a wavelength of 32 cm (900 MHz band) or 16 cm (1800 MHz band). Theoretically, the dipole 'arms' have a length of 80 mm (900 MHz) or 40 mm (1800 MHz). However, because the conductors (1) have a finite thick-ness, (2) are not suspended in free space but etched on a printed circuit board and (3) are close to other objects, the actual length is not given by theory. Observing all of the above factors, a 'velocity factor' of about 0.85 was found by trial and error. Fortunately, a dipole is wideband in nature, and all we need is a relative indication of the transmitter fieldstrength. In

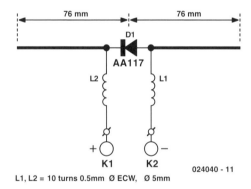

L1, L2 = 10 turns 0.5mm Ø ECW, Ø 5mm

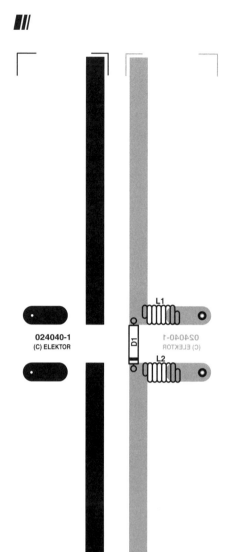

COMPONENTS LIST

Capacitors:
C1,C2 = 1nF ceramic

Semiconductors:
D1 = AA117, AA119

Inductors:
L1,L2 = choke, 10 turns, 0.5 mm ECW, internal diameter 5 mm

practice, good results were obtained from a dipole with a track length of 150 mm (75 mm). A half-wave dipole has its feeder point in the middle, so that element lengths of At resonance, a half-wave dipole is marked by minimum voltage between the feeder points and maximum current at the end of the elements. The current distribution along the dipole is exactly the opposite. In principle, all we need to do is insert a rectifier diode between the feeder points and shunt it with a sensitive micro-ampèremeter. Because of their low threshold voltage of about 0.2 V, germanium (and some Schottky) diodes are preferred over their silicon counterparts. Suitable diodes include the AA119 and AA117. The dipole should have the same polarisation (usually vertical) as the phone you are checking. The RF signal is decoupled from

the measured antenna current by two chokes inserted in the lines to the micro-ampèremeter.

Both chokes consist of about 10 turns of enamelled copper wire. Their internal diameter is about 5 mm. The ammeter may be connected at some distance from the antenna and to the chokes by a (long) length of twisted wire. To prevent erroneous readings owing to the connecting cable picking up RF signals, 1 nF decoupling capacitors are connected between across the wires at both ends of the cable. In principle, the fieldstrength meter should also be capable of locating active DECT handsets and base stations operating at frequencies between 1880 and 1930 MHz.

T. Bulla   (024040-1)

025
RMS to DC Converter

In order to measure the RMS value of an alternating voltage an accurate converter is required to produce the true RMS value of its alternating input as a DC output. With simple sinewave inputs the RMS voltage can simply be calculated as 0.707 times the peak AC voltage, but with complex waveforms the calculation is not nearly as straightforward. The RMS value is defined as the DC voltage that would give the same heat-ing effect in a resistor as the alternating voltage. The LTC 1966 from Linear Technology (www.linear-tech.com) uses a new form of delta-sigma conversion and is designed for battery operation, drawing only 170 µA from the supply. The new technique is accurate to 0.02 % between 50 mV and 350 mV and is highly linear. It can operate from 50 Hz to 1 kHz (with an error of 0.25 %) and up to 6 kHz with a 1 % error. The input voltage range on the differential inputs IN1 and IN2 extends to the supply rails, and so in the non-symmetrical circuit shown here the voltage on IN1 can swing between 0 V and the supply voltage. If the signal to be measured is AC only, then another coupling capacitor will be required. The input impedance is many megohms. The output voltage at the OUT pin can be offset by applying a DC voltage to the OUTRTN pin. This is particularly

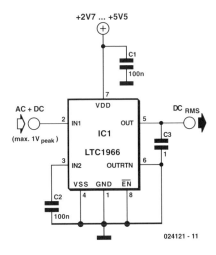

helpful when using the device with LCD multime ter ICs such as the C3 7106. A further capacitor is connected to the output which is charged up to the required voltage by the switched-capacitor circuit in the converter. The capacitor required is ten times smaller than that demanded by previous RMS to DC converter designs. The LTC1966 is not temperature sensitive and is available in an 8-pin MSOP package. It allows a tiny RMS to DC converter to be constructed using just four components.

G. Kleine   (024121-1)

026
Rolling Shutter Control

An electrically operated rolling shutter usually has a standard control panel with a three-position switch: up, down and stop. If you would like to automate the opening and closing with a time controlled switch, a few additional wires will have to be connected. Typically, the controls are implemented as indicated in the schematic 'Normal Situation'.

If this is indeed the case, then you can see in 'New Situation' how the shutter can be automated with a timer. There is only one method to determine the actual schematic of your control circuit, and that is to open the control box and using an ohmmeter, pencil and paper to check out and draw the circuit. Make sure you turn the power off first though! Connect a 230 V relay (with both the contacts and the coil rated 230 VAC) to the timer. The changeover switch between automatic and manual control needs to be rated 230 VAC as well and may not be a hazard for the user. The relay and switch are preferably fitted in a

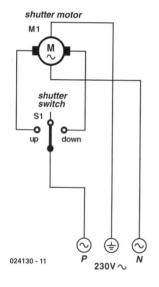

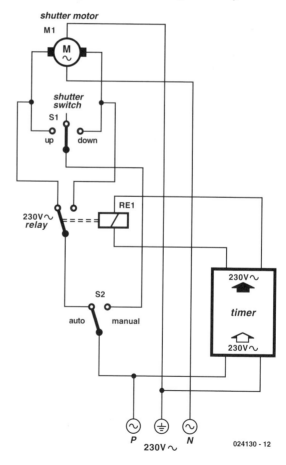

plastic mains adapter enclosure with built-in plug, which is plugged into the timer. It is a good idea to check first if this will actually fit. Because of the manual/automatic-switch, the operation is the same time. completely fail-safe and mis-understandings are out of the question.

The switch prevents the issue of conflict-ing commands (with disastrous consequences) when, for example, the shutter is being automatically raised and manually lowered at the same time.

T. Knipa

(024130-1)

027
Electronic Die

The simplicity of a traditional die makes it exceptionally difficult to create a fully equivalent electronic version, if only be-cause an electronic version requires a power supply and a collection of electronic components that occupy a much larger vol-ume than a normal die. This article de-scribes an electronic die that can be built using normal components or SMDs as de-sired, and which comes very close to hav-ing the same format as a traditional die in the latter case. Despite its simplicity, this

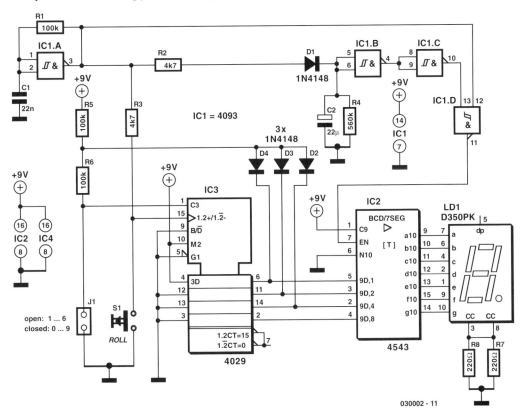

030002 - 11

electronic die incorporates several interesting features. For instance, the range of 'spots' can be increased from 1–6 to 0–9 using a jumper, and it has standby function that disables the display approximately 8 seconds after the die has been 'thrown', in order to save energy. The electronic die also uses energy efficiently by driving the display in pulsed mode. As a result of the latter two features, the current consumption of the circuit is approximately 25 mA in use and 12 mA in standby. This means that it can easily be powered by a 9 V battery.

The circuit consists of the following parts: a free-running oscillator (IC1a), additional logic for driving the display (IC1c & IC1d), a timer (IC1b), a counter (IC3) and a display decoder (IC2). The oscillator is very simple. Its frequency, which is determined by R1 and C1, is approximately 225 Hz, with a duty cycle of around 50–60 percent. The signal from the oscillator acts as a clock signal for the counter (via R2) and a blanking signal for the display decoder (via IC1d). However, the counter will not count as long as the 'throw' switch (S1) remains closed, since the clock input of IC3 is grounded by S1. The blanking input of the display decoder is driven by a pulse waveform, so the display is in principle illuminated only around 50 percent of the time, but it appears to be constantly illuminated due to the high clock frequency. The standby mode works as follows. As long as there is a signal on the clock input of the counter (S1 pressed), the output of gate IC1b is low and the display is enabled. If S1 is released, the counter stops and a number will be shown on the display. How-ever, the clock pulses will have charged C2 via D1, and C2 will slowly discharge via R4. After approximately 8 seconds, the out-put of gate IC1b will go high, causing the display to be blanked. The design of the counter is relatively simple. It is wired as an up counter by connecting the U/D pin to VCC. The preset inputs (pins 4, 12, 13 and 3) are configured to binary

'0001', and the counter normally has a counting range of 0–9 (pin 9 connected to ground). Diodes D2, D3 and D4, in combination with resistor R5, act as a logic AND gate, so if the value of the counter is greater than 6, the preset value of 1 is latched into the counter and it starts to count again from 1 to 6. This only happens if jumper J1 is open. If it is closed, the preset pulse on PREN is suppressed and the counter range is 0–9. The A, B, C and D inputs of the decoder IC (IC2) are driven directly by the counter. The series resistors normally used for the individual segments of the display are instead placed in the common-cathode lead (R7 & R8). This has the advantage of allowing the number of resistors to be reduced, although it has the drawback that the brightness of the display depends on the displayed number. If the segment current is sufficiently large, (light) saturation occurs and this brightness variation is no longer noticeable. The Blank input (BL) controls whether the display is enabled. If you choose to build this circuit using SMD technology, that will not affect the schematic diagram, but it will naturally affect the choice of components. In this case, SMD components must be used for the resistors and C1, the diodes must be replaced by BAS32 types, and BT versions of ICs IC1– IC3 must be used instead of conventional types.

An SMD version of C2 was not used in the prototype, since SMD electrolytic capacitors are expensive, and normally they are only sold in lots of 10, just like other passive components. It is also recommended to use a socket for the display of the SMD version of the electronic die, to allow the space under the display to also be used and the dimensions of the circuit board to be further reduced. Any desired DC power source providing a voltage of 5 to 15 V can be used as a power supply. Due to the low current consumption of the circuit, a 9 V battery will last quite a long time.

L. de Hoo   (030002-1)

028
Isolated Fuse Fail Indicator

This circuit uses standard components and shows a method of indicating the fuse status of mains powered equipment while providing electrical isolation from the mains supply. A standard miniature low power mains transformer (e.g. with an output of around 6 V at 1.5 VA) is used as a 'sense' transformer with its primary winding (230 V) connected across the equipment's input fuse so that when the fuse blows, mains voltage is applied to the transformer and a 6 V ac output voltage appears at the secondary winding. The 1N4148 diode rectifies this voltage and the LED lights to indicate that the fuse has failed. The rectified voltage is now connected to an RC low-pass filter formed by the 10 kW resistor and 100 nF capacitor. The resulting positive signal can now be used as an input to an A/D converter or as a digital input to a microcontroller (make sure that the signal

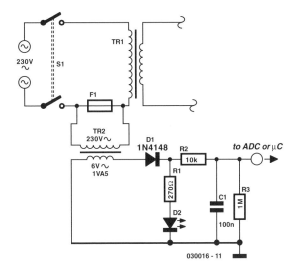

level is within the microcontroller input voltage level specification). The 1 MΩ resistor is used to discharge the capacitor if the input impedance of the connected equipment is very high. As long as the fuse remains intact it will short out the primary winding of the 'sense' transformer so that its secondary output is zero.

G. Kleine   (030016-1)

029
Overvoltage Protection

When a sensitive circuit must under no circumstances have too high a supply voltage applied, then some means of disconnecting the supply must be provided. One way to achieve this is to trigger a thyristor to blow a fuse. A less destructive alternative possibility is to use a MOSFET to disconnect the supply. An overvoltage protection IC, the

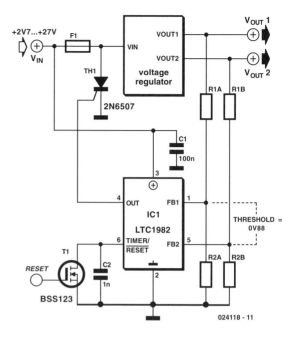

LTC1696 from Linear Technology (www.linear-tech.com), has recently become available, which is suitable for triggering and driving such a device. It operates from a power supply in the range 2.7 V to 27 V and can be connected to the unregulated input of a voltage regulator. Two voltages can be monitored using feedback pins FB1 and FB2, suitably divided down using potential dividers. The trigger threshold for both FB1 and FB2 is +0.88 V. The value of the upper resistor in the potential divider can be calculated using the following formula:

$$R_1 = 33\,k\Omega \cdot \left(\frac{V_{LIMIT} - 0.88\,V}{0.88\,V} \right)$$

The value of the capacitor connected to the TIMER/RESET pin sets the delay before the protection is triggered. The charging current for this capacitor depends non-linearly on the amount by which the voltage exceeds the threshold value. The greater the overvoltage, the faster the IC triggers.

Once triggered the IC remains in that state until either the input voltage is removed or the internal latch is cleared using the MOSFET connected to the TIMER/RESET input.

G. Kleine   (024118-1)

030
Floating 9 V Supply for DVM Modules

Most commercial DVM modules with an LCD readout are 9 V powered and based on an ICL7106 or similar A-D converter chip. These modules are typically used in laboratory power supplies and other test and measurement equipment where a drop-in solution needs to be found to realize a voltmeter readout. Particularly in power supply units, the LCD module will need to 'float' relative to the PSU supply rails, and this inevitably requires a separate 9 volt power supply. In some cases, batter-

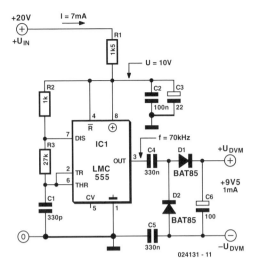

ies may be used but these have distinct advantages. The alternative, a 9 V converter effectively powered by the PSU and yet floating, is shown here. It is built from the ubiquitous TLC555, LMC555 or 7555)

timer IC acting in a stable multivibrator configuration producing a 70 kHz square wave fed into a simple rectifier. In essence, capacitors C5 and C6 afford the above mentioned electrical isolation between the PSU supply rails and the LCD module. The old, bipolar NE555 IC should not be used here because it presents a too heavy loads on the converter's own supply voltage. Depending on the exact type and brand of the CMOS 555 you're using, resistor R6 may need to be re-dimensioned a bit to ensure a supply voltage of about 10 volts at pins 8 and 4 of the chip. At an output voltage of 9.5 V, the maximum output current of the converter s about 1 mA.

D. Ponting   (024131-1)

031
Case Modding

The aesthetics of 'case modding' (modifying a PC's enclosure) offer plenty of scope for debate and plenty of scope for original circuits. In the April 2003 issue of Elektor Electronics we presented a PC temperature indicator; and the circuit shown here falls into the same category. Low current LED indicators are usually fitted in PC enclosures. Although this certainly saves energy, the LEDs do not light particularly brightly. It is not completely straightforward to replace the low-current types with high-brightness types since the latter draw a current of 20 mA rather than 2 mA. This can – whatever people might tell you to the contrary – in some instances lead to excessive load on the LED drivers on the motherboard. The problem can be solved using a

small external driver stage: two resistors and a transistor mounted on a small piece of perforated board, connected in place of the original LED. The new high-brightness

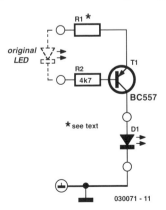

LED is then connected between the output of the current source and a spare motherboard ground connection (for example on the infrared port) or to a grounded screw in the enclosure.

R1 is responsible for the constant current. High-brightness red LEDs are driven at 20 mA (R1 = 150 Ω), whereas high-brightness blue LEDs require 10mA (R1 = 75 Ω). In view of the large number of different PC motherboards available, it is not certain that this driver can be used in every PC. It is easy to check whether the circuit will work: use a DC voltmeter to measure the voltage between the positive connection on the LED (generally a red wire or a pin marked with an arrow on the LED connector) and ground (not the other pin of the connector). If this reads +5 V independent of whether the LED is on or not, then the driver can be used.

S. Büching   (030071-1)

032
Audio Tester

Wouldn't it be nice to have an oscilloscope, spectrum analyser and function generator in one instrument without doing enormous damage to your wallet? And in addition, this instrument will not take up space on your crowded desk or workbench and does not require a separate power supply? This sounds too good to be true! But it is possible, provided you already own a Windows-PC with a sound card. Granted, the frequency range is limited to the audio range and the accuracy depends on your sound card, but for 18 pounds or so this is certainly a very handy and desirable measuring instrument. You have probably realised by now that this is not a build-it-yourself circuit, but a program that is available as share-ware on the Internet. You can register the program for 25 euros (approx. £18) and use it is much as you desire. The (2-channel!) oscilloscope is rather Spartan by design, but that is logical, since the input of a sound card does not possess as many bells and whistles as a normal oscilloscope. It is possible however, to adjust the time base and change the sensitivity of the input channels. This oscilloscope is also fitted with a trigger function, which allows a nice 'stationary' picture to be displayed on the screen. The oscilloscope is quite usable in practice, provided the limited bandwidth and maximum input voltage don't become a problem. The function generator has the option of generating various waveforms such as sine, rectangle and triangle. White noise and pink noise can also be produced with this generator. A final option permits a wave-file to be played, so that, in principle, any arbitrary waveform is possible (provided it can be generated by the sound card). The generator also has a second function, namely as a sweep generator. In this case the generator produces different sine waves of increasing frequency. At each frequency, the level at

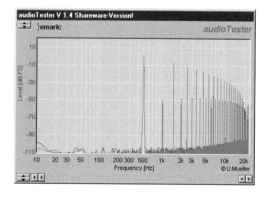

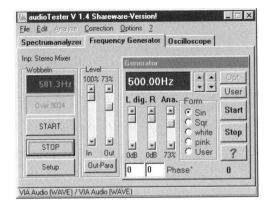

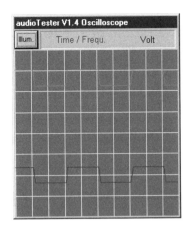

the input is measured. These values are then displayed in a nice Bode-plot. This is very useful when measuring filters, amplifiers, etc. The final instrument is a spectrum analyser. This makes use of FFT-analysis, which gives the analyser a fast update rate.

The adjustments provide, among other things, different windowing options that can be applied to the input signal. The number of samples that are used for analysis is also adjustable. Being able to carry out a good measurement with this program is entirely dependent on the quality of the sound card. Currently, the average sound card is good enough to make reasonable

measurements possible, but they are not really ideal, of course. It should be clear, that the inside of a PC enclosure is hardly a desirable environment for analogue signals. There are manufacturers who, for these reasons, make sound cards available where the A/D and D/A converters are fitted in a separate enclosure from the PC. To reiterate, a standard soundcard is good enough to perform measurements, but you should not expect miracles from it! The program can be downloaded from the designer's   website: http://www.sumuller.de/audiotester/.

P. Goossens   (030027-1)

033
Programming Tool for ATtiny15

A small microcontroller like the ATtiny15 is ideal for many simple tasks not requiring too much I/O activity. The ATtiny15 has two I/O connections, offers a 1024-byte Flash memory and a 64-byte EEPROM, all in an 8-pin case. The programming tool shown here is connected to the RS232 port on your PC. The programming clock signal is generated using a byte sent across the serial interface. The programmer employs

the so-called low-power programming algorithm under Windows ME. Identify which allows manufacturer data as well as calibration register contents to be read from the ATtiny15 chip.

Flash which expects a file in HEX or binary format for programming into the on-chip Flash memory. When the HEX format is used, the checksum is ignored.

The associated software offers four tabs:
```
.equ     lastflash       = 0x1ff

Reset:
Ldi      ZL,LOW(lastflash*2)+1
Ldi      ZH,HIGH(lastflash*2);//FLASHEND
Lpm      ;in last memory location of calibration register
Out      OSCCAL,r0
```

1

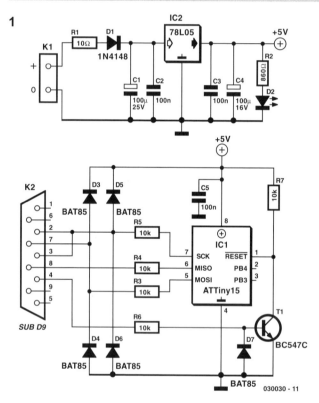

Read and write access to the Flash memory is only possible using the HEX format. The calibration byte is read with the aid of PutCallbyte and then written to the indicated buffer address. The internal oscillator is controlled as shown in the listing. EEPROM does essentially the same as Flash function, only the PutCallbyte is not present. Fuses allows the chip fuses to be read and programmed. Note, however, that the Fuse and Lockbit programming is not tested because afterwards only the High Voltage mode will work. A small PCB, order code 030030-1, was designed for the programmer.

The PCB artwork may be obtained from the Free Downloads section of the

2

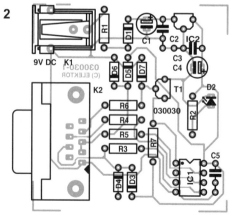

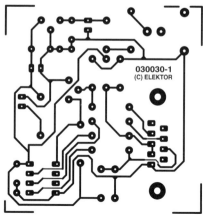

COMPONENTS LIST

Resistors:
R1 = 10Ω
R2 = 820Ω
R3-.R7 = 10kΩ

Capacitors:
C1,C4 = 100µF 25V radial
C2,C3,C5 = 100nF

Semiconductors:
D1 = 1N4148
D2 = LED, green
D3...D7 = BAT85
IC1 = 78L05
T1 = BC547C

Miscellaneous:
IC1 = 8-way DIL programming socket
K1 = mains adaptor socket, PCB mount
K2 = 9-way sub-D socket (female), PCB mount, angled pins
PCB, order code 030030-1
Disk, project software, order code 030030-11

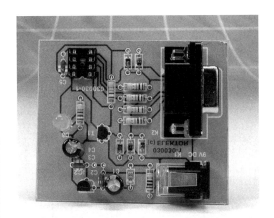

Publishers' website. The software is also found there. For those without access to the Internet the software is also available on diskette, order code 030030-11.

H. Volmer   (030030-1)

034
1.2 GHz VCO with Linear Modulation

Since high frequency voltage-controlled oscillators, or VCOs, are not easy to construct, Maxim (www.maxim-ic.com) has produced an integrated 1.2 GHz oscillator, the MAX2754. The centre frequency is set using the TUNE input, and a linear modulation input allows the frequency to be modulated. The IC is available in an 8-pin µMAX package, operates from a supply of between 2.7 V and 5.5 V, and draws a current of less than 20 mA. Both TUNE and MOD operate over control voltage range of +0.4 V to +2.4 V. TUNE allows the VCO frequency to be adjusted from 1050 MHz to 1270 MHz. In some applications a PLL control voltage will be applied here, allowing the centre frequency to be set exactly to a desired value. For simplicity in the circuit

diagram we have shown a potentiometer. The MOD input allows the VCO to be modulated in a digital or analogue fashion, with a transfer slope of –500 kHz/V. In the circuit we have shown an example where MOD is used for frequency shift keying (FSK) modulation. Resistors R1 to R4 shift the level of the data signal so that it has a centre value of +1.4 V and an amplitude corresponding to the desired frequency deviation. One example set of values, suitable for use with a 5 V power supply, is as follows: R1 = 480 Ω, R2 = 100 Ω, R3 = 220 Ω und R4 = 270 Ω. The input impedance is about 1 kΩ. The output level of the MAX2754 at OUT is around –5 dBm into 50 Ω. A coupling capacitor is not required here: the IC already contains one. The

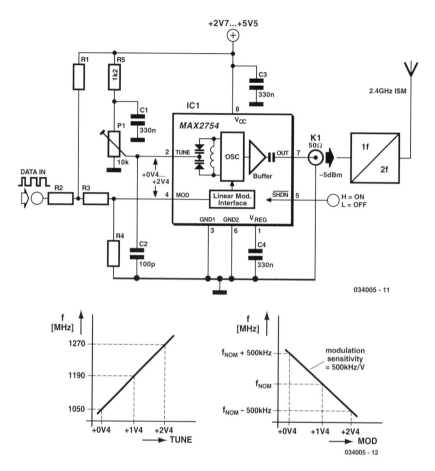

034005 - 11

034005 - 12

MAX2754 is designed for use in transmitters in the 2.4 GHz ISM (industrial, scientific and medical) band. This requires the addition of a frequency doubler, which, along with the 2.4 GHz antenna, is shown symbolically in the circuit diagram.

G. Kleine

(034005-1)

035
FM Wireless Microphone

Here is a very simple, inexpensive and interesting project which provides lot of fun to a home experimenter or hobbyist. This simple transmitter can transmit speech over a short range. It can be used as a simple cordless microphone. The circuit uses two integrated circuits from Maxim. IC1 a MAX4467, is an amplifier raising the microphone signal to a level suitable for frequency modulation (FM). IC2 is a

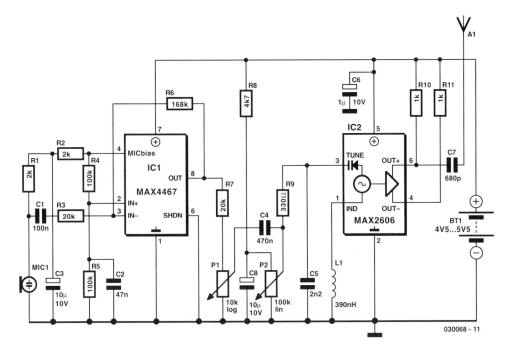

030068 - 11

voltage-controlled oscillator (VCO) with integrated varactor (a.k.a. varicap diode). Its nominal frequency of oscillation is set by inductor L1. The inductor value 390 nH provides an oscillation frequency of about 100 MHz. For best performance, L1 should be a high-Q component. L1 may consist of 4 turns of silver-plated wire wound around a 10 mm drill bit, and stretched to a length of about 1.5 cm. The wire diameter can be anything between 26 SWG (0.5 mm) and 20 SWG (1 mm). No core is used. The MAX4467 is a micropower opamp for low voltage operation and providing 200 kHz gain bandwidth at a supply current of just 24 µA. When used with an electret microphone, some form of DC bias for the microphone capsule is necessary. The MAX4467 has the ability to turn off the bias to the microphone when the device is in shutdown mode.

This can save several hundred microamps of supply current, which can be significant in low power applications particularly for battery powered applications like cordless microphones. The MICBias pin provides a switched version of Vcc to the bias components. Resistor R1 resistor limits the current to the microphone element. The output impedance of the MAX4467 is low and well suited to driving cables over distances up to 50 m. The MAX2606 intermediate-frequency (IF) voltage-controlled oscillators (VCO) has been designed specifically for portable wireless communication systems. The IC comes in a tiny 6-pin SOT23 package. The low-noise VCO features an on-chip varactor and feedback capacitors that eliminate the need for external tuning elements. Only an external inductor (here, L1) is required to set the oscillation frequency and produce a properly operating VCO. To minimize the effects of parasitic elements, which degrade circuit performance, place L1 and C5 close to their respective pins. Specifically, place C5 directly across pins 2 (GND) and 3 (TUNE). Potentiometer P2 then lets you select a free channel by tuning over the FM band of 88 MHz to 108 MHz. Output power is about –21dBm (approx. 10 µW) into 50 Ω. P1 serves as a volume control by modulating the RF frequency. Signals above 60 mV introduce distortion, so the pot attenuates from that level. To decrease stray capacitance, minimize trace lengths

57

by placing external components close to IC1's pins. Using a wire antenna of about 75 cm the transmitter should have a range of about 35 m. Try to keep all leads as short as possible to prevent stray capacitance. The transmitter operates on a single supply voltage in the range 4.5 V to 5.5 V from

any standard battery source. The transmitter must be housed in a metal case, with shielding installed between the two stages (AF and RF). Try to keep all leads as short as possible to prevent stray capacitance.

D. Prabakaran   (030068-1)

036
Battery Switch with LDO Regulator

In the form of the LT1579 Linear Technology (www.lineartech.com) has produced a practical battery switch with an integrated low-dropout regulator. In contrast to previous devices no diodes are required.

The circuit is available in a 3.3 V version (LT1579-CS8-3.3) and in a 5 V version (LT1579CS8-5), both in SO8 SMD packages. There is also an adjustable version and versions in an SO16 package which offer a greater range of control and drive signals.

The main battery, whose terminal voltage must be at least 0.4 V higher than the desired output voltage, is connected to pin IN1. The backup battery is connected to pin IN2. The regulated output OUT can deliver a current of up to 300 mA. The LDO regulator part of the IC includes a pass transistor for the main input voltage IN1 and another for the backup battery on IN2.

The IC will switch over to the backup battery when it detects that the pass transistor for the main voltage input is in danger of no longer being able to maintain the required output voltage. The device then smoothly

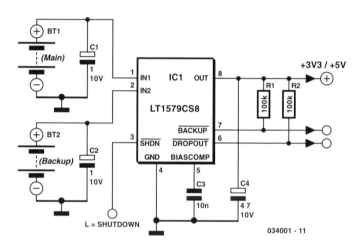

switches over to the backup battery. The open-drain status output BACKUP goes low to indicate when this has occurred. When neither battery is able to maintain the output voltage at the desired level the open-drain output DROPOUT goes low.

The LT1579 can operate with input voltages of up to +20 V from the batteries. The regulator output OUT is short-circuit proof. The shutdown input switches off the output; if this feature is not required, the input can simply be left open.

G. Kleine   (034001)

037
Resistance Calculator

This is a project that does not require any soldering. It is a resistance calculator made from cardboard, which can be used to determine the value of a resistor from the coloured bands and vice versa. Building the calculator couldn't be simpler. First scan and print this page (or cut it out), glue everything to a reasonably light piece of cardboard and then cut out the individual discs and the main body. Cut out the small windows in the body piece and then fold it along the heavy line, so that the markings are to the outside. Secure the discs in the body using paper fasteners. The locations of the fasteners are marked on the body by dots, and the crosshatched areas must be cut out. The first two numbers must be multiplied by the third number. The fourth ring gives the tolerance of the resistance in percent. The first two discs have coloured segments with the following colours (in the order listed): black, brown, red, orange, yellow, green, blue, violet, grey and white. Disc 3 has the following colours: black, brown, red, orange, yellow, green, blue, violet, silver and gold. The fourth disc has green, brown, red, gold, and silver. To finish everything off, fasten the edges using

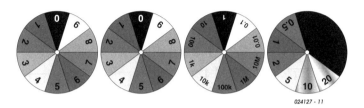

024127 - 11

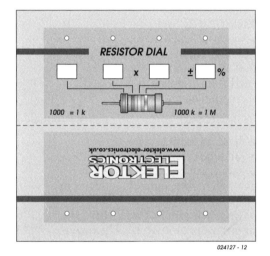

024127 - 12

024127 - 13

Sellotape, and your calculator is ready for use.

A. Rossius   (024127-1)

038
Isolated 1-Hz ClockM. Myo

One of the author's physics projects required an accurate 1-Hz (seconds) clock signal. Unfortunately, precision 10 MHz quartz crystals are expensive, while another problem was found in the inability of most common or garden 40xx CMOS logic chips to work at such a high frequency. However, a typical CMOS counter like the 4017 has such a high input resistance that its clock input has 'radio' properties. The effect is exploited here to convert the stray magnetic field picked up from a mains transformer into a clock signal. Here, the signal is induced in a short piece of wire (approx. 5 cm) connected to the clock input of a CD4017 decade counter for division by 10. The resulting 5-Hz signal is then divided by 5 by a second 4017 (IC2) to give an output of 1 Hz. LED D1 flashes to indicate the presence of a sufficiently strong magnetic field.

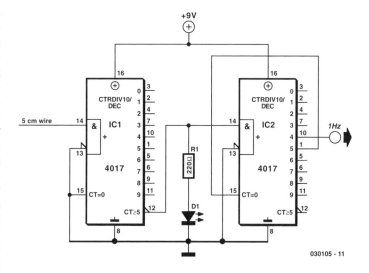

030105 - 11

The pickup wire should be placed close to the mains transformer, without compromising electrical safety. Always use the greatest distance at which a clock signal is reliably generated.

For 1-Hz output from 60-Hz power systems, use output 6 of IC2 (pin 5).

S. Büching   (030105)

039
Multisound for Guitars

Electric guitars use coils (guitarists call them pickups or elements) to convert the vibrations of the strings into an electrical signal. Usually, a guitar has more than one element built in, so that the musician can select with a switch which element or elements are used to generate the signal. Because of the differences in construction of

original

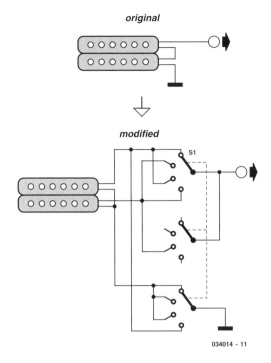

modified

034014 - 11

ements that contain one core and coil for each string. Humbuckers can be regarded as two elements that are connected in series. Many humbuckers have four connections (actually two single-coils with two connections each). These two individual coils are usually interconnected with fixed wiring so that they are always used in series. The circuit proposed here offers the possibility of using a hum-bucker with four connections in no less than four different modes, each of which having its own sound. The only things that have to be changed on the guitar are the wiring and the addition of a four-position switch. The latter requires drilling holes in the guitar of course, but if there is a control cover plate (along the lines of a Fender Stratocaster, for example) then it makes sense to put the switch there. This avoids the need for drilling holes in the wood while keeping an (expensive) guitar reasonably unmarred. The schematic shows what the various things look like, electrically speaking, before and after the multisound modification.

the elements and the varying positions of where they are mounted, each element sounds different. The elements can be roughly divided into two categories. There are the so-called 'single-coils' and 'humbuckers'. Single coil elements are el-

P. Goossens (034014)

040
Flip-Flop Using CMOS NAND Gates

Using just two NAND or inverter gates its possible to build a D type (or 'toggle') flip-flop with a pushbutton input. At power-up the output of gate N2 is at a logical '1', ensuring that transistor T2 is switched off. When the pushbutton is pressed the output of N2 changes to a logical '0' and transistor T2 conducts. The coupling between N1 and N2 ensures that the output of N1 will always be the inverse of N2 so T1 and T2 will always be in opposite states and will flip each time the pushbutton is pressed. In some cases it is possible to omit T1 and T2 and use the outputs of N2 and N1 to drive external circuitry directly but only if the loading on these outputs is low enough. The 4000 series CMOS family can source/sink a maximum of 0.5 mA (at 5 V) so for the sake of safety its best to use these open-collector configured transistor buffers. This circuit is particularly useful if

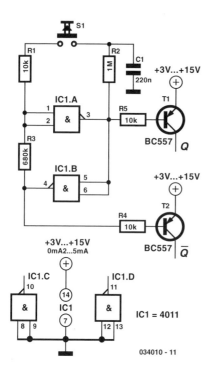

034010 - 11

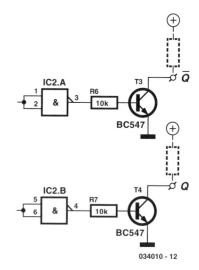

034010 - 12

you have some unused gates left over in a circuit design, avoiding the need to add a dedicated flip-flop IC. It is worth remem-

bering that all other unused CMOS gates must have their inputs connected to either the positive or ground rail. The supply voltage can be in the range 3 V to 15 V for CMOS ICs and the current taken by this circuit is between 0.2 mA and 5 mA (no load).

L. Libertin   (034010-1)

041
Optical Serial Interface

It is sometimes necessary to install microcorollers quite far away from data processing systems, such as in temperature measurement systems, security systems and other situations.

An interference-free connection between the two parts of the system is thus an absolute 'must'. This simple optical interface is excellent for this sort of application. The system consists of two parts: a transceiver (Figure 1) using a MAX323, which is connected to the serial bus of a PC, and an even simpler transceiver (Figure 2) at the microcontroller end. Both circuits can eas-

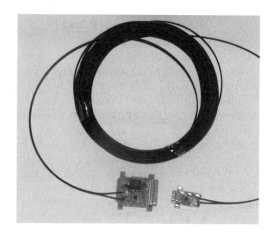

1

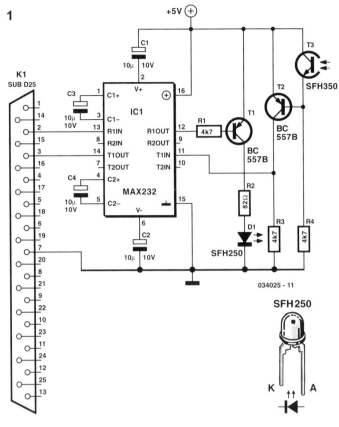

034025 - 11

SFH 250

2

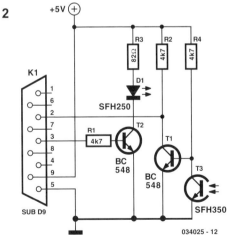

034025 - 12

The author had no difficulty bridging distances of up to 25 metres using this optical interface. It will probably be relatively easy to power the circuit shown in Figure 2 from the microcontroller power supply. In many cases, the required 5 V voltage will already be present on pin 9 of the 9-way Sub-D connector.

For the circuit in Figure 1, it may be possible to take the supply voltage from the PC with a bit of fiddling. If this proves not to be possible, a simple external supply with a 5 V regulator can be used. Now for a few practical remarks. First, almost all types of optical transmitters and receivers can be used in place of the SFH250 and SFH350 devices shown here, including the popular TOTX173 and TORX173. A 9-way Sub-D connector can naturally also be used instead of a 25-way connector for connecting the circuit of Figure 1 to the PC, but in this case, it will not be possible to fit the circuit into the connector shell.

If you want to build the circuit so it can be connected directly to the PC, as with the author's prototype, you will have to use a female connector. Note that the wiring of the Rx and Tx leads (pins 2 and 3) is not always the same; there may be differences between 9-way and 25-way connectors and between male and female connectors. The only solution is to measure every-thing using a multimeter.

The pin bearing a voltage of –10 to –12 V in the quiescent state is the Tx-connection of the PC. The same problem also arises with interconnecting cables. There are extension cables with a male connector at one

ily be built so small that they can be fitted into the shell of a Sub-D25 or Sub-D9 connector, respectively.

The link between the two optical modules is provided by two standard optical cables.

end and a female connector at the other end, which are connected 1:1, but there are also null-modem cables with two female connectors, which usually (!) have pins 2 and 3 cross-connected. Here again, the answer is to check everything with a meter.

R. Schüler

(034025-1)

042
US-Style Siren

The circuit described here can create three different 'US-style' siren sounds: police, ambulance and fire engine. The desired sound can be selected using switch S1. The circuit can be used in toys (such as model vehicles), as part of an alarm system, and in many other applications. For use in a toy, a BC337 is an adequate device for driver T5, since it is capable of directly driving a 200 mW (8 Ω) loud-speaker. In this case the current consumption from a 9 V power supply is around 140 mA. If a louder sound is required, a BD136 is recommended: this can drive a 5 W (8 Ω) loudspeaker. The current consumption from a 12 V supply will then be about 180 mA. If still more volume is desired, then T5 (a BD136) can be used as a first driver stage, and a 15 W (8 Ω) consumption of the circuit will now be about 500 mA with a loudspeaker can be connected via output transistor T6. Here 12 V power supply. Capacitor C1 is not required for battery an AD162 or an

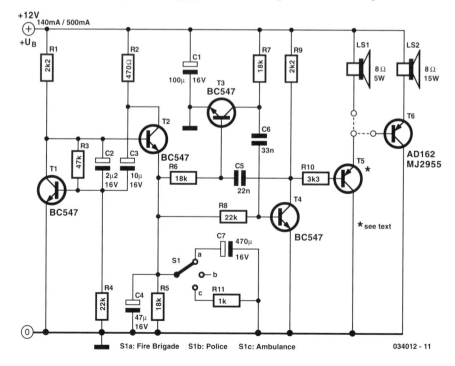

MJ2955 can be used, which, for continuous operation, must be provided with cooling. The peak current consumption of the circuit will now be about 500 mA with

a 12 V power supply. Capacitor C1 is not required for battery operation.

L. Libertin   (034012-1)

043
Low-cost Inductance Meter

This inductance meter consists, all in all, of four components that can be mounted on a small piece of prototype board in a jiffy. In order to use this meter the following are also necessary: a power supply, a pulse generator and an oscilloscope. So this is not really a complete measuring instrument, but more of a handy tool for the hobby laboratory.

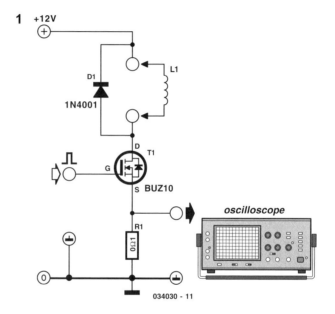

034030 - 11

Inductors

An inductor does with current what a capacitor does with voltage, namely: store and release. A constant voltage across an ideal inductor causes a linearly increasing current. Or mathematically: $dI/dt = U/L$. This formula may look a little strange, but it is actually very simple. dI/dt indicates how quickly the current changes in one second. U is the voltage across the inductor and L is the inductance (value) of the coil. From the above equation we can deduce that the larger the value of the coil, the slower will be the rate of increase of current.

The Circuit

If we apply a constant voltage for a short period of time across the inductor, we can determine the value of the coil by measuring how quickly the current increases. In our circuit (Figure 1), an external pulse

generator provides a pulse, which drives the gate of FET T1. This causes the FET to conduct, which in turn causes the entire power supply voltage to be applied across the inductor.

We make the assumption here that there is no current flowing in the inductor the moment the FET is switched on. The voltage across this (unknown) coil results in an increasing current through it. This current flows via the FET through resistor R1. The current will cause a voltage drop across the resistor. If you examine this voltage with an oscilloscope, you will see that it increases linearly at first. The rate at which this occurs is a measure of the value of the inductor. The instant the FET blocks, the inductor will continue to supply the current. That is because a negative voltage is required across the inductor for the current to fall. This is why a diode is connected in parallel with the inductor. The inductor will cause current to flow through the diode once the FET switches off. Because there is now a negative voltage across the inductor, the current in the coil will decrease. Without the diode, the coil would generate a very high voltage in an attempt to maintain constant current flow.

The ignition coil in a car makes use of this property to generate a spark between the spark plug electrodes. But with electronics, sparks, smoke and fire are usually an indicator that something is wrong, hence the diode.

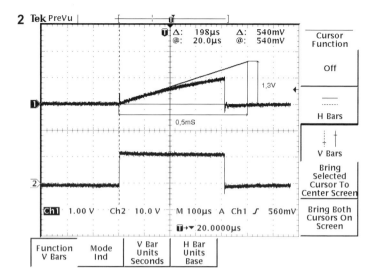

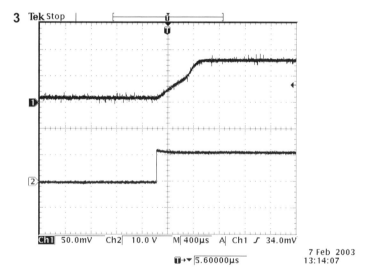

Calculating

If everything is correct, there will now be a nice picture on the oscilloscope. From this we should be able to calculate the value of the unknown coils, since that's what we were after in the first place! The necessary calculations are fortunately very straightforward. As mentioned before, $dI/dt = U/L$. We also remember Ohm's Law, namely $U_R = I_R \times R$. Since the current through the resistor is equal to the current through the coil we may replace dI_L/dt with $0.1 \times dU_R/dt$. Also, the voltage across the coil (U_l) is known as 12 V. Substituting this in the first formula leads to:

$0.1 \times dU_R/dt = 12/L$. Finally, rearranging the terms gives: $L = 120/(dU_R/dt)$.

Example

An oscilloscope trace of a measurement of an unknown inductor is shown in Figure 2. The sloping line has been drawn in order to determine the increase of voltage per time unit. This line intersects the measurement at exactly the beginning of the measurement, since the rate of change is most accurate here. As the picture shows, the voltage initially rises at a rate of 1.3 V/0.5 ms. This corresponds with a rate of change $dU_R/dt = 2600$ V/s. The inductance is therefore: $L = 120/2600 \approx 46$ mH. In the example measurement you can see that the current does not increase linearly. This can be explained as follows: as soon as the current starts to increase, the voltage drop across T1 and R1 will also increase, reducing the effective voltage across L1. The rate of change of current is directly proportional to the voltage across the inductor. If the voltage across the inductor drops, then the rate of increase of current will also be smaller. A

second effect that can occur is when the graph curves upward. This can be clearly seen in the next measurement (refer Figure 3). This effect only occurs with inductors that have a core. Once the current increases to the point where the magnetic field causes the core to saturate, the core does not contribute to the magnetic field any further. From this point onwards, the inductance reduces with the consequence that the current will increase faster.

P. Goossens   (034030-1)

044
RIAA Phono Preamplifier

Since modern sound systems usually lack inputs for record players, separate MD preamplifiers are becoming increasingly popular. They are needed not only by people who still regularly like to listen to vinyl records, but also by those who want to finally transcribe their LP collections to CD using a CD recorder. The author has built innumerable phono preamplifiers for friends and acquaintances. In many cases, for the sake of simplicity these were based on an old circuit design with two µA741s, which was originally described by B. Wolfenden in a 1976 issue of Wireless World. In that simple design, the first 741

simply amplified the full range of the frequency spectrum, while the second one was fitted with RIAA frequency compensation – a fairly common configuration at that time. However, a variant on this classic design was recently born after a bit of experimenting. It also uses two opamps, with the difference that the RIAA frequency compensation is distributed over both opamps. The accompanying figure shows the schematic diagram of this preamplifier. The first opamp attenuates the signal at 6 dB/octave starting at 2.2 kHz, while the second opamp looks after the other corner frequency. The objective of the new design was to keep the feedback factor as high as possible in both stages. To the considerable surprise of the developer, this modification turned out to have an unexpected side effect: when records were played, certain scratches were no longer audible! The difference between the new and old preamplifiers could be

034023 - 11

clearly heard; it was certainly not just imagination. What could be the cause of this?

A quick calculation showed that a 0.05 mm scratch in a record groove moving past a needle at a speed of 0.5 m/s produces a square-wave pulse with a frequency of 10 kHz. Evidently, there is a lot to be gained by attenuating such pulses with a low-pass filter as early as possible, which means in the first stage, in order to prevent them from overdriving the rest of the circuit.

H. Riegstra   (034023-1)

045
78xx Voltage Regulators

The voltage regulators from the 78xx-series are found in many analogue power supplies. It seems, then, somewhat superfiuous to say much more about them. But on the other hand, it can be very useful to highlight the important points, just because they are so ubiquitous. The 78xx is almost always used 'bare', because additional components are almost unnecessary. In fact, only one additional component is required, and that is capacitor C2. Based on the manufacturer's recommendation, this capacitor should be 220 nF in order to prevent oscillatory behaviour. In practice, you will almost always see that a 100 nF capacitor used here. This is a value that does not cause problems. C1 is the smoothing (reservoir) capacitor, its purpose is to reduce the ripple of the rectified AC voltage and is not actually related to voltage regulation. If the DC voltage is provided by a mains adapter, then this electrolytic capacitor is usually already part of the adapter, although the value is rather small sometimes. C2 may only be omitted if C1 is close to the 78xx and C1 is of good quality (low ESR). But there is nothing wrong by playing it safe and always fitting C2. Rule of thumb: always place a 100 nF capacitor on the input as close to the regulator as is practicable. Strictly speaking, there is no need for a capacitor on the output. However, a capacitor of at least 100 nF (C3) ensures much-improved regulation with fast (several microseconds) changes in load current. In practice, a decoupling-capacitor is placed close to the power supply pins of many ICs. These can provide the same function, provided they are placed not too far away. An electrolytic capacitor (C4) can be added for similar reasons: to catch slow (and fast, if it is a good capacitor) variations in load current. There is actually no compelling reason to deal with slow variations, because the IC is fast enough of regulating these on its own.

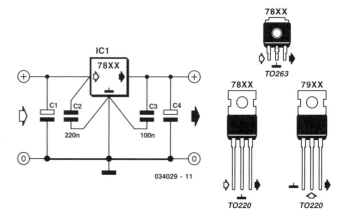

Rule of thumb: always place an output capacitor of at least 100 nF preferably as close as possible to the IC with the greatest current consumption (read: greatest changes in current consumption).

When building the circuit, it is important to connect the capacitors via the shortest possible path. So don't use long wires or make large loops. Obtain the input voltage for the regulator directly from the connections of the smoothing capacitor, because the ripple is smallest there. Finally, a few remarks about the temperature that a 78xx may run at. As a first approximation, if seem to burn your fingers when touching the regulator, the temperature of the regulator is above 60°C and a (small) heatsink is definitely recommended. It is not really a

problem when the IC gets too hot, because it was designed in such a way that it will turn itself off when the temperature is too high. It doesn't actually turn off, but the output current reduces when the temperature increases. When the internal temperature has reached 150°C, the output current will be only a little more than half the current delivered at 25°C. That is why it is possible that the out-put voltage of the regulator has dropped even though the out-put current is less than the rated current for the IC. A heatsink is the obvious solution. Rule of thumb: you should be able to touch the regulator (or the heatsink) without burning yourself. If not, then the heatsink has to be larger.

K. Walraven   (034029-1)

046
Flight Simulator Extensions

The flight simulators from Microsoft have improved much over the years. A great deal of attention has been paid to realistic flight behaviour and the graphical representation of the aircraft and its surroundings. Planning of the flight, flight lessons, weather circumstances and even radio contact are currently part of the simulation. The only thing that is missing is a cockpit with all the bells and whistles, so that we won't have to use the keyboard, and display the relevant information (height, speed, etc.) on a real instrument panel surrounding us. There are companies that manufacture these kind of cockpits or parts of cockpits, but do-it-yourself construc-

tion is much more fun of course… It is quite a job to build a beautiful cockpit, but this is not usually the issue that hobbyists have difficulties with. The big problem is

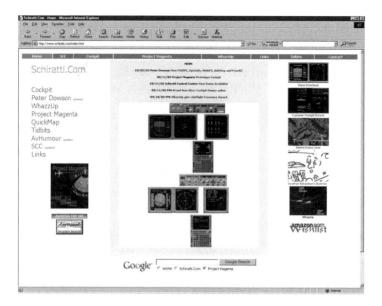

the communication between the flight simulator and your own cockpit. The control of the stick (via a joystick) is probably not going to be too much of a problem, but how do you deal with such things as tuning the radio to a different channel or adjusting the landing flaps to 10%? And another big question is how to obtain the relevant information for the instrument panel. Fortunately, the flight simulator can be ex-

panded with your own functions through DLLs. On the Internet you can find legions of people who enthusiastically occupy themselves with these problems. They often will make their software available to others at no charge or for a very nominal sum. A summary of the available software would never be complete. It is therefore much better if you take a look on the Internet for yourself. A good place to start is: www.schiratti.com/index.html. A diverse number of things can be found here, from DLLs to write your own communications programs, to complete programs that perform a particular function. For information, examples and help when building your own cockpit, http://www.simpits.org/ is a good starting point.

(034034-1)

047
DirectX for Delphi (DelphiX)

It is hard to believe that it's only been a few short years since the game 'Wolfenstein' made its debut and caused a revolution in the PC game world. This was because the game gave a realistic, for the time, impression of the 3D-world, something that used to be the realm of very expensive CAD software only.

Many other games have subsequently been released that also present a 3D-world through which the player can move freely

about (only obstructed by various monsters, soldiers, aliens, etc. who are all out to get you). With the rise of Windows, Microsoft has recognised the importance of migrating these kinds of games to the Windows platform. It was time for a new API, so that the operating system became responsible for the display of the 3D-world, because it was not desirable that programs accessed the hardware directly under Windows. This new API has been

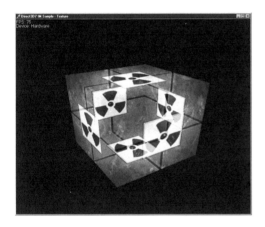

ful for 'normal' programs. It is possible, for example, to use the network interface for other purposes. Plus, it looks very professional if one or more 3D-objects enhance the appearance of your program. But above all: it is fun and educational to use DirectX. The Japanese programmer Hiroyuki Hori has apparently made an extensive study of the DirectX-material, with the result a package of Delphi-components that make programming of DirectX-applications easier. The installation of these components is done with a setup-program. After installation, an additional tab with the name 'DelphiX' has been added to the components (Figure 1). It is a good idea to study the many examples that are included to see what can be done with all these new components. The examples vary from simple applications (that make it easier to examine the various functions of DelphiX) to very beautiful graphical effects. To illustrate these, Figure 2 shows the screen dumps of two of the examples: Samples/d3drm/Alph-aBlending and Samples/d3dim/Texture.

called DirectX, a complicated piece of interface technology. Later on, the API was further expanded with other functions for the PC game industry. These days there are DirectX interfaces for the joystick (including force-feedback), keyboard and other input devices as well a network interface optimised for game play, sound card interface and much more. Although this API is primarily intended for games, it is also use-

The file can be downloaded from www.yks.ne.jp/~hori/ DelphiX-e.html. We suggest that you download the three tools on this page as well. These are very handy when converting graphics images or audio to a file format that is suitable for DirectX.

P. Goossens   (034031)

048
AT90S2313 Programmer

In January 2003, we published an article about programming AVR microcontrollers using Bascom AVR. In that article and in the manual for this Basic compiler, a circuit from Sample Electronics is described as one of the simplest ways to program the microcontroller. This circuit provided the basis for the practical implementation described here. For this purpose, the parallel port of the PC is directly connected to the 'Serial Interface for In-system Program-

ming' (SPI) via resistors that protect the parallel port. The serial link to the internal Flash memory is enabled when the Reset signal (pin 1) is low. The three connections for the SPI interface are SCK, MISO (output) and MOSI (input), which are located on pins 19, 18 and 17 of the microcontroller, respectively. The 100-pF capacitor (C1) is intended to avoid potential timing problems. It can be connected between the SCK line and ground using

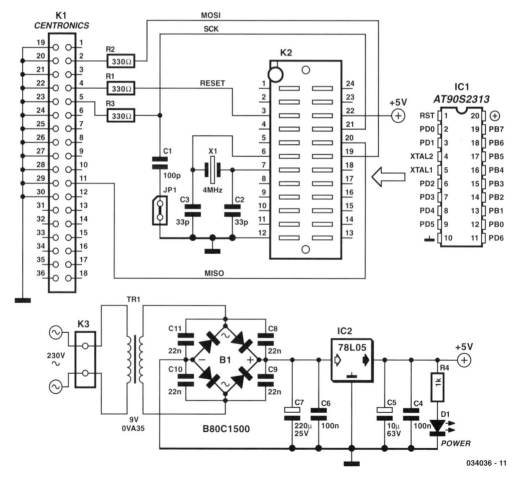

034036 - 11

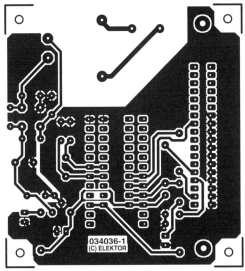

jumper JP1. However, during our tests this proved not to be necessary. To make building and connecting this programmer as easy as possible, a Centronics connector is located on the associated printed circuit board to allow a standard parallel printer cable to be used to connect the programmer to a PC. If you want to program multiple microcontrollers, it is highly recommended to use a zero insertion force (ZIF) socket. If you want to build a less expensive version, you can also use a normal IC socket.

The design of the printed circuit board allows only one socket to be fitted. The circuit has a small 5 V supply, consisting of a tiny 0.35 VA transformer, a bridge rectifier and a 78L05 voltage regulator (IC2). Capacitors C4, C6 and C8–C11 suppress any interference that may be present.

COMPONENTS LIST

Resistors:
R1,R2,R3 = 330Ω
R4 = 1kΩ

Capacitors:
C1 = 100pF
C2,C3 = 33pF
C4,C6 = 100nF ceramic
C5 = 10µF 63V radial
C7 = 220µF 25V radial
C8-C11 = 22 nF ceramic

Semiconductors:
B1 = B80C1500, round case (80V piv, 1.5A)
D1 = LED

IC1 = AT90S2313
IC2 = 78L05

Miscellaneous:
JP1 = 2-way pinheader with jumper
K1 = Centronics socket (female), PCB mount, angled pins
K2 = 24-pin ZIF-socket (e.g., Farnell # 178-235) or 20-pin IC socket with turned pins

K3 = 2-way PCB terminal block, lead pitch 7.5 mm
X1 = 4MHz quartz crystal
TR1 = mains transformer 9 V/0.35 VA (e.g., Hahn BV201 0136)
PCB, order code 034036-1

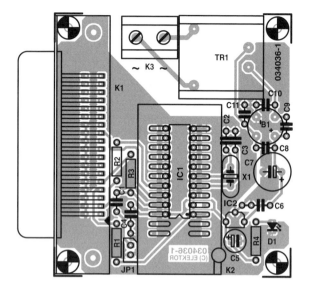

the PC is switched off. This is in any case always a good idea when connecting equipment to the parallel port of a PC, in order to avoid damage to the hardware. After the programmer is connected, switch on the PC and the programmer and start Bas-com AVR. The program should select the Sample Electronics programmer by default. If this doesn't happen, look under Options/Programmer to verify that the address of the parallel port (LPT address) is properly configured.

The power indicator LED D1 also acts as a minimum load for the voltage regulator, in order to prevent the voltage from rising excessively in the absence of a load, which is a problem with some types. The programmer should preferably be connected while

T. Giesberts   (034036-1)

(based on an idea from Sample Electronics)

049
Quad Car Amp

This quad final amplifier is actually intended to be used in a car, but it can naturally also be used for a variety of other medium-power applications. The TDA7375A can be successfully used in all situations in which a reasonable amount of audio power is desired and only a relatively low supply voltage is available. This IC is the successor to the TDA7374B, which forms the heart of the active loudspeaker

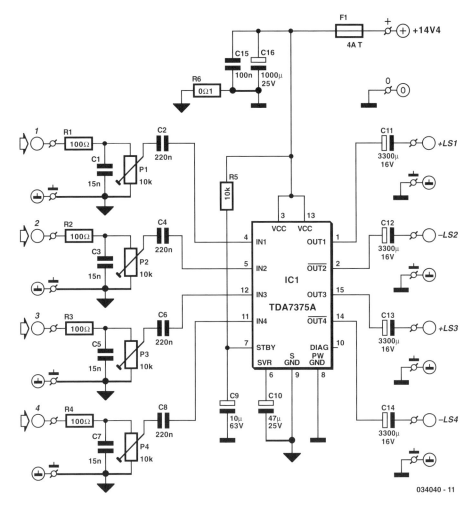

034040 - 11

system described earlier this year. Such a quad IC amplifier is naturally an excellent choice for this application, especially since the individual amplifiers can be connected in pairs in the bridge configuration, which allows them to provide approximately four times as much power. An example of such a bridge configuration is shown elsewhere in this issue.

The new IC can handle a peak voltage of 50 V (10 V more than the TDA7374B), but what is more important is that it is also truly intended to be used for single-ended operation. It includes all imaginable types of protection in order to avoid the premature demise of the four amplifiers, and in fact it is ideally suitable for a 'no-non-

sense' mini surround-sound system. For more information about the TDA7375A, we refer you to its data sheet, which can be found at www.st.com. The circuit shown here has four trim-pots for individually setting the output levels of the amplifiers. In addition, all inputs have RC networks (R1/C1, etc.) to block possible RF interference. The function of R6 is to separate the grounds of the input and output stages, in order to avoid possible ground loops that might arise with the use of multiple modules. A 5 W type is used for this resistor, in order to prevent it from going up in smoke if the ground connection of the power supply comes loose. C10 decouples the internal voltage divider, which biases the inter-

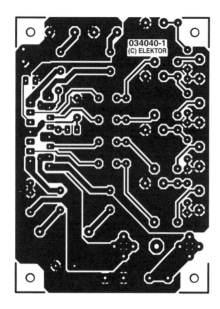

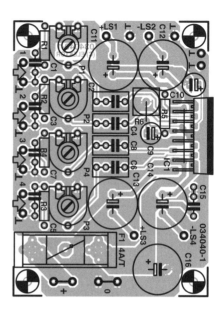

nal amplifier stages to half of the supply voltage. RC network R5/C9 provides a delayed, plop-free switch-on. C15 and C16 are local bypass capacitors for the supply voltage. The power supply ripple rejection of the TDA7375A is approximately 50 dB. If you want to use only a transformer, bridge rectifier and smoothing capacitor for the power supply, the minimum requirement is a transformer rated at 12 V / 30 VA in combination with a 10,000 µF electrolytic capacitor (remember that the maximum allowable supply voltage is 18 V). One of the few drawbacks of this quad amplifier is that two of the channels are inverted with respect to the other two.

For this reason, the polarity of each loudspeaker terminal is marked on the circuit board layout (e.g., +LS1 and –LS4) to indicate which terminal of the loudspeaker should be connected where. Radial electrolytic capacitors rated at 3300 µF / 16 V and having a diameter of only 12 mm are used for the output capacitors, which allows the circuit board to remain relatively compact. Our preferred type of electrolytic capacitor is a member of the Rubycon ZL series, which can handle no less than 3.4 A of ripple current. The maximum current consumption of the circuit with all four channels driven to the clipping level (with 4 Ω loads) is approximately 2.1 A.

COMPONENTS LIST

Resistors:
R1-R4 = 100Ω
R5 = 10kΩ
R6 = 0Ω1, 5W
P1-P4 = 10 k preset

Capacitors:
C1,C3,C5,C7 = 15nF
C2,C4,C6,C8 = 220nF
C9 = 10µF 63V radial
C10 = 47µF 25V radial
C11-C14 = 3300µF 16V 2 fast-on (spade) terminal, radial, max. diameter 13mm, e.g.,

Panasonic (Iripple 2500 mA) or Rubycon (Iripple 3400 mA)
C15 = 100nF
C16 = 1000µF 25V radial, max. diameter 13mm

Semiconductors:
IC1 = TDA7375A (ST)

Miscellaneous:
F1 = fuse, 4A/T (time lag), with PCB mount holder male, vertical, solder type (2-pin version)

The TDA7375A can also be used with 2 Ω loads. However, in this case the internal temperature rises considerably, since the Multiwatt 15V package has a rather large thermal impedance of 1.8 °C/W. In the interest of the service life of the IC, it is thus a good idea to use a somewhat larger heat sink. A 4 A/T fuse has been selected in consideration of possible 2 Ω operation. If you limit the load to 4 Ω, the fuse value can be reduced to 2 A/T. The output terminals of the amplifiers can be found on the circuit board next to the associated electrolytic capacitors. The related ground connections for LS1 and LS2 are located next to the LS1 and LS2 terminals, but the ground connections for LS3 and LS4 are located on the left, next to the IC, since this gives the best current paths on the circuit board and the least distortion. Vertical car connectors (spade terminals) are used for the power supply connections.

T. Giesberts   (034040-1)

050
Operating Life Extender for Incandescent Lamps

One way to extend the operating life of incandescent lamps is to reduce the voltage across the lamp by connecting something in series with it. Naturally, a series resistor is the simplest solution, but this produces unnecessary power dissipation. In order to reduce the voltage on a 100 W lamp by approximately 10 V, slightly more than 4 W must be dissipated in the resistor, and that is naturally a waste of energy. A better approach is to use a phase control circuit with a triac controller to reduce to voltage on the lamp. To reduce the voltage on the lamp by 10 V, a phase cut of approximately 44 degrees is necessary.

Three ways in which this can be done are presented here. The simplest version is a triac with a resistor in the gate circuit (Figure A). If the voltage on T2 is sufficiently great, the triac will start to conduct. With the indicated type of triac and a 33 kΩ resistor, the voltage is reduced by approximately 10 V. However, the drawback of this approach is that the gate sensitivity of most triacs is different in the various quadrants. For the type shown here, the gate sensitivity is typically 5 mA with T2 positively biased and a positive gate current, while with T2 negatively biased and a neg-

ative gate current, it is 11 mA! This results in a DC voltage across the lamp and a 50-Hz component that is just noticeable in the form of a slight flickering. With the tested circuit, the positive triggering threshold was approximately 30 degrees, while the negative triggering threshold was approximately 54 degrees. A better solution is to add a Zener diode to compensate for the difference in sensitivity. Naturally, this depends on the specific type of triac used. With a positive mains voltage, the di-

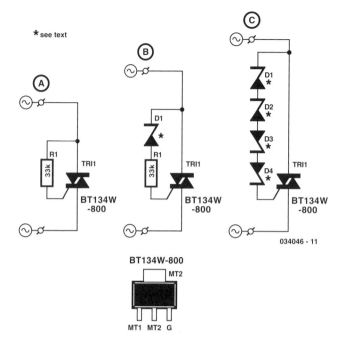

*see text

BT134W-800

034046 - 11

ode acts as a Zener, so a higher voltage is required to cause sufficient current to flow through the gate. When the mains voltage is negative, the Zener diode acts as a normal diode, so a specific trigger threshold is reached more quickly. There are various families of Zener diodes with rated voltages up to 100 V (such as the Vishay ZPY100) or even 200 V (the Vishay BZX85C series). If it is difficult to obtain a 100 V type, it is also possible to connect two 47 V Zeners in series.

The 50-Hz and DC components can be nicely reduced using this approach. However, there is still a possibility of asymmetric triggering. An example of this circuit is shown in Figure B. In a prototype, the voltage was decreased by around 12 V using a 75 V Zener and a 33 kΩ resistor. With the tested circuit, the positive triggering threshold was approximately 48 degrees, while the triggering threshold for negative voltages was approximately 54 degrees. If you want to obtain a voltage that is as symmetric as possible, you can connect two or four Zeners in reverse series. With this approach, triggering depends only on the tol-

erances of the Zener diodes. The circuit shown in Figure C illustrates what is meant. Naturally, this can only be used with an incandescent lamp. A DC component is not healthy for fluorescent lamps, and in the case of low-voltage halogen lamps, any DC current flowing through the DC resistance of the transformer could be rather large and even cause saturation of the transformer, with all of the associated nasty consequences.

Now for a few remarks about interference. Naturally, a switched-voltage circuit can generate interference signals. It may be necessary to use an interference-suppression choke, along with a snubber network to protect the triac against transients. Building the circuit is of course dead easy. If desired, you can design a small printed circuit board to fit in an existing lighting fixture or plug. However, the circuit can also be built in a separate (tiny) enclosure fitted 'in line' with the mains cable. Be sure to provide adequate isolation separation. Here we have used a type BT134W-800 V triac, which has an SOT223 SMD package and can handle voltages up to 800 V. In the SOT223 package, the tab is intended to be used for cooling. If it is soldered to the printed circuit board, the copper surface of the circuit board acts a heat sink. With this triac, the load must be limited to 100 W. Of course, you are free to experiment with other types of triacs, but remember that when the cable is plugged in, the entire circuit is connected to the mains, and the same applies to any measurement equipment you may be using!

T. Giesberts   (034046-1)

051
Quad Bridge Car Amp

If you want to obtain more power at a given supply voltage (such as the battery voltage of a car) and you do not want to resort to a switched-mode power supply or class-H amplifier, your only choice is a traditional bridge amplifier. Various types of integrated versions have been available for years already. New types also appear regularly, such as the ST TDA7375A, which has the advantage that it requires very few external components. This IC is a pin-compatible successor to the TDA7374A, which can also be used in the circuit shown here. Since many people nowadays are already used to surround-sound amplifiers, and it is common for more than two loudspeakers to be fitted in cars, four channels per module is just about the minimum. This is why we have fitted two ICs on a single circuit board for this design. The two ICs share common

Measurement results

Supply voltage	+14.4 V
Quiescent current	200 mA
Pmax (THD+N = 0.1 %)	4 x 17.9 W
	(1 kHz, 4 x 4 Ω)
Sensitivity (P= 17.9 W)	0.46 V
THD+N	
(B = 80 kHz, 1 kHz, 1 W/4 Ω)	<0.03 %
Bandwidth	24 Hz to 55 kHz

input and power grounds. These two grounds are connected by a 0.1 Ω resistor (R7), in order to limit the undesirable effects of ground loops when two ore more modules are used together. The inputs have RC networks to suppress RF interference. The four channels can be aligned to each other using the four trimpots. The values of

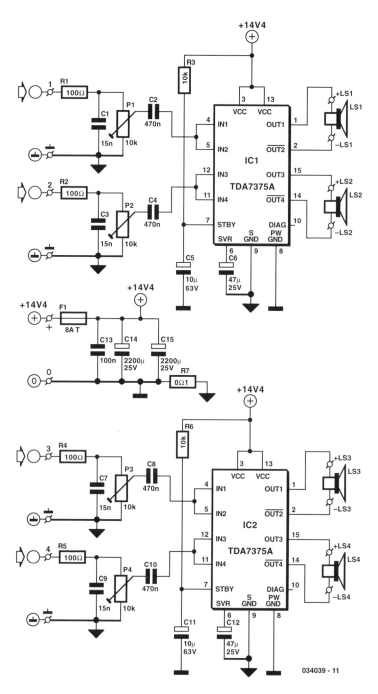

034039 - 11

C2, C4, C8, and C10 are twice as great as in a single-ended configuration (see the 'Quad Car Amp' article elsewhere in this issue). R3/C5 and R6/C11 ensure plop-free switch-on and switch-off. C6 and C12 de-couple the internal voltage dividers, which bias the output stages at half the supply voltage. Both ICs are also well decoupled using separate electrolytic capacitors (C4 and C15). RF decoupling capacitor C13 is

COMPONENTS LIST

Resistors:
R1,R2,R4,R5 = 100Ω
R3,R6 = 10kΩ
R7 = 0Ω1, 5W
P1-P4 = 10kΩ preset

Capacitors:
C1,C3,C7,C9 = 15nF
C2,C4,C8,C10 = 470nF
C5,C11 = 10µF 63V radial
C6,C12 = 47µF 25V radial

C13 = 100nF
C14,C15 = 2200µF 25V radial, max. diameter 16mm.

Semiconductors:
IC1,IC2 = TDA7375A (ST)

Miscellaneous:
F1 = fuse, 8A/T (time delay) with PCB mount holder 2 fast-on (spade) terminal, male, vertical, solder type (2-pin version)
PCB, order code 034039-1

shared by the two ICs and located between them. The printed circuit board has been designed to be as compact as possible, with the result that some of the connections are a bit cramped. The loudspeaker connections are located on both sides of the ICs. When connecting the loudspeakers, it is a good idea to cut the cables to the proper length and solder them directly to the circuit board. The power supply connections (which use car connectors) are located at the ends of the fuse holder ('0' and '+'). Don't forget the three wire links, and use insulated wire for the two links underneath the ICs to avoid shorts. The third wire link is right next to R7. If the amplifier is driven to its maximum output level using four sine-wave signals, it draws 7.7 A! This is why an 8 A fuse is used to protect the circuit. In addition, a heat sink with a maximum thermal resistance of 1 K/W is recommended to provide adequate cooling. When fitting the circuit board, be careful with the corner holes next to he ICs. The copper extends to the holes here, in or-

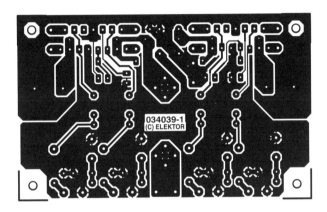

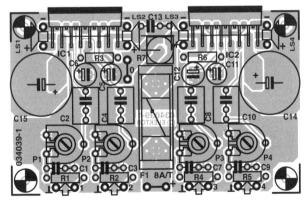

der to keep the circuit board as small as possible, so you must use plastic standoffs to fasten the board. Use insulating washers and rings to attach the ICs to the heat sink, in order to electrically isolate them from the heat sink.

T. Giesberts   (034039-1)

052
LCD Status Panel

PCs are currently used for just about anything. E-mail has become a standard communications tool and a network at home is no longer the exclusive domain of geeks. In many households, the stereo is connected to the PC in order to listen to MP3s, etc., etc. There is such a huge choice of software that everyone should be able to find something to their liking and be able to put it to good use.

The disadvantage of much software is that it is sometimes hard to see the windows for the screen (well... trees, forest...). Granted, it is very easy to check whether you have email, to look which track Winamp is playing or keep an eye on the current speed of your internet connection. But it often takes a little bit of hunting around to find this information and it is often obscured when another window is in focus. The end result is that displaying this information inevitably requires several mouse clicks. Not difficult really, but rather cumbersome. Here we present a solution to this problem: an LCD screen built into the PC enclosure that displays the desired information continuously. This display can easily be mounted on a spare blanking panel of the PC. The power requirements can conveniently be fulfilled by the PC's own power supply, thanks to the low current consumption. In this case, the plug for a

3.5"-drive has been used, of which there is usually a spare one available. To keep things cheap, we can use the parallel port for communication between the PC and the LCD. This connector is being used less and less anyway, due to the rise of USB and network printers. The schematic (Figure 1) is very simple, since the parallel port has sufficient inputs and outputs to drive an LCD directly.

Although the circuit requires very little current, it is much better of the power is not obtained from the parallel port directly. As mentioned before, we used a power connector that is normally reserved for 3.5"

1

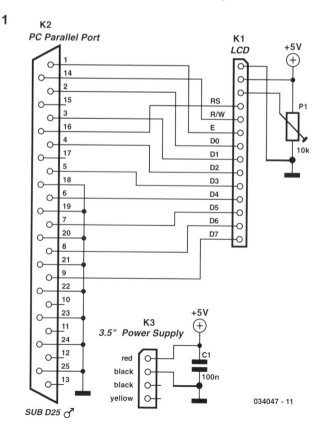

floppy drives (K3). That leaves potentiometer P1 and capacitor C1. P1 is used to adjust the contrast of the LCD. C1 helps provide a clean power supply without unnecessary interference. You can choose the type of LCD according to your own requirements, provided only that the display contains an HD44780 controller. The software supports the various standard dimensions (from 1 × 10 characters to 4 × 40 characters). The entire circuit is easily built on a piece of prototype board. In our prototype we used a 14-pin header to connect the LCD module.

The connection to the power supply is made with a simple 4-way pin header. Make absolutely sure you connect the power supply connector the correct way around, otherwise you run the chance that you use +12 V instead of +5 V, and the LCD is unlikely to survive that! If the LCD module has a backlight, you may like to connect that as well. It is best to consult the

datasheet for the LCD module to check how this is done. Some models require a series resistor whereas other implementations have this resistor built in. To mount the LCD in a PC, an unused blanking panel is eminently suitable. Once everything has been built, we need the software to drive the circuit. Several programs are available on the Internet, complete with all the relevant information. The nicest program we have come across so far is 'LCD Smartie 5.1'. The program can be downloaded from:

http://backupteam.gamepoint.net/smartie/

The capabilities of the program are so extensive that it is a hopeless task to detail them all. The text messages on the display are fully adjustable. The messages can include up-to-date information, such as current username, temperature of the processor, speed of the Internet connection, alerts when new emails arrive, etc. Fortunately, the program is very user friendly, so that everyone will be up-and-running quickly. Figure 2 shows the Smartie 5.1 emulation of an LCD on the screen, while Figure 3 shows the program's configuration screen.

P. Goossens   (034047-1)

2

3

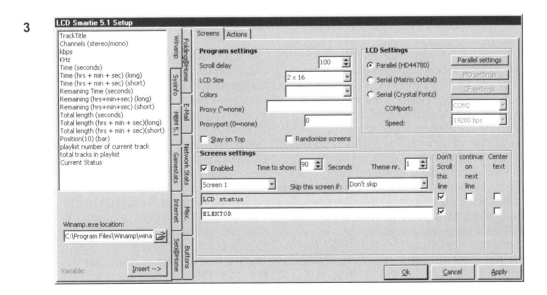

053
Hard Disk Selector

In the last few years, the available range of operating systems for PCs has increased dramatically. Various free (!) operating systems have been added to the list, such as BeOS, OpenBSD and Linux. These systems are also available in different colours and flavours (versions and distributions). Windows is also no longer simply Windows, because there are now several dif-

ferent versions (Windows 95, 98, ME, NT and XP). Computer users thus have a large variety of options with regard to the operating system to be used. One problem is that not all hardware works equally well under the various operating systems, and with regard to software, compatibility is far from being universal. In other words, it's difficult to make a good choice.

Switching from one operating system to another – that's a risky business, isn't it? Although this may be a bit of an exaggeration, the safest approach is still to install two different operating systems on the same PC, so you can always easily use the 'old' operating system if the new one fails to meet your needs (or suit your taste). A software solution is often used for such a 'dual system'. A program called a 'boot manager' can be used to allow the user to choose, during the start-up process, which hard disk will be used for starting up the computer. Unfortunately, this does not always work flawlessly, and in most cases this boot manager is replaced by the standard boot loader of the operating system when a new operating system is installed. In many cases, the only remedy is to rein-

stall the software. The solution presented here does not suffer from this problem. It is a hardware solution that causes the primary and secondary hard disk drives to 'swap places' when the computer is started up, if so desired. From the perspective of the computer (and the software running on the computer), it appears as though these two hard disks have actually changed places. This trick is made possible by a feature of the IDE specification called 'CableSelect'. Every IDE hard disk can be configured to use either Master/Slave or CableSelect. In the latter case, a signal on the IDE cable tells the hard disk whether it is to act as the master or slave device. For this reason, in every IDE cable one lead is interrupted between the connectors for the two disk drives, or the relevant pin is omitted from the connector. This causes a low level to be present on the CS pin of one of the drives and a high level to be present on the CS pin of the other one (at the far end of the cable). The circuit shown here is connected to the IDE bus of the motherboard via connector K1. Most of the signals are fed directly from K1 to the other connectors (K2 and K3). An IDE hard disk is connected to K2,

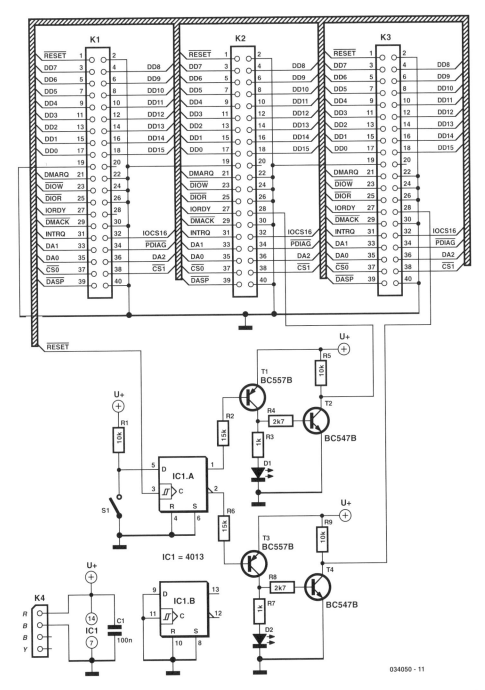

034050 - 11

and a second one is connected to K3. When the computer is switched on or reset, a pulse will appear on the RESET line of the IDE interface. This pulse clocks flip-flop IC1a, and depending on the state of switch S1, the Q output will go either high or low. The state on the Q out-through T1 will cause LED D1 to light up and transistor T2 to yellow conduct. The hard disk attached to connector K2 will thus see a low level on

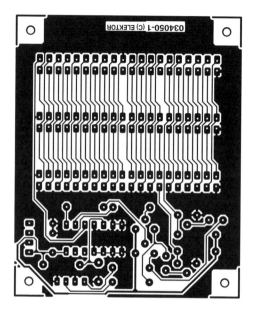

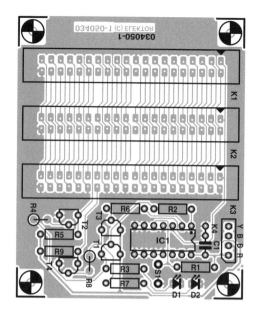

COMPONENTS LIST

Resistors:
R1 = 10kΩ
R2,R6 = 15kΩ
R3,R7 = 1kΩ
R4,R8 = 2kΩ7
R5,R9= 10kΩ

Capacitors:
C1 = 100 n

Semiconductors:
D1 = LED, low-current, yellow

D2 = LED, low-current, green
IC1 = 4013
T1,T3 = BC557B
T2,T4 = BC547B

Miscellaneous:
K1,K2,K3 = 40-way boxheader
K4 = 4-way SIL pinheader
S1 = switch, 1 make contact
PCB, order code 034050-1

its CS pin, which will cause it to act as the master drive and thus appear to the computer as the C: drive. A high level will appear on the Q output following the reset pulse. This will prevent T3 and T4 from conducting, with the consequence that LED D2 will be extinguished and the hard disk attached to connector K3 will see a high level on its CS pin. For put is naturally always the opposite of that on the Q output. If we assume that the switch is closed during start-up, a low level will be present on D input of IC1a, so the Q output will be low following the reset pulse. This low level on the Q output will cause transistor T1 to conduct. The current flowing this disk, this

indicates that it is to act as a slave drive (D: drive). If S1 is open when the reset pulse occurs, the above situation is of course reversed, and the hard disk attached to connector K2 will act as the D: drive, while the hard disk attached to connector K3 will act as the C: drive. Flip-flop IC1a is included here to prevent the hard disks from swapping roles during use. This could have disastrous consequences for the data on the hard disks, and it would most likely cause the computer to crash. This means that you do not have to worry about affecting the operation of the computer if you change the switch setting while the computer is running. The state of the flip-flop, and thus the configuration of the hard disks, can only be changed during a reset. The circuit is powered from a power connector for a 3.5-inch drive. This advantage of using this connector is that it easily fits onto a standard 4-way header. However, you must observe the correct polarity when attaching the connector. The red lead must be connected to pin 1. Constructing the hard disk selector is easy if the illustrated printed circuit board is used.

You will need three IDE cables to connect the circuit. The best idea is to use short cables with only two connectors, with all pins connected 1:1 (no interruption in the CS line). The IDE connector on the mother-board is connected to K1 using one cable. A cable then runs from K2 to first hard disk, and another cable runs from K3 to the second hard disk. This means that it is not possible to connect more than two hard disks to this circuit. You must also ensure that the jumpers of both disk drives are configured for CableSelect. To find out how to do this, refer to the user manual(s) for the drives.

P. Goossens (034050-1)

054
Longwire Match for SW Receivers

Most shortwave receivers for use 'in the shack' have a 50 Ω coaxial input (usually a SO239 socket) which is not directly suitable for the high impedance of a typical longwire antenna. The problem is usually overcome by inserting a balun (balanced to unbalanced) transformer whose primary purpose is to step down the antenna impedance from 'high' to to 50 Ω and not, as would be expected, to effect a change from balanced to unbalanced (note that a longwire is an unbalanced antenna). Unfortunately, such a balun may be difficult to obtain, make yourself, or both. The circuit shown here is a transistorised (i.e. inductor-free) equivalent of the wire balun. The grounded-collector configuration is used because a relatively high input impedance (the longwire antenna) has to be stepped down to 50 Ω (the receiver input impedance). Voltage amplifica-

tion is not required here. The two anti-parallel diodes at the antenna input prevent damage to the circuit as a result of static discharges or extremely strong signals. Like an active antenna, the circuit receives its supply voltage (in this case, 9 V) via the downlead coax cable. Current consumption will be of the order of 20 mA. The coax cable should be earthed at the receiver side. The length of the antenna wire will depend on local conditions and what you

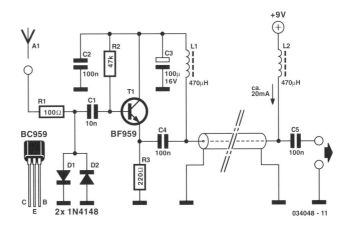

034048 - 11

hope to be able to receive. For most SW Broadcast service and amateur radio listening, a wire of about 3 m will be sufficient but bear in mind that the longwire antenna is prone to pick up electrical interference.

S. Delleman

(034048-1)

055
Keypad Extension

Bascom AVR application note no. 105 (www.mcselec.com/an_105.htm) describes a simple circuit for using an AT90S2313 (an 8-bit AVR microcontroller) to transmit RC5 or Sony remote control codes. In this circuit, a small 3 × 4 matrix is used to connect 12 pushbutton switches to only 7 pins of the microcontroller (see Figure 1). If you need more pushbuttons, the matrix will have to be extended, but quite a bit can be achieved using a simple auxiliary circuit. If

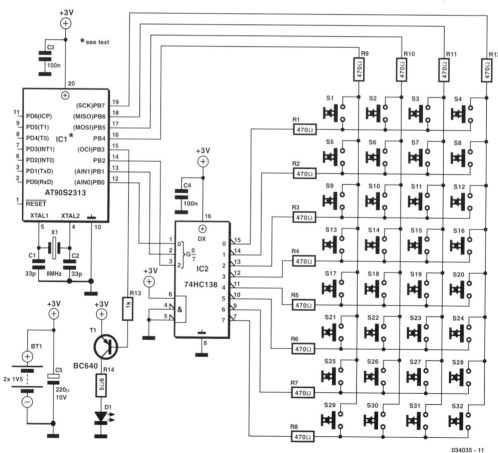

034035 - 11

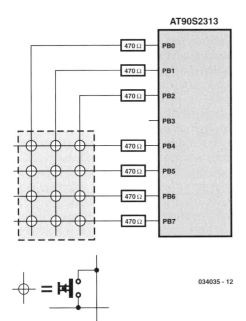

AT90S2313

034035 - 12

the three column bits are used to drive a de-multiplexer, up to 32 pushbuttons can be connected. Figure 2 shows and example of such an approach. Here a 74HC138 has been chosen for the de-multiplexer. This IC has inverted outputs. If active-high outputs are more convenient for software reasons, the pin-compatible 74HC238 can be used instead.

Naturally, the software will have to be modified, since it is unlikely that a standard statement or instruction is present in Bascom AVR for this auxiliary circuit. It may be necessary to write a new assembler routine. This is not a particularly difficult task, and moreover, it is a pleasant challenge for would-be users of the extended keypad.

T. Giesberts   (034035-1)

056
Cross-linking with Two Patch Cables

In networks, the supremacy of coax cable is a thing of the past. Nowadays, Ethernet connections are made using UTP cables. The BNC plug has yielded to the 8-way RJ45 plug. Previously, coax cables were daisy-chained from computer to computer and terminated at the two ends using 50 Ω resistors, but modern networks use central 'socket boxes' (switches and/or hubs) to interconnect everything. The connections between the hubs and the computers are made using patch cables having the same sequence of leads in the RJ45 connectors at each

end. For making a direct connection between two computers without using a hub or switch, a 'crossover cable' is used. Such

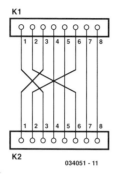

034051 - 11

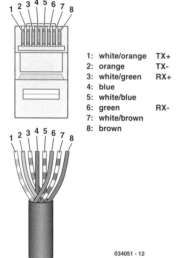

1: white/orange   TX+
2: orange   TX-
3: white/green   RX+
4: blue
5: white/blue
6: green   RX-
7: white/brown
8: brown

034051 - 12

a cable has the leads cross-linked in order to allow the two computers to directly communicate with each other. If there are problems with the network, it can be handy to be able to directly interconnect two computers, or directly connect a computer to a cable or ADSL modem without using a hub or switch. A long crossover cable is not always available, and shoving around computers is not an attractive alternative. Consequently, we can use a dual RJ45 wall outlet box to construct an adapter, which can be used to interconnect the two patch cables coming from the equipment in question. This outlet box must be wired to create a cross-linked connection. This is done by making the following internal connections:

$$1 \rightarrow 3$$
$$2 \rightarrow 6$$
$$3 \rightarrow 1$$
$$4 \rightarrow 4$$
$$5 \rightarrow 5$$
$$6 \rightarrow 2$$
$$7 \rightarrow 7$$
$$8 \rightarrow 8$$

L. Lemmens   (034051-1)

057
Lambda Probe Readout for Carburettor Tuning

A lambda probe (or oxygen sensor) can be found on the exhaust system of most cars running on unleaded fuel. Having reached its normal operating temperature (of about 600 degrees Celcius!) the lambda probe supplies an output voltage proportional to the amount of residual oxygen measured in the exhaust gas. This information is indicative of, among others, the air/fuel ratio supplied by the carburettor(s) and hence the combustion efficiency. In modern car (and

motorcycle) engines, this information is used to (electronically) adjust engine parameters like ignition timing and fuel injection. The indicator described here is intended for permanent installation on a motorcycle of which the air/fuel ratio needed to be watched, with the obvious aim engine power tuning after fitting a different set of carburettors. Apart from this obvious technical use the unit's bright LEDs will no doubt attract the attention of

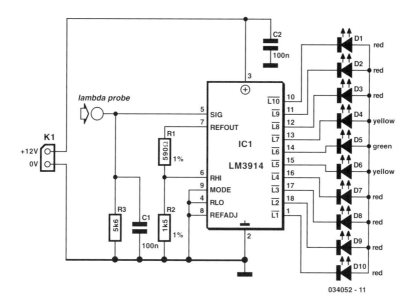

034052 - 11

curious motorcyclists. At the local junk-yard a single-wire lambda probe may be salvaged from a wrecked car. Once a suitable nut has been found, the probe can screwed into the exhaust pipe of the motorcycle, at about 30 cm from the cylinders. Since we're talking of welding and drilling in an expensive (chrome-plated) exhaust pipe, you may find that actually fitting the probe is best left to specialists! The starting point for the design of a suitable electronic indicator is that in the noble art of carburettor tuning an air/fuel ratio of 14.7 to 1 is generally considered 'perfect', the range covering 16.2 to 1 ('lean') to 11.7 to 1 ('rich'). The perfect ratio typically corre-

sponds to a probe output voltage of 0.45 V. Referring to the circuit diagram, that is the input level at which 5 of the 10 LEDs will light, including the green one, D5. If one of the red LEDs lights, the mixture is definitely too rich.

Note that in general it is better to have a mixture that is a little to rich than one that's on the lean side, hence a yellow LED lights between the green LED and the first red one. Also note that the engine needs to be at its normal operating temperature before a meaningful indication is obtained.

P. Goossens   (034052-1)

058
Egg Timer

This egg timer, which is both simple and functional, shows once again that it is not essential to use a microcontroller for everything these days. The circuit consists of only two ICs from the standard 4000 logic

family, a multi-position rotary switch and a few individual components. The combination of a 4040 oscillator/counter and a 4017 decimal counter is certainly not new, but it is an ideal combination for timers that are

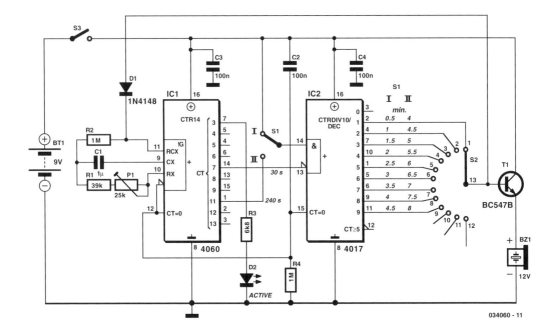

034060 - 11

required to generate long intervals that can be programmed in steps. The circuit can be directly powered from a 9 V battery, without using a voltage regulator. The signalling device is a 12 V buzzer, which generally works quite well even at a much lower voltage. We won't explain the operation of the two ICs here; if you would like to know more about this, we recommend consulting the device data sheets. The RC configuration has been selected for the oscillator circuit of the 4060, since the frequencies of standard crystals and resonators would be too high (even 32.768 Hz is much too high), making it impossible to achieve the desired times. With an RC oscillator, it's

also easier to modify the times to suit our purposes. For instance, if the oscillator frequency is reduced by a factor of two, we obtain a range of 1 to 16 minutes in steps of 1 minute. The range is split into two by taking advantage of the fact that the 4017 has an AND gate at its input (with an inverted input). The two ranges overlap by two steps. The oscillator has been dimensioned such that the 23 divider output (pin 14) has a period of 30 seconds, so IC2 receives a clock pulse every 30 seconds. This means that the oscillator frequency must be set to 8.5333 Hz. The first output of IC2 is active after a reset, so it cannot be used. If S1 is in position I, pin 14 of IC2 is connected to the

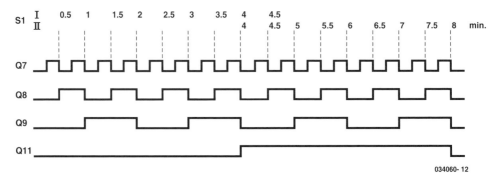

034060- 12

positive supply line. This input is used as an enable input. Directly after the first pulse from the 4060, the second output of IC2 goes high (which means after exactly half a minute).

The subsequent outputs become active in turn at intervals of one clock pulse, and thus generate the states for 1 to 4.5 minutes. In the second range (II) of S1, the 'enable' pin of IC2 is connected to the 212 divider output of the 4060 (pin 1). This output goes high 4 minutes after the reset (which is why it is labelled '240 s', instead of the period time of 480 s). Since the 4060 is an asynchronous counter, this output goes high a short time after the 23 output goes low. This delay provides the proper condition for an extra clock pulse for the 4017. The outputs of the 4017 will thus count upwards once. This means that the second output will become active after 4 minutes, with the rest of the outputs becoming active after 4.5 to 8 minutes. The desired timing interval is selected using switch S2. The output of S2 is connected directly to emitter follower T1, which energises the buzzer when the level on the wiper of the switch is high. At the same time, the counter of IC1 is disabled via diode D1 by forcing the oscillator input high. The buzzer thus remains active until the circuit is switched off. The first counter output of the 4060 is connected to an LED (D2), which indicates that the circuit is active and the battery not yet exhausted. The blinking rate is approximately 0.5 Hz. The current through the LED is set to a modest 1 mA, since this current represents the majority of the current drawn by the circuit. This ranges from 0.5 to 1.5 mA, with the average current consumption being approximately 1 mA while the timer is running.

The buzzer used in our prototype increases the current to around 13 mA when it is energised, but this naturally depends on the actual type used. In principle, the circuit will work with any supply voltage between 3 and 16 V. However, the actual supply voltage should be taken into account in selecting the buzzer. The value of the supply voltage also has a small effect on the time interval, but in practice, the deviation proved to be less than 5 percent – which is not likely to matter too much to the eggs.

T. Giesberts (034060-1)

059
12 V Glow Plug Converter

Most small internal-combustion engines commonly used in the model-building world use glow plugs for starting. Unfortunately, glow plugs have an operating voltage of 1.5 V, while fuel pumps, starter motors, chargers and the like generally run on 12 V. This means that a separate battery is always needed to power the glow plug. The standard solution is to use an additional 2 V lead storage battery, with a power diode in series to reduce the voltage by approximately 0.5 V.

However, this has the annoying consequence that more than 30 percent of the energy is dissipated in the diode. Naturally, this is far from being efficient. The converter presented here allows glow plugs to be powered from the 12 V storage battery that is usually used for fuelling, charging, starting and so on. A car battery can also be used as a power source. Furthermore, this circuit is considerably more efficient than the approach of using a 2 V battery with a series power diode. The heart of the

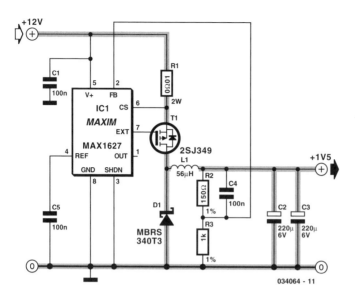

034064 - 11

DC/DC converter is IC1, a MAX 1627. The converter works according to the well-known step-down principle, using a coil and an electrolytic capacitor. Here the switching stage is not integrated into the IC, so we are free to select a FET according to the desired current level. In this case, we have selected a 2SJ349 (T1), but any other type of logic-level FET with a low value of RDSon would also be satisfactory. Of course, the FET must be able to handle the required high currents. Diode D1 is a fast Schottky diode, which must be rated to handle the charging currents for C2 and

C3. This diode must also be a fairly hefty type. The internal resistances of coil L1 and capacitors C2 and C3 must be as low as possible. This ensures efficient conversion and prevents the components from becoming too warm. The resistor network R2/R3 causes 87 percent of the output voltage to be applied to the FB pin of IC1. This means that an output voltage of 1.5 V will cause a voltage of approximately 1.3 V to be present at the FB pin. The IC always tries to drive the switching stage such that it 'sees' a voltage of 1.3 V on the FB input. If desired, a different output voltage can be provided by modifying the values of R2 and R3.

When assembling the circuit, ensure that C5 and C1 are placed as close as possible to IC1, and use sufficiently heavy wiring between the 12 V input and the 1–5 V output, since large currents flow in this part of the circuit. A glow plug can easily draw around 5 A, and the charging current flowing through the coil and into C2 and C3 is a lot higher than this!

P. Goossens   (034064-1)

060
Extended Simulation Package

Simulation programs are often used in developing electronic circuits. They allow a variety of versions to be quickly tested and enable design errors to be detected at an early stage, even before a prototype has been built. Depending on the particular simulation package, the results can be

viewed in the form of diagrams or even using virtual instruments. However, most simulation programs have trouble simulating microcontrollers. If a microcontroller (with its associated software) is present in the circuit, most simulation packages fail. A pleasant exception to this situation is

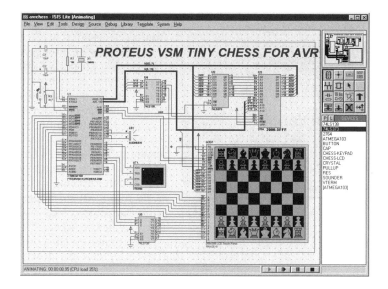

tested in the simulation package! GALs and PALs are also supported. Naturally, the program can also simulate 'normal' electronic circuits. The supplied examples are more than worth looking at. They range from simple, standard electronic circuits, such as a transistor amplifier and oscillators, to microcontroller applications, such as a clock and even a chess computer with a touch screen! One of the noteworthy features of the simulations is their execution speed. Most of the microcontroller simulations could be run at real time on a PC with an AMD Athlon 900 MHz processor! An evaluation version of the program can be downloaded from http://www.labcenter.co.uk.

provided by the ISIS simulation package from our advertiser Labcenter Electronics. This program can simulate a variety of microcontrollers in electronic circuits. In addition to the tools commonly found in simulation packages, it thus includes debuggers for several types of microcontrollers, including the 8051/AVR/HC11 family and PIC microcontrollers. This software package also includes several components typically found in microcontroller applications, such as 7-segment displays, a keypad and (graphic) LCDs. These components can be operated using the mouse during the simulation, allowing the user interface to also be

The restrictions of the evaluation version are that the circuits cannot be saved or printed. The 'schematic in schematic' option is also not available in the evaluation version. Otherwise, it is fully usable.

(034055-1)

061
Flight Simulator Display

Elsewhere in the Small Circuits Collection there is an article about connecting an LCD to a PC (see 'LCD Status Display' on page 82). There is also an article about creating your own options for the Microsoft Flight Simulator (see 'Flight Simulator Exten-

sions' on page 69). We couldn't resist the temptation to combine these two subjects. The result is an LC display that shows essential information while you are flying with the Microsoft Flight Simulator. The software for this article is very simple,

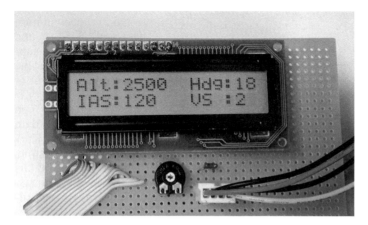

config.cfg belongs in the main folder of Smartie51. Once everything is properly installed, you're ready to have fun. First, start the flight simulator. After it is fully loaded, you have to start Smartie51. If everything is OK, the LCD will now continuously display the following information: Alt (altitude), Hdg (heading), IAS (indicated air speed) and VS (vertical speed). The necessary files, including the source code, can be downloaded from the Elektor Electronics website (http://www.elektor-elektronics.co.uk) under number 034056-11, see month of publication.

The design of the software is simple, so people with a bit of experience with Delphi can easily extend the program to display other flight simulator data. If you want to do this, it's a good idea to also download the FSUIPC SDK, which can be found at the following site: http://www.schiratti.com/dowson.html.

since most of the work has already been done for us by the programmers of FSUIPC and Smartie51. The Smartie51 LCD program can be extended using DLLs, and FSUIPC allows us to read data from the flight simulator. The only thing we had to do was to write a DLL that can read the desired data from the flight simulator and pass them to Smartie51 as necessary.

We did this using the FSUIPC SDK. Naturally, to get everything to work properly you will have to install the necessary software, as described in the previously mentioned articles. You will also have to place the DLL for this article (FSSSmart.dll) in the Smartie plugins folder. The file

P. Goossens   (034056-1)

062
CMOS Crystal Frequency Multiplier

Crystals usually operate at fundamental frequencies up to about 15 MHz. Whenever higher frequencies are required a frequency multiplier is placed after the crystal oscillator. The resulting output signal is then a whole multiple of the crystal frequency. Other frequency multipliers often use transistors, which produce harmonics due to their non-linearity. These are

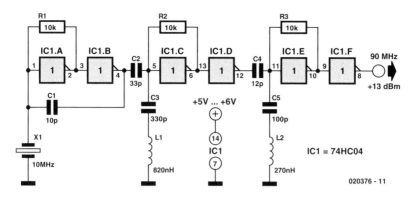

020376 - 11

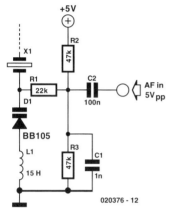

020376 - 12

subsequently filtered from the signal. One way of doing this is to put a parallel L-C filter in the collector arm. This filter could then be tuned to three times the input frequency. A disadvantage is that such a circuit would quickly become quite substantial. This circuit contains only a single IC and a handful of passive components, and has a complete oscillator and two frequency triplers. The output is therefore a signal with a frequency that is 9 times as much as that of the crystal. Two gates from IC1, which contains six high-speed CMOS inverters, are used as an oscillator in combination with X1. This works at the fundamental frequency of the crystal and has a square wave at its output. A square wave can be considered as the sum of a fundamental sine wave plus an infinite number of odd multiples of that wave. The second stage has been tuned to the first odd multiple (3 ×). We know that some of our readers will have noticed that the filter used

here is a band-rejection (series LC) type. Worse still, when you calculate the rejection frequency you'll find that it is equal to the fundamental crystal frequency! The fundamental frequency is therefore attenuated, which is good. But how is the third harmonic boosted? That is done by the smaller capacitor of 33 pF in combination with the inductor. Together they form the required band-pass filter. (The same applies to the 12 pF capacitor in the next stage.) Through the careful selection of components, this filter is therefore capable of rejecting the fundamental and boosting the third harmonic! Clever, isn't it? The output in this example is a signal of 30 MHz. The inverter following this stage heavily amplifies this signal and turns it into a square wave. The same trick is used again to create the final output signal of 3 times 30 MHz = 90 MHz. At 5 V this circuit delivers about 20 milliwatt into 50 Ω. This corresponds to +13 dBm and is in theory enough to drive a diode-ring balanced mixer directly. The circuit can be used for any output fre-+5V quency up to about 100 MHz by varying the component values. When, for example, an 8 MHz crystal is used to obtain an output frequency of 72 MHz (9 × 8 = 72), the frequency determining inductors and capacitors have to be adjusted by a factor of 10/8. You should round the values to the nearest value from the E12 series. Another application is for use in an FM transmitter; if you connect a varicap in series with the crystal, you can make an FM modulator. An added bonus here is that the relatively small modulation level is also increased by a factor of 9. Crystals with frequencies near 10 MHz are relatively easy to find and inexpensive, so you should always be able to find a suitable frequency within the FM band. A crystal of 10.245 MHz for instance the border of the HC specifications. If this would give you a frequency of causes any problems you should increase 92.205 MHz and

10.700 MHz results in the supply voltage a little to 6 V. an output of 96.300 MHz. (020376-1) You may find that the circuit operates on the border of the HC specifications. If this causes any problems, you should increase the supply voltage a little to 6 V.

Gert Baars (020376)

063
Inductorless 3-to-5 Volts Converter

By configuring a comparator and a transistor to control the oscillator in a charge pump circuit, you enable the pump to generate a regulated output of – in principle – any desired value. Charge pump ICs can either invert or double an input voltage (for example, 3 V to –3 V or 3 V to 6 V). The charge pump itself does not regulate the output voltage and one running off 3 V is not normally capable of generating intermediate output voltage levels like 5 V. However, by adding a comparator and a reference device, you can create arbitrary output levels like 5 V and regulate them as well. Charge pump IC1 (a MAX660) has an internal oscillator whose 45 kHz operation transfers charge from C1 to C2, causing the regulated output to rise. When the feedback voltage (pin 3 of IC2) exceeds 1.18 V, the output of comparator IC2 (a MAX921) goes high, turning off the oscillator via T1. The comparator hysteresis (easily added on IC2) is zero here simply because no hysteresis is required in the control loop. The oscillator when enabled generates two cycles, which is sufficient to drive VOUT slightly above the desired level. Next, the feedback turns the oscillator off again.

The resulting output ripple will depend mainly on the input voltage and the output load current. Output ripple may be reduced at the expense of circuit efficiency by adding a small resistor (say, 1 Ω) in series with C1. You'll find that ripple also depends on the value and ESR associated with C1 – smaller values of C1 transfer less charge to C2, producing smaller jumps in VOUT.

D. Prabakaran (020230)

020230 - 11

064
DMM Fuse Protector

Typically the input protection fuse of your DMM will blow in the middle of a demonstration or an exciting phase of your construction work. Spare fuses are always hard to find, and if available take a lot of time to install. This circuit replaces the fuse by a 500 mA current limiter. When resistor R1 passes about 500 mA, it will drop 0.75 V which is sufficient to switch on T2. With the buzzer acting as a pull-up resistor (and, of course, as a very loud acoustic warning device), the voltage at the gate of power FET T1 will drop to a level at which the drain-source current is limited to a safe value of about 500 mA. Of course, the excess energy caused by the current limiting action is dissipated by the FET. Cooling is required in all cases where the dissipation can be expected to exceed about 1-2 watts. After all, without cooling, the voltage allowed to occur across the FET will be just 4 V (2 W = 0.5 A × 4 V). Although an IRF740S is indicated in the circuit diagram, almost any power FET may be used. The popular BUZ10, for example, is a good choice when a lot of power has to be

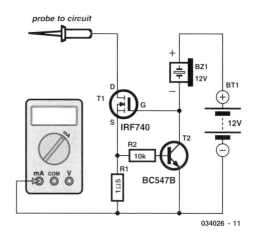

034026 - 11

dissipated. If a 12 V mini battery is used then the buzzer should also be a 12 V device. However the circuit will also work fine from the more commonly found (and certainly less expensive) 9 V battery and a matching buzzer. If the latter is not required it is simply replaced by a 10 kΩ resistor.

P. Sicherman   (034026-1)

065
SSB Add-On for AM Receivers

Given favourable radio wave propagation, the shortwave and radio amateur band are chock-a-block with SSB (single-sideband) transmissions, which no matter what language they're in, will fail to produce intelligible speech on an AM radio. SSB is

transmitted without a carrier wave. To demodulate an SSB signal (i.e. turn it into intelligible speech) it is necessary to use a locally generated carrier at the receiver side. As most inexpensive SW/MW/LW portable radios (and quite a few more ex-

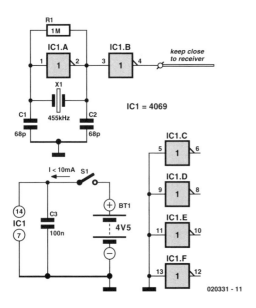

frequency (IF), adding SSB amounts to no more than allowing the radio's IF to pick up a reasonably strong 455 kHz signal and let the existing AM demodulator do the work.

The system is called BFO for 'beat frequency oscillator'. The heart of the circuit is a 455 kHz ceramic resonator or crystal, X1. The resonator is used in a CMOS oscillator circuit supplying an RF output level of 5 Vpp. which is radiated from a length of insulated hookup wire wrapped several times around the receiver.

The degree of inductive coupling needed to obtain a good beat note will depend on the IF amplifier shielding and may be adjusted by varying the number of turns. All unused inputs of the 4069 IC must be grounded to prevent spurious oscillation.

pensive general coverage receivers) still use plain old 455 kHz for the intermediate

D. Prabakaran (020331-1)

066
Irregular Flasher

Two multivibrators with different frequencies can be built using the NAND gates of a 4011 IC. If the output of IC1.B is positive with respect to IC1.C, LED D1 is on. As the levels of IC1.A and IC1.D are exactly opposite, D2 is always on when D1 is off, and the other way around. The two oscillators have different frequencies, which are determined by the values of R2/C2 and R5/C5 respectively according to the formula:

$$f_0 = \frac{1}{1,4 \cdot R \cdot C}$$

With the given component values, the frequencies are 2.2 Hz and 7.2 Hz. Low-current LEDs should be used, since the CMOS IC cannot sink or source sufficient

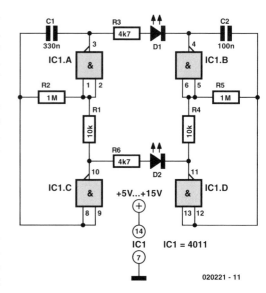

current for 'normal' LEDs. The values of series resistors R3 and R6 are suitable for a supply voltage of 12 V, in which case the current consumption of the circuit is around 5 mA.

However, in principle the 4011 can be operated over a supply voltage range of 5 – 15 V. Higher currents can be provided by the HC family (supply voltage 3 – 6 V) or the HCT family (5 V). Incidentally, the part number of the quad gate IC in the HC family is HC7400.

Ludwig Libertin (020221)

067
Slave Switch

A peculiarity of this slave switch is that it has a delay when it turns off. This can be very useful. The author has used this design in his workshop to turn on a vacuum cleaner automatically. The delayed turn-off lets the vacuum cleaner continue for a while after the sawing stops. With the help of a double-wound suppression

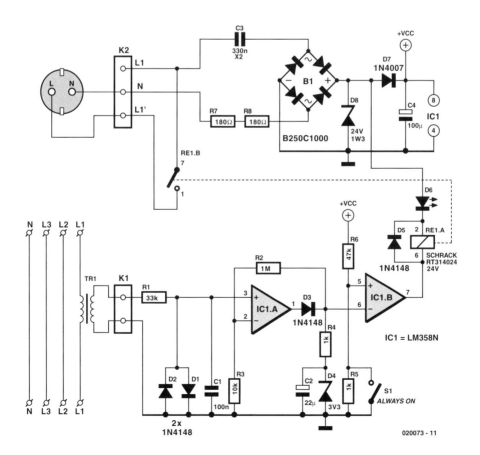

020073 - 11

choke (16 amps) the current through one phase is detected when the equipment is turned on. This signal is passed through a clipping filter network and amplified/rectified by opamp IC1.A. Capacitor C2 then charges up, with the voltage across it limited to 3.3 V by zener diode D4. The capacitor is discharged via R4, R2 and R3, which takes around 2 minutes. IC1.B is used to drive relay RE1. This is a 24 V type that requires just 20 mA and can switch up to 16 A. The dual opamp used here (LM358N) is similar to a LM324, but only has two opamps instead of four. These opamps can sink 50 mA, so can easily drive the relay. This isn't self-evident, as many other types of opamp won't work at all! S1 can be used to turn on the relay permanently. The circuit has been designed such that the relay remains off when the mains is turned on or off. The circuit can be built on prototyping board and mounted into a combined power

socket with a 3-phase socket and an ordinary mains socket. Conrad, for example, has a suitable item in their range of products. The workshop equipment plugs into the 3-phase socket and the vacuum cleaner plugs into the ordinary socket. Keep in mind during construction of the circuit that all cables carrying mains voltage to the board have to have a cross-section of at least 1 mm^2. The unused holes should be drilled out to avoid flashovers. The completed board should be isolated with a plastic spray, or better still immersed in potting compound! The LED and the suppression choke have to be fixed firmly in place. A piece of transparent plastic should be placed in front of the LED. Make sure that all connections to the sockets and mains are reliably made. The connection pins on the board should be separated by at least 3 mm (Class I).

H. Tempelman   (020073-1)

068
Very Low Power 32 kHz Oscillator

The 32 kHz low-power clock oscillator offers numerous advantages over conventional oscillator circuits based on a CMOS inverter. Such inverter circuits present problems, for example, supply currents fluctuate widely over a 3 V to 6 V supply range, while current consumption below 250 µA is difficult to attain. Also, operation can be unreliable with wide variations in the supply voltage and the inverter's input characteristics are subject to wide tolerances and dif-

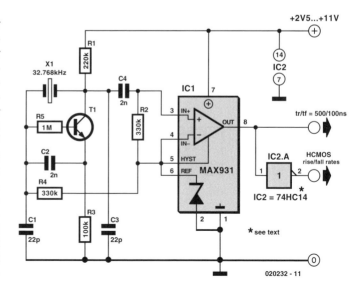

ferences among manufacturers. The circuit shown here solves the above problems. Drawing just 13 µA from a 3 V supply, it consists of a one-transistor amplifier/oscillator (T1) and a low-power comparator/reference device (IC1). The base of T1 is biased at 1.25 V using R5/R4 and the reference in IC1. T1 may be any small-signal transistor with a decent beta of 100 or so at 5 µA (defined here by R3, fixing the collector voltage at about 1 V below Vcc). The amplifier's nominal gain is approximately 2 V/V. The quartz crystal combined with load capacitors C1 and C3 forms a feedback path around T1, whose 180 degrees of phase shift causes the oscillation. The bias voltage of 1.25 V for the comparator inside the MAX931 is defined by

the reference via R2. The comparator's input swing is thus accurately centred around the reference voltage. Operating at 3 V and 32 kHz, IC1 draws just 7 µA. The comparator output can source and sink 40 mA and 5 mA respectively, which is ample for most low-power loads. However, the moderate rise/fall times of 500 ns and 100 ns respectively can cause standard, high-speed CMOS logic to draw higher than usual switching currents. The optional 74HC14 Schmitt trigger shown at the circuit output can handle the comparator's rise/fall times with only a small penalty in supply current. Further information on the MAX931 from: www.maxim-ic.com.

D. Prabakaran   (020232-1)

069
Solar-Powered High
Efficiency Charger

This is a simple NiCd battery charger powered by solar cells. A solar cell panel or an array of solar cells can charge a battery at more than 80 % efficiency provided the available voltage exceeds the 'fully charged' battery voltage by the drop across one diode, which is simply inserted between the solar cell array and the battery. Adding a step-down regulator enables a solar cell array to charge battery packs with various terminal voltages at optimum rates and with efficiencies approaching those of the regulator itself. However, the IC must then operate

in an unorthodox fashion (a.k.a. 'Elektor mode') regulating the flow of charge current in such a way that the solar array output voltage remains near the level required for peak power transfer. Here, the

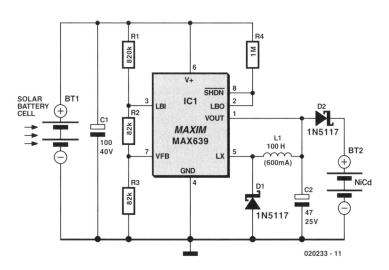

020233 - 11

MAX639 regulates its input voltage instead of its output voltage as is more customary (but less interesting).

The input voltage is supplied by twelve amorphous solar cells with a minimum surface area of 100 cm2. Returning to the circuit, potential divider R2/R3 disables the internal regulating loop by holding the V-FB (voltage feed-back) terminal low, while divider R1/R2+R3 enables LBI (low battery input) to sense a decrease in the solar array output voltage. The resulting deviation from the solar cells' peak output power causes LBO (low battery output) to pull SHDN (shutdown) low and consequently disable the chip. LBI then senses a rising input voltage, LBO goes high and

the pulsating control maintains maximum power transfer to the NiCd cells. Current limiting inside the MAX639 creates a 'ceiling' of 200 mA for Iout. Up to five NiCd cells may be connected in series to the charger output. When 'on' the regulator chip passes current from pin 6 to pin 5 through an internal switch representing a resistance of less than 1 ohm. Benefiting from the regulator's low quiescent current (10 micro-amps typical) and high efficiency (85 %), the circuit can deliver four times more power than the single-diode configuration usually found in simple solar chargers. Coil L1 is a 100 µH suppressor choke rated for 600 mA.

D. Prabakaran   (020233-1)

070
Luxury Car Interior Light

This circuit belongs to the 'car modding' category. This is similar to the popular case modding in the computer world and has found its way into a substantial proportion of cars. The modifications vary from light effects to complete movie playback systems. This circuit is much more modest, but certainly still worth the effort. It provides a high quality interior light delay. This is a feature that is included as standard with most modern cars, although the version with an automatic dimmer is generally only found in the more expensive models. With this circuit it is possible to upgrade second hand and mid-range models with an interior light delay that slowly dims after the door has been closed. The dimming of the light is implemented by means of pulse-width modulation. This requires a triangle wave oscillator and a comparator. Two opamps are generally required to generate a good triangle wave, but because the waveform doesn't have to be accurate, we

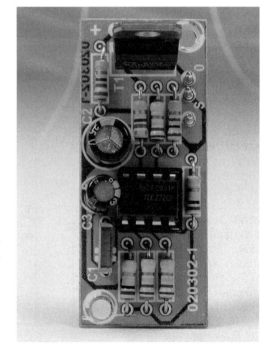

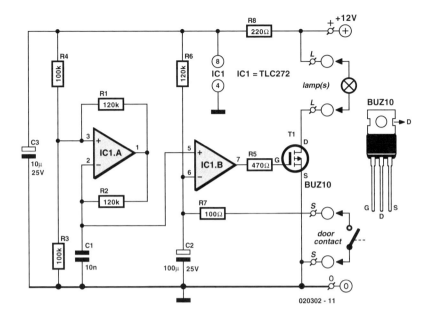

020302 - 11

can make do with a single opamp. This re-sults in the circuit around IC1.A, a relax-ation oscillator supplying a square wave output. The voltage at the inverting input has more of a triangular shape. This signal can be used as long as we do not put too much of a load on it. The high impedance input of IC1.B certainly won't cause prob-lems in this respect. This opamp is used as a comparator and compares the voltage of the triangular wave with that across the door switch. When the door is open, the switch closes and creates a short to the chassis of the car. The output of the opamp will then be high, causing T1 to conduct

and the interior light will turn on. When the door is closed the light will continue to burn at full strength until the voltage across C2 reaches the lower side of the triangle wave (about 5 V).

The comparator will now switch its output at the same rate of the triangle wave (about 500 Hz), with a slowly reducing pulse width, which results in a slowly reducing brightness of the interior light. R8 and C3 protect the circuit from voltage spikes that

Components list

Resistors:
R1,R2,R6 = 120 kΩ
R3,R4 = 100 kΩ
R5 = 470 Ω
R7 = 100 Ω
R8 = 220 Ω

Capacitors:
C1 = 10 nF
C2 = 100 µF 25V
C3 = 10 µF 25V

Semiconductors:
T1 = BUZ10
IC1 = TLC272CP

Miscellaneous:
PCB available from ThePCBShop

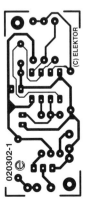

may be induced by the fast switching of the light. The delay and dimming time can be adjusted with R6 and C2. Smaller values result in shorter times. You can vary the dimming time on its own by adjusting R1, as this changes the amplitude of the triangle wave across C1. R7 limits the discharge current of C2; if this were too big, it would considerably reduce the lifespan of the capacitor. There is no need to worry about reducing the life of the car battery. The circuit consumes just 350 µA when the lamp is off and a TLC272 is used for the dual opamp. A TL082 will take about 1 mA. These values won't discharge a normal car battery very quickly; the self-discharge is probably many times higher. It is also possible to use an LM358, TL072 or TL062 for IC1. R8 then needs to have a value between 47 Ω and 100 Ω. Since T1 is always either fully on or fully off, hardly any heat is generated. At a current of 2 A the voltage drop across the transistor is about 100 mV, giving rise to a dissipation of 200 mW. This is such a small amount that no heatsink is required.

The whole circuit can therefore remain very compact and should be easily fitted in the car, behind the fabric of the roof for example.

Cuno Walters   (020302-1)

071
Shortwave Monitor

This broadband AM receiver enables you to 'monitor' the shortwave radio band. The circuit has been deliberately designed to have low selectivity and is most sensitive in the range from 6 to 20 MHz. This frequency range contains most of the shortwave broadcast stations. In this configuration, whichever station has the strongest signal will be the easiest to hear. An interesting fact is that the signal strength of stations in this band changes quite a lot. This is because the ionosphere reflects the radio signals. Because this layer of the atmosphere is in constant motion, the received signal strengths from different directions are subject to continuous variation. During testing of our prototype Radio Netherlands World Ser-

vice, Radio Finland and Deutsche Welle alternated as the strongest station at regular intervals. This receiver not only gives a good indication of the myriad of stations on offer in the short-wave band but is also an excellent tool for monitoring the state of

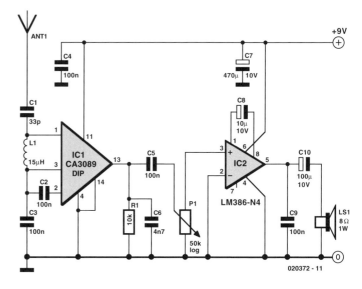

020372 - 11

the ionosphere. The circuit actually consists of no more than an RF and an AF amplifier. The high-frequency amplification is carried out by the IF stage of a CA3089. This IC is actually intended for FM receivers, but the FM section is not used here. The internal level detector provides a signal of sufficient strength to drive an audio amplifier directly. An LM386 was selected for this task. This IC can directly drive an 8 Ω loudspeaker or headphones without any difficulty. The power supply voltage is 9 V. Because of the modest power consumption a 9 V battery is very suitable. In addition, the circuit will work down to a voltage of about 5.5 V, so that the battery life will be extra long. The antenna will require a little experimentation. We obtained reasonable results with a piece of wire 50 cm long. A length of wire in the range of 5 to 15 meters should provide even better results at these frequencies.

Gert Baars (020372-1)

072
Tuned Radio Frequency (TRF) Receiver

Super-heterodyne receivers have been mass-produced since around 1924, but for reasons of cost did not become successful until the 1930s. Before the second world war other, simpler receiver technologies such as the TRF receiver and the regenerative receiver were still widespread. The circuit described here is based on the old technology, but brought up-to-date. The most important part of the circuit is the input stage, where positive feedback is used to achieve good sensitivity and selectivity. The first stage is adjusted so that it is not quite at the point of oscillation. This increases the gain and the selectivity, giving a narrow bandwidth. To achieve this, the potentiometer connected to the drain of the FET must be adjusted very carefully: opti-

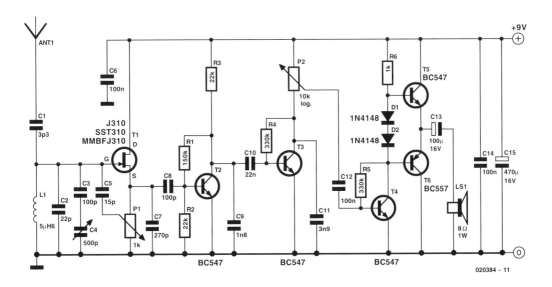

020384 - 11

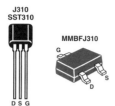

mal performance of the receiver depends on its setting. In ideal conditions several strong stations should be obtainable during the day using a 50 cm antenna. At night, several times this number should be obtainable. The frequency range of the receiver runs from 6 MHz to 8 MHz. This range covers the 49 m and the 41 m shortwave bands in which many European stations broadcast. Not bad for such a simple circuit! The circuit employs six transistors. The first stage is a selective amplifier, followed by a transistor detector. Two

low-frequency amplifier stages complete the circuit. The final stage is a push-pull arrangement for optimal drive of the low-impedance loudspeaker. This circuit arrangement is sometimes called a '1V2 receiver' (one preamplifier, one detector and two audio-frequency stages). Setting-up is straightforward. Adjust P1 until the point is reached where the circuit starts to oscillate: a whistle will be heard from the loudspeaker. Now back off the potentiometer until the whistle stops. The receiver can now be tuned to a broadcaster. Occasional further adjustment of the potentiometer may be required after the station is tuned in. The receiver operates from a supply voltage of between 5 V and 12 V and uses very little current. A 9 V PP3 (6F22) battery should give a very long life.

Gert Baars   (020384-1)

073
Digital Isolation up to 100 MBit/s

When it is necessary to send a digital signal between two electrically isolated circuits you would normally choose an optoisolator or some form of transformer coupling. Neither of these solutions is ideal; optocouplers run out of steam beyond about 10 MHz and transformers do not have a good low frequency (in the region of Hertz) response.

The company NVE Corporation (www.nve.com) produces a range of coupler devices using an innovative 'IsoLoop' technology allowing data rates up to 110 Mbaud. The example shown here uses the IL715 type coupler providing four TTL or CMOS

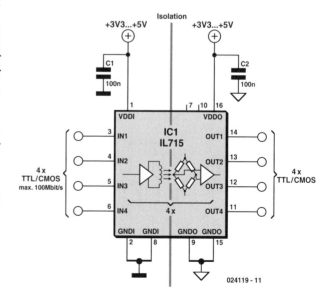

compatible channels with a data rate of 100 Mbit/s. Inputs and outputs are compatible with 3.3 V or 5 V systems. The maximum isolation voltage is 2.5 kV and the device can cope with input transients up to 20 kV/µs.

The company produce many other configurations including bi-directional versions that would be suitable for RS485 interfacing. The IsoLoop coupler is based on relatively new GMR (GiantMagnetoResistive) technology. The input signal produces a current in a planar coil. This current generates a magnetic field that produces a change in resistance of the GMR material.

This material is isolated from the planar coil by a thin film high voltage insulating layer. The change in resistance is amplified and fed to a comparator to produce a digital output signal. Differences in the ground potential of either the input or output stage will not produce any current flow in the planar coil and therefore no magnetic field changes to affect the GMR material. Altogether the circuit provides a good electrical isolation between input and output and also protects against input signal transients (EMV).

Gregor Kleine   (024119-1)

074
Buck/Boost Voltage Converter

Sometimes it is desired to power a circuit from a battery where the required supply voltage lies within the discharge curve of the battery. If the battery is new, the circuit receives a higher voltage than required, whereas if the battery is towards the end of its life, the voltage will not be high enough. This is where the new LTC 3440 buck/boost voltage converter from Linear Technology (www.linear.com) can help. The switching regulator in Figure 1 converts an input voltage in the range +2.7 V to +4.5 V into an output voltage in the range +2.5 V to +5.5 V using one tiny coil. The level of the output voltage is set by the voltage divider formed by R2 and R3. The device switches as necessary between step-up (or 'boost') operation when Vin is less than Vout, and step-down

(or 'buck') operation when Vin is greater than Vout. The maximum available output current is 600 mA. The IC contains four MOSFET switches (Figure 2) which can connect the input side of coil L1 either to Vin or to ground, and the output side of L1 either to the output voltage or to ground. In

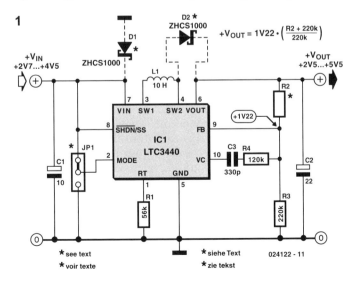

$$+V_{OUT} = 1V22 \cdot \left(\frac{R2 + 220k}{220k} \right)$$

* see text
* voir texte
* siehe Text
* zie tekst
024122 - 11

2

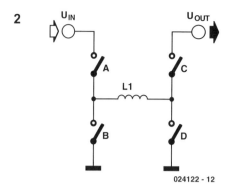

024122 - 12

step-up operation switch A is permanently on and switch B permanently off. Switches C and D close alternately, storing energy from the input in the inductor and then releasing it into the output to create an output voltage higher than the input voltage. In step-down operation switch D is permanently closed and switch C permanently open. Switches A and B close alternately and so create a lower voltage at Vout in proportion to the mark-space ratio of the switching signal. L1, together with the output capacitor, form a low-pass filter. If the input and output voltages are approxi-

mately the same, the IC switches into a pulse-width modulation mode using all four switches. Resistor R1 sets the switching frequency of the IC, which with the given value is around 1.2 MHz. This allows coil L1 to be very small. A suitable type is the DT1608C-103 from Coilcraft (www.coilcraft.com).

The IC can be shut down using the SHDN/SS input. A 'soft start' function can also be implemented by applying a slowly-rising voltage to this pin using an RC network. The MODE pin allows the selection of fixed-frequency operation (MODE connected to ground) or burst mode operation (MODE=Vin). The latter offers higher efficiency (of between 70 % and 80 %) at currents below 10 mA. At currents of around 100 mA the efficiency rises to over 90 %. A further increase in efficiency can be obtained by fitting the two Schottky diodes shown dotted in the circuit diagram. These operate during the brief period when both active switches are open (break-before-make operation).

Gregor Kleine   (024122-1)

075
Simple Infrared Control Extender

Lots of consumer electronic equipment like TV sets, VCRs, CD and DVD players employs infrared remote control. In some cases, it is desirable to extend the range of the available control and this circuit fits the bill, receiving the IR signal from your remote control and re-transmitting it, for example, around a corner into another room. Photodiode D4 is connected to the inverting input of a 741 opamp through resistor R2 and capacitor C1. Since the BPW41 photodiode (from Vishay/Telefunken) needs to be reverse-biased to turn light energy into a corresponding voltage, it is also

connected to the positive supply rail via R1. The non-inverting input of the '741 is held at half the supply voltage by means of equal resistors R3 and R4. The opamp is followed by a BD240 after-burner transistor capable of supplying quite high current pulses through IR LEDs D2 and D3. However, the pulsed current through the LD274s should not exceed 100 mA or so, hence a fixed resistor is used in series with preset P1. D1 is an ordinary visible-light LED that flashes when an IR signal is received from the remote. With regard to the setting of P1, do not make the IRED cur-

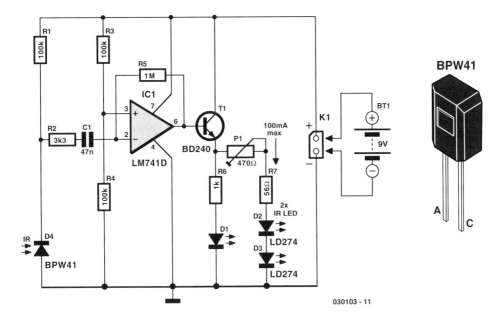

030103 - 11

rent higher than necessary to reliably reach the final destination of the IR signal. Also, the currents mentioned above are peak levels – due to the small duty factor of the IR pulses, the average current drawn from the battery will be much smaller. The directivity of the IR LEDs – and consequently the range of the control extender – may be increased by fitting the devices with reflective caps.

Raj. K. Gorkhali   (030103-1)

076
Pseudo Track Occupancy Detector for Märklin Digital Systems

Track occupancy detectors are needed for hidden yards and other sections of track that are hard to see, but they are also necessary for block operation. The circuit described here uses an LED to indicate track occupancy for digitally controlled Märklin HO model train systems (including Delta Control). In contrast to a real track occupancy detector, which detects all vehicles, it only responds to vehicles that draw traction current. This means it can be used without making complicated modifications to the rolling stock and track, since it is only necessary to gap the third rail. This circuit is thus especially suitable for retrofitting to existing installations, and it is equally well suited to M, K and C tracks. The basic idea of the circuit is simple. If a locomotive enters the monitored track section, a current flows through the motor. This current is sensed and generates an indication. With a Märklin Digital system, power is provided to the locomotive via a controller or a booster in the form of a

1

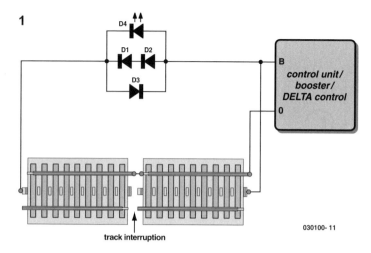

track interruption

030100- 11

2

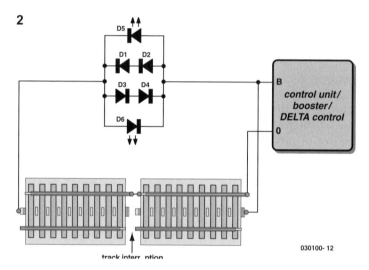

track interr ption

030100- 12

circuit. If a locomotive travels over the monitored track section, the positive component of the drive current flows through diodes D1 and D2, while the negative component flows through D3. With a motor current of approximately 250 mA, the voltage drop across a single diode (1N4001 types are used here) is a good 1 V. The voltage drop across the two diodes connected in series (D1 and D2) is sufficient to illuminate LED1. Although the locomotive will travel somewhat slower due to the voltage drop, this will not cause any problems. A second detector can be obtained by connecting an additional diode to the circuit as shown in Figure 2.

This causes a second LED to illuminate for negative drive current. Due to the pulse trains and fluctuations in the traction current, the LED illumination is not constant, but instead flickers more or less strongly. Other traction-power loads, such as coach lighting or taillights, will also generate an 'occupied' indication. In such cases, the LED will remain illuminated even if the locomotive is standing still with no current flowing through the motor. Sometimes the quiescent current through a decoder is sufficient to cause the LED to illuminate (a little bit) even if the locomotive is standing still. Another possibility is to use an optocoupler instead of an LED. This would allow the circuit to be connected to an s88 detection module.

square-wave voltage. The voltage levels on the rails are approximately –15 V and +15 V. Digital control information is transferred by a continuous sequence of alternating plus and minus levels. This means that the detector circuit must be able to respond to AC signals. In Figure 1, the monitored track section on the left is connected to the ground terminal '0' via the rails. The third rail, with conducts the traction current to the locomotive, is isolated from the rest of the system (special third-rail insulators are available for this purpose), and it is connected to the 'B' terminal of the controller or booster via the detector

Nils Körber

(030100-1)

077
Symmetrical-Out from Mains Adapter

Mains adapters (also known by other popular names like 'wall cubes' or 'battery eliminators') come in many different guises and are now used in almost every household and office. However, try to find one with a symmetrical output, for example, ±12 VDC, and you will come a cropper. If and when such a rare beast exists, it would have an odd-looking output voltage connector, namely one with three pins! In this article we propose an alternative in the form of a mains adapter of the 'AC-out' variety and combine it with a simple add-on circuit fitted in the equipment requiring a symmetrical supply voltage. A 15 VAC wall cube is used to power a standard symmetrical supply consisting of single-phase rectifiers (D1 and D2) and two

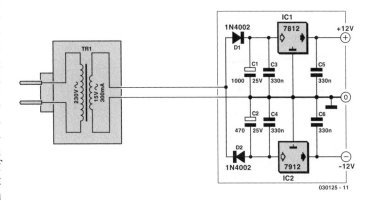

12 volt voltage regulators with the usual reservoir (C1 and C2) and decoupling capacitors (C3-C6) strewn around them.

As you will require just two wires to connect the cube to the supply input, an ordinary two-pin plug/socket combination may be used.

Flemming Jensen   (030125-1)

078
Bluish Flasher

This circuit is innovative in more than one way and therefore belongs per se in Elektor's Small Circuits Collection. Firstly, it demonstrates how the combination of a blue and a white LED can be used to give a realistic imitation of a camera flashlight. Secondly, the good old 555 IC is used in a way many of you may never have

seen before – alternately mono-stable / astable – without too much in the way of external parts. Initially C3 will be empty, pulling output pin 3 to +12 V and causing the blue LED, D1, to light via R3. Next, C3 will charge up via R2. Meanwhile C1 has been building up charge through R1 and D3. If the voltage on C3 reaches about 8 V

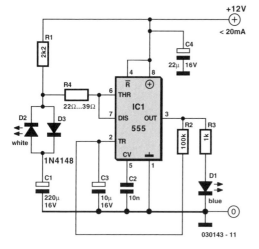

030143 - 11

the 555's output has dropped Low, the voltage on C3 will decrease as well. Ad soon as a level of 4 V is reached (one third of 12 V), the above cycle is repeated. Resistor R4 limits the current through the 555 to safe levels.

You may want to experiment with the latest hyper-bright white LEDs. SDK's AlInGaP LEDs, for example, are claimed to light three times as brightly as regular white LEDs. A number of blue LEDs may be connected in series instead of just one as shown in the circuit diagram. Unfortunately, that is not possible at the 'white' side. For the best visual effect, the white blue LEDs should be mounted close together. When fitted close to the extra brake light in your car, the bluish white flash is sure to make even persistent tailgaters back off. Note however that this use of the circuit may not be legal in all countries.

Myo Min   (030143-1)

(two-thirds of 12 V), pin 3 of the 555 will drop Low. So does pin 7, causing the white LED to light, pulling its energy from C1. This energy drops quickly, causing D2 to dim in an exponentially decaying fashion, just like a camera flashlight. Now, because

079
Adjustable Zener Diode

A Zener diode is the simplest known type of voltage limiter (Figure 1) As soon as the voltage exceeds the rated voltage of the Zener diode, a current can flow through the diode to limit the voltage. This is exactly the right answer for many protection circuit applications. However, if it is necessary to limit a signal to a certain voltage in a control circuit, Zener diodes do not pro-

vide an adequate solution. They are only available with fixed values, which are also subject to a tolerance range. What we are looking for is thus an 'adjustable' Zener di-

1

030150 - 11

2

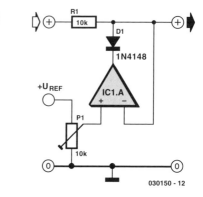

030150 - 12

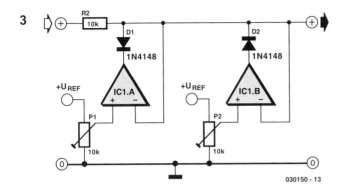

030150 - 13

ode. Such a component would be useful in a heating controller with a preheat temperature limiting, for example, or in a battery charger to provide current limiting. The answer to our quest is shown in Figure 2. Assume for example that the output voltage must not exceed 6.5 V. The control voltage on the non-inverting input is thus set to 6.5 V. Now assume that 4.2 V is present at the input. The result is that the maximum positive voltage is present at the opamp output, but the diode prevents this from

having any effect on the signal. However, if the voltage rises above 6.5 V, the output of the opamp goes negative and pulls the voltage back down to 6.5 V. The current is limited by R3. Another example is a situation in which exactly the opposite is required. In this case, the voltage must not drop below a certain value. This can be easily achieved by reversing the polarity of the diode. Another option is a voltage that is only allowed to vary within a certain voltage window. It must not rise above a certain value, but it also must not drop below another specific value. In the circuit shown in Figure 3, the left-hand opamp provides the upper limit and the right-hand opamp provides the lower limit. Each opamp is wired as a voltage follower.

Dieter Bellers   (030150-1)

080
SMPSU with a Relay

Switched mode power supply units (SMP-SUs) are popular but difficult to build oneself as well problematic when it comes to understanding their design principles. Building your own SMPSU typically requires a lot of expertise, hard to find components and time.

The circuit shown here is educational only and devised to demonstrate the principle of the step-up switch mode circuits. It is not intended to be incorporated in a 'real' design. Relay Re1 has a normally-closed (NC) change-over contact and is connected to act as a vibrator. When power is applied to the circuit, the relay is energized and actuates its contact. This action may appear

to break the circuit. However, the energy stored in the relay coil will produce an induced voltage which is fed to D1 and C1 for rectification and smoothing. The output voltage will be of the order of 150 V and

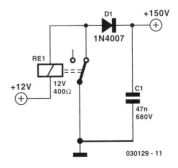

030129 - 11

strongly dependent on the type of relay used. In general, the faster the relay, the higher the output voltage. The circuit will oscillate at a low frequency (100-200 Hz), and a buzzing sound will be heard from the relay.

Myo Min   (030129-1)

081
Simple Audio Peak Detector

This audio peak detector allows a pair of stereo channels to be monitored on a single LED. Identical circuitry is used in the left and right channels. Use is made of the switching levels of Schmitt trigger NAND gates inside the familiar 4093 IC. The threshold level for gate IC1.A (IC1.B) is set with the aid of preset P1, which supplies a high-impedance bias level via R2 (R1). When, owing to the instantaneous level of the audio signal superimposed on the bias voltage by C3 (C2), the dc level at pins 1 and 2 (5 and 6) of the Schmitt trigger gate drops below a certain level, the output of IC1.A (IC1.B) will go High. This level is copied to the input of IC1.C via D2 (D1) and due to the inverting action of IC1.C, LED D3 will light. Network R3-C1 provides some delay to enable very short audio peaks to be reliably indicated. Initially turn the wiper of P1 to the +12 V extreme – LED D3 should remain out. Then apply 'line' level audio to K1 and K3, preferably music with lots of peaks (for example, drum 'n bass). Carefully adjust P1 until the peaks in the music are

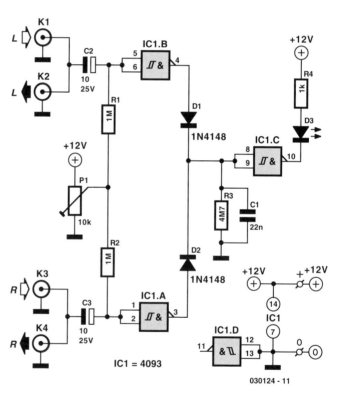

those rare and expensive audio-indicated

by D3. splitter ('Y') cables. The circuit has double RCA connectors for the left and right channels to obviate the use of those rare and extensive audio splitter ('Y') cables.

left and right channels to obviate theadjust
use of

Flemming Jensen   (030124-1)

082
Zero Gain Mod for Non-Inverting Opamp

Electronics textbooks will tell you that a non-inverting opamp normally cannot be regulated down to 0 dB gain. If zero output is needed then it is usual to employ an inverting amplifier and a buffer amp in front of it, the buffer acting as an impedance step-up device. The circuit shown here is a trick to make a non-inverting amplifier go down all the way to zero output.

The secret is a linear-law stereo potentiometer connected such that when the spindle is turned clockwise the resistance in P1a increases (gain goes up), while the wiper of P1b V+ moves towards the opamp output (more signal). When the wiper is turned anti-clockwise, the resistance of P1a drops, lowering the gain, while P1b also supplies a smaller signal to

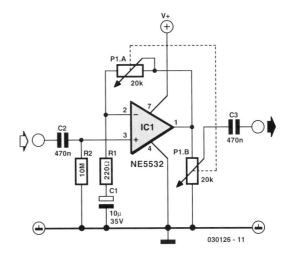

the load. In this way, the output signal can be made to go down to zero.

Flemming Jensen   (030126-1)

083
Push Off / Push On

The ubiquitous 555 has yet another airing with this bistable using a simple push-button to provide a push-on, push-off action. It uses the same principle of the stored charge in a capacitor taking a Schmitt trigger through its dead-band as previously published as 'Pushbutton Switch' (038) in the Small Circuits collection of 2002. Whereas the Schmitt trigger in that reference was made from discrete components, the in-built dead-band arising from the two

comparators, resistor chain and bistable within the 555 is used instead. The circuit demonstrates a stand-by switch, the state of which is indicated by illumination of either an orange or red LED, exclusively driven by the bipolar output of pin 3. Open-collector output (pin 7) pulls-in a 100 mA relay to drive the application circuit; obviously if an ON status LED is provided elsewhere, then the relay, two LEDs and two resistors can be omitted, with pin 3 being used to

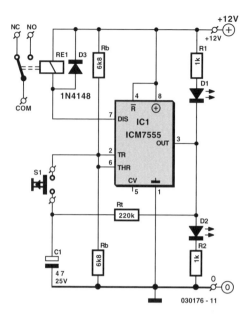

030176 - 11

drive the application circuit, either directly or via a transistor. The original NE555 (non-CMOS) can source or sink 200 mA from / into pin 3. Component values are not critical; the 'dead-band' at input pins 2 and 6 is between 1/3 and 2/3 of the supply voltage. When the pushbutton is open-circuit, the input is clamped within this zone (at half the supply voltage) by two equal-value resistors, Rb. To prevent the circuit powering-up into an unknown condition, a power-up reset may be applied with a resistor from supply to pin 4 and capacitor to ground. A capacitor and high-value resistor (Rt) provide a memory of the output state just prior to pushing the button and creates a dead time, during which button contact bounce will not cause any further change. When the button is pressed, the stored charge is sufficient to flip the output to the opposite state before the charge is dissipated and clamped back into the neutral zone by resistors Rb. A minimum of 0.1 µF will work, but it is safer to allow for button contact-bounce or hand tremble; 10 µF with 220 kΩ gives approximately a 2-second response.

Trevor Skeggs   (030176-1)

084
Single-Chip VHF RF Preamp

Here is a high performance RF amplifier for the entire VHF broadcast and PMR band (100-175 MHz) which can be successfully built without any special test equipment. The grounded-gate configuration is inherently stable without any neutralization if appropriate PCB layout techniques are employed. The performance of the amplifier is quite good. The noise figure is below 2 dB and the gain is over 13 dB. The low-noise figure and good gain will help car radios or home stereo receivers pick up the lower-power local or campus radio stations, or distant amateur VHF stations in the 2-metres band. Due to the so-called threshold effect, FM receivers

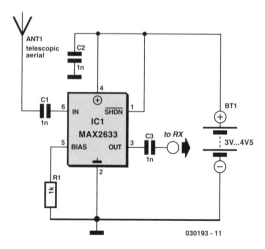

030193 - 11

loose signals abruptly so if your favourite station fades in and out as you drive, this amplifier can cause a dramatic improvement. The MAX2633 is a low-voltage, low-noise amplifier for use from VHF to SHF frequencies. Operating from a single +2.7 V to +5.5V supply, it has a virtually flat gain response to 900 MHz. Its low noise figure and low supply current makes it ideal for RF receive, buffer and transmit applications. The MAX2633 is biased internally and has a user-selectable supply current, which can be adjusted by adding a single external resistor (here, R1).

This circuit draws only 3 mA current. Besides a single bias resistor, the only external components needed for the MAX2630 family of RF amplifiers are input and output blocking capacitors, C1 and C3, and a VCC bypass capacitor, C2.

The coupling capacitors must be large enough to contribute negligible reactance in a 50 Ω system at the lowest operating frequency. Use the following equation to calculate their minimum value:

$$C_C = \frac{53000}{f_{laag}} \, [pF]$$

Further information:
www.maxim-ic.com.

D. Prabakaran   (030193-1)

085
LED Light Pen

Physicians and repair engineers often use small light pens for visual examination purposes. Rugged and expensive as these pens may be, their weak point is the bulb, which is a 'serviceable' part. In practice, that nearly always equates to 'expensive' and / or 'impossible to find' when you need one.

LEDs have a much longer life than bulbs and the latest ultra bright white ones also offer higher energy-to-light conversion efficiency. On the down side, LEDs require a small electronic helper circuit called 'constant-current source' to get the most out of them. Here, T1 and R1 switch on the LED. R2 acts as a current sensor with T2 shunting off (most of) T1's base bias current when the voltage developed across R2 exceeds about 0.65 V.

The constant current through the white LED is calculated from

$$R2 = \frac{0.65}{I_{LED}}$$

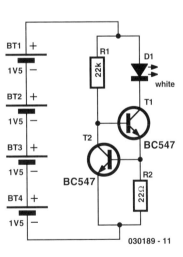

030189 - 11

With some skill the complete circuit can be built such that its size is equal to an AA battery. The four button cells take the place of the other AA battery that used to be inside the light pen.

Myo Min   (030189-1)

086
Monitor Life Xtender

This circuit was designed to protect a computer monitor from overheating. It is recommended to attach this circuit to power users' monitors! Most computer monitors of the CRT type fail owing to overheating. After one or two hours of use, the rear of a monitor may become as hot as 45 degrees C, or 20 degrees above ambient temperature. Most heat comes from the VGA gun drivers, the horizontal circuit, vertical circuit and power supply. The best possible way to extract heat and so prolong monitor life (and save the environment) is to add a brushless fan, which is lighter, energy-wiser and more efficient than a normal fan. In the diagram, diodes D2, D3 and D4 sense the monitor's temperature. These diodes have a total negative temperature coefficient of 6 mV per degree Celsius. To eliminate noise, shielded wire should be used for the connection of the temperature

sensor to the circuit sensor. The +12 V supply voltage is borrowed from the computer's power supply. Alternatively, a mains adapter with an output of 12 VDC may be used. C1 and C2 are decoupling capacitors to eliminate the ripple developed by switching or oscillation. R1 provides bias current to D1, a 6 V zener diode acting as a reference on the non-inverting pin of opamp IC2.B. IC1, a 'precision shunt regulator' raises the sensor diodes' voltage to just over 6 V depending on the adjustment of P1. C4 is the decoupling capacitor with the sensor network. Integrator network R4-C5 provides a delay of about 3 seconds, transforming the on/off output signal of IC2.B into an exponentially decreasing or increasing voltage. This voltage is fed to pin 3 of the second opamp, IC2.A. The hard on/off technique would produce a good amount of noise whenever the load is

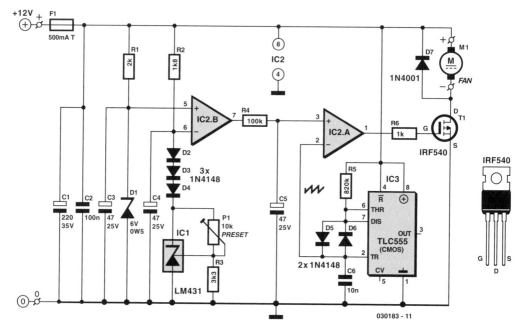

030183 - 11

switched, hence an alternative had to be found. IC3, a TLC555, is used as an astable multivibrator with R5 and C6 controlling the charging network that creates a sawtooth voltage with a frequency of about 170 Hz. This sawtooth is coupled to pin 2 of IC2.A, which compares the two voltages at its input pins and produces a PWM (pulsewidth modulated) output voltage. The sawtooth wave is essential to the PWM signal fed to power output driver T1 by way of stopper resistor R6. The power FET will switch the fan on and off fan according to the PWM drive signal. The back emf pulses that occur when T1 switches on

and off are clamped by a high-speed diode, D7. Initially, turn P1 to maximum resistance. Blow hot air from a hair dryer onto the sensor-diodes for a minute or so, then get the temperature meter near the sensor diodes and adjust P1 slowly towards the minimum resistance position with a digital meter hooked up on pin 7 of IC2.B. Roughly calibrate the temperature to 40 degrees C. At this temperature, the meter will show approximately 12 V. The circuit will draw about 120 mA from its 12 V supply.

Myo Min   (030183-1)

087
Whistling Kettle

Most electric kettles do not produce a whistle and just switch off when they have boiled. Fitting a box of electronics directly onto an electric kettle (or even inside!) to detect when the kettle has boiled is obviously out of the question. The circuit shown here detects when the kettle switches off, which virtually all kettles do when the water has boiled. In this way, the electronics can be housed in a separate box so that no modification is required to the kettle. The box is preferably a type incorporating a mains plug and socket. In this application, the current flowing in coil L1 provides a magnetic field that actuates reed switch S1. Since the current drawn by the kettle element is relatively large (typically 6 to 8 amps), the coil may consist of a few turns of wire around the reed switch. The reed switch is so fast it will actually follow the AC current flow through L1 and produce a 100-Hz buzz. The switching circuit driven by the reed switch must, therefore, disregard these short periods when the contacts open, and respond only when they remain open for a relatively long period

when the kettle has switched off. The circuit is based on a simple voltage controlled oscillator formed around T2 and T3. Its operation is best understood by considering the circuit with junction R4/R5 at 0 V and C4 discharged. T2 will receive base current through R5 and turn on, causing T3 to turn on as well. The falling collector voltage of T3 is transmitted to the base of T2 by C4 causing this transistor to conduct harder. Since the action is regenerative, both transistors will turn on quickly and conduct heavily. C4 will therefore charge quickly through T2's base-emitter junction and T3. Once the voltage across C4 exceeds about 8.5 V (leaving less than 0.5 V across T2's be junction), T2 will begin to turn off. This action is also regenerative so that soon both transistors are switched off and the collector voltage of T3 rises rapidly to +9 V. With C4 still charged to 8.5 V, the base of T2 will rise to about 17.5 V holding T2 (and thus T3) off. C4 will now discharge relatively slowly via R5 until T2 again begins to conduct whereupon the cycle will repeat. The voltage at the collector of T3

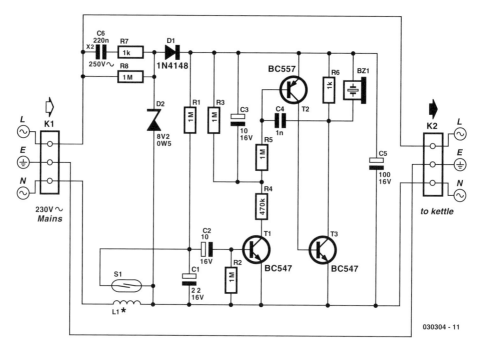

030304 - 11

will therefore be a series of short negative going pulses whose basic frequency will depend on the value of C4 and R5. The pulses will be reproduced in the piezo sounder as a tone.

The oscillation frequency of the regenerative circuit is heavily dependent on the voltage at junction R4/R5. As this voltage increases, the frequency will fall until a point is reached when the oscillation stops altogether. With this in mind, the operation of the circuit around T1 can be considered. In the standby condition, when the kettle is off, S1 will be open so that C1 and C2 will be discharged and T1 will remain off so that the circuit will draw no current. When the kettle is switched on, S1 is closed, causing C1 and C2 to be discharged and T1 will remain off. C3 will remain discharged so that T2 and T3 will be off and only a small current will be drawn by R1. Although S1 will open periodically (at 100 Hz), the time constant of R1/C1 is such that C1 will have essentially no voltage on it as the S1 contacts continue to close. When the kettle switches off, S1 will be permanently open and C1/C2 will begin to charge via R1, causing T1 to switch on. C3 will then begin

to charge via R4 and the falling voltage at junction R4/R5 will cause T2/T3 to start oscillating with a rising frequency. However, once T1 has switched off, C3 will no longer be charged via R4 and will begin to discharge via R3 and R5 causing the voltage at R4/R5 to rise again. The result is a falling frequency until the oscillator switches off, returning the circuit to its original condition. As well as reducing the current drawn by the circuit to zero, this mimics the action of a conventional whistling kettle, where the frequency rises as more steam is produced and then falls when it is taken off the boil.

The circuit is powered directly by the mains using a 'lossless' capacitive mains dropper, C6, and zener a diode, D2, to provide a nominal 8 V dc supply for the circuit. A 1-inch reed switch used in the prototype required about 9 turns of wire to operate with a 2 kW kettle element. Larger switches or lower current may require more turns. In general, the more turns you can fit on the reed switch, the better, but do remember that the wire has to be thick enough to carry the current. It is strongly recommended to test the circuit using a

9 volt battery instead of the mains-derived supply voltage shown in the circuit diagram. A magnet may be used to operate S1 and so simulate the switching of the kettle. Warning.

This circuit is connected directly to the 230 V mains and none of the components must be touched when the circuit is in use. The circuit must be housed in an approved ABS case and carry the earth connection to the load as indicated. Connections and solder joints to components with a voltage greater than 200 volts across them (ac or dc) must have an insulating clearance of least 6 mm. An X2 class capacitor must be used in position C6.

Bart Trepak   (030304-1)

088
Relay Coil Energy Saver

Some relays will become warm if they remain energised for some time. The circuit shown here will actuate the relay as before but then reduce the 'hold' current through the relay coil current by about 50%, thus considerably reducing the amount of heat dissipation and wasted power. The circuit is only suitable for relays that remain on for long periods. The following equations will enable the circuit to be dimensioned for the relay on hand:

$$R3 = \frac{0.7}{I}$$

Charge time $= 0.5 \times R2 \times C1$

Where I is the relay coil current. After the relay has been switched off, a short delay should be allowed for the relay current to return to maximum so the relay can be energised again at full power. To make the delay as short as possible, keep C1 as small as possible. In practice, a minimum delay of about 5 seconds should be allowed but this is open to experimentation.

The action of C2 causes the full supply voltage to appear briefly across the relay coil, which helps to activate the relay as fast as possible. Via T2, a delay network consisting of C1 and R2 controls the relay coil current flowing through T1 and R3, effectively reducing it to half the 'pull in'

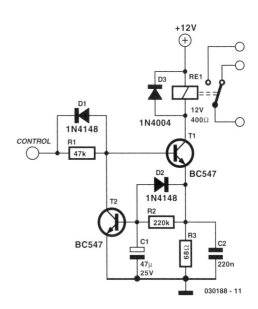

current. Diode D2 discharges C1 when the control voltage is Low. Around one second will be needed to completely discharge C1. T2 shunts the bias current of T1 when the delay has elapsed. Diode D1 helps to discharge C1 as quickly as possible. The relay shown in the circuit was specified at 12 V / 400 ohms. All component values for guidance only.

Myo Min   (030188-1)

089
Voltage LevelsControl Relays

This circuit proves that microcoprocessors, PCs and the latest ultra-accurate DACs are overkill when it comes to controlling four relays in sequence in response to a rising control voltage in the range 2.4 V – 12 V. By using equal resistors in ladder network R1-R5, equal intervals are created between the voltages that switch on the relays in sequence. Each resistor drops 1/5th of the supply voltage or 2.4 V in this case, so we get +2.4 V =Re1, +4.8 V = Re2, +7.2 V = Re3, +9.6 V = Re4. Obviously, these switching levels vary along with the supply voltage, hence the need to employ a stabilised power supply. Looking at the lowest level switching stage, when the control voltage exceeds 2.4 V, IC1 will flip its output to (nearly)the supply level. The resulting current sent into the base of T1 is limited to about 1 mA by R6.

With T1 driven hard, relay Re1 is energised by the collector current. Because the BC548 has a maximum collector current spec of 100 mA, the relaycoil resistance must not be smaller than120 ohms.

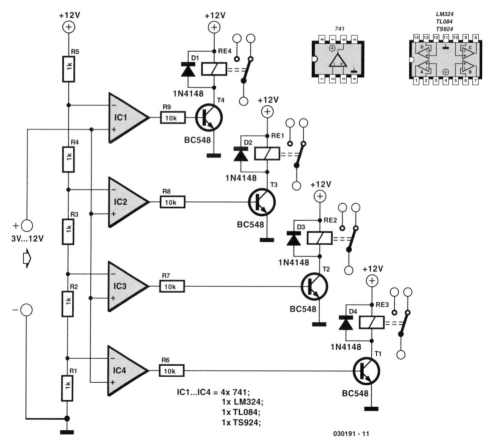

IC1...IC4 = 4x 741;
1x LM324;
1x TL084;
1x TS924;

030191 - 11

Nearly all current consumed by the circuit goes on account of the relaycoils, so depending on your relays a pretty hefty power supply of up to 500 mA may be required. When dimensioning the ladder network to create the desired switching levels, it is good to remember that the 741 will not operate very well with input voltages below 1.5 V or above 10.5 V, while voltage levels outside the supply range (i.e., negative or above +12 V) are out of the question. If you do need a switching level in the range 0-1.5 V, consider using an LM324, which contains four opamps in one package.

For the high side of the range (10.5 to 12 V), a TL084 or a 'rail-to-rail' opamp like the TS924 is required. However, the TS924 cannot be used with supply voltages above 12 V.

Raj. K. Gorkhali   (030191-1)

090
Power Flip-Flop Using a Triac

Modern electronics is indispensable for every large model railroad system, and it provides a solution to almost every problem. Although ready-made products are exorbitantly expensive, clever electronics hobbyists try to use a minimum number of components to achieve optimum results together with low costs. This approach can be demonstrated using the rather unusual semiconductor power flip-flop described here. A flip-flop is a toggling circuit with two stable switching states (bistable multi-vibrator). It maintains its output state even in the absence of an input pulse. Flip-flops can easily be implemented using triacs if no DC voltage is available. Triacs are also so inexpensive that they are often used by model railway builders as semiconductor power switches. The decisive advantage of triacs is that they are bi-directional, which means they can be triggered during both the positive and the negative half-cycle by applying an AC voltage to the gate electrode (G). The polarity of the trigger voltage is thus irrelevant. Triggering with a DC current is also possible. Figure 1 shows the circuit diagram of such a power flop-flop. A permanent magnet is fitted to the model train, and when it travels from left to right, the magnet switches the flip-flop on and off via reed switches S1 and S2. In order for this to work in both directions of travel, another pair of reed switches (S3 and S4) is connected in parallel with S1 and S2. Briefly closing S1 or S3 triggers the triac.

The RC network C1/R2, which acts as a phase shifter, maintains the trigger current. The current through R2, C1 and the gate electrode (G) reaches its maximum value when the voltage across the load passes through zero. This causes the triac to be triggered anew for each half-cycle, even though no pulse is present at the gate. It remains triggered until S2 or S4 is closed, which causes it to return to the blocking state. The load can be incandescent lamps in the station area (platform lighting) or a solenoid-operated device, such as a crossing gate. The LED connected across the output (with a rectifier diode) indicates the state of the flip-flop. The circuit shown here is designed for use in a model railway system, but there is no reason why it could not be used for other applications. The reed switches can also be replaced by normal pushbutton switches. For the commonly used TIC206D triac, which has a maximum current rating of 4 A, no heat sink is necessary in this application unless a load

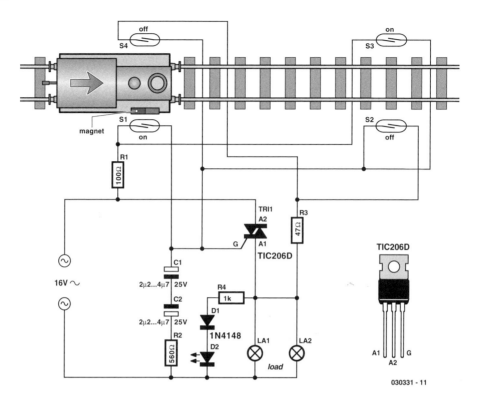

030331 - 11

current exceeding 1 A must be supplied continuously or for an extended period of time. If the switch-on or switch-off pulse proves to be inadequate, the value of electrolytic capacitor C1 must be increased slightly.

R. Edlinger   (030331-1)

091
Acoustic Sensor

This acoustic sensor was originally developed for an industrial application (monitoring a siren), but will also find many domestic applications. Note that the sensor is designed with safety of operation as the top priority: this means that if it fails then in the worst-case scenario it will not itself generate a false indication that a sound is detected. Also, the sensor connections are protected against polarity reversal and short-circuits. The supply voltage of 24 V is suitable for industrial use, and the output of the sensor swings over the supply voltage range. The circuit consists of an electret microphone, an amplifier, attenuator, rectifier and a switching stage. MIC1 is supplied with a current of 1 mA by R9. T1 amplifies the signal, decoupled from the supply by C1, to about 1 Vpp. R7 sets the collector current of T1 to a maximum of 0.5 mA. The operating point is set by feedback resistor R8. The sensitivity of

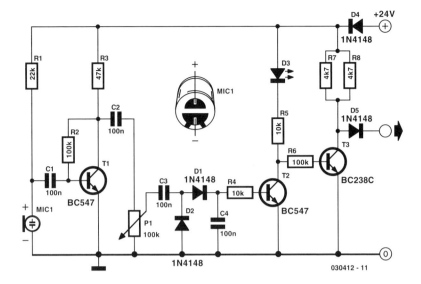

030412 - 11

the circuit can be adjusted using potentiometer P1 so that it does not respond to ambient noise levels. Diodes D1 and D2 rectify the signal and C4 provides smoothing. As soon as the voltage across C4 rises above 0.5 V, T2 turns on and the LED connected to the collector of the transistor lights.

T3 inverts this signal. If the microphone receives no sound, T3 turns on and the output will be at ground. If a signal is detected, T3 turns off and the output is pulled to +24 V

by R4 and R5. In order to allow for an output current of 10 mA, T3's collector resistor needs to be 2.4 kΩ. If 0.25 W resistors are to be used, then to be on the safe side this should be made up of two 4.7 kΩ resistors wired in parallel. Diode D4 protects the circuit from reverse polarity connection, and D3 protects the output from damage if it is inadvertently connected to the supply.

Engelbert Göpfert (030412-1)

092
Model Railway Short-Circuit Beeper

Short circuits in the tracks, points or wiring are almost inevitable when building or operating a model railway. Although transformers for model systems must be protected against short circuits by built-in bimetallic switches, the response time of such switches is so long that is not possible to immediately localise a short that occurs while the trains are running, for example. Furthermore, bimetallic protection

switches do not always work properly when the voltage applied to the track circuit is relatively low. The rapid-acting acoustic short-circuit detector described here eliminates these problems. However, it requires its own power source, which is implemented here in the form of a GoldCap storage capacitor with a capacity of 0.1 to 1 F. A commonly available reed switch (filled with an inert gas) is used for the cur-

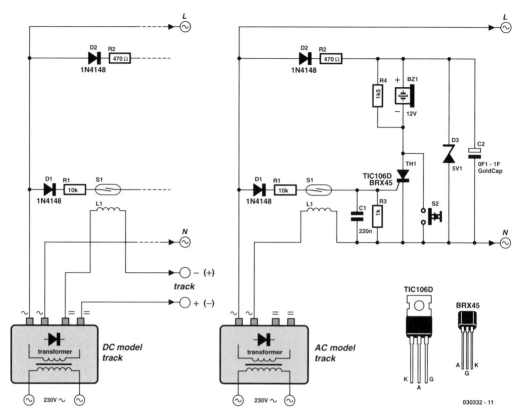

rent sensor, but in this case it is actuated by a solenoid instead of a permanent magnet.

An adequate coil is provided by several turns of 0.8–1 mm enamelled copper wire wound around a drill bit or yarn spool and then slipped over the glass tube of the reed switch. This technique generates only a negligible voltage drop. The actuation sensitivity of the switch (expressed in ampère-turns or A-t)) determines the number of turns required for the coil. For instance, if you select a type rated at 20–40 A-t and assume a maximum allowable operating current of 6 A, seven turns ($40 \div 6 = 6.67$) will be sufficient. As a rule, the optimum number of windings must be determined empirically, due to a lack of specification data. As you can see from the circuit diagram, the short-circuit detector is equally suitable for AC and DC railways. With Märklin transformers (HO and I), the track and lighting circuits can be sensed together, since both circuits are powered

from a single secondary winding. Coil L1 is located in the common ground lead ('O' terminal), so the piezoelectric buzzer will sound if a short circuit is present in either of the two circuits. The (positive) trigger voltage is taken from the lighting circuit (L) via D1 and series resistor R1. Even though the current flowing through winding L1 is an AC or pulsating DC current, which causes the contact reeds to vibrate in synchronisation with the mains frequency, the buzzer will be activated because a brief positive pulse is all that is required to trigger thyristor Th1. The thyristor takes its anode voltage from the GoldCap storage capacitor (C2), which is charged via C2 and R2. The alarm can be manually switched off using switch S1, since although the thyristor will return to the blocking state after C2 has been discharged if a short circuit is present the lighting circuit, this will not happen if there is a short circuit in the track circuit. C1 eliminates any noise pulses that may be generated. As a continuous tone does not attract as much attention as an intermittent beep, an intermittent piezoelectric generator is preferable. As almost no current flows during the intervals between beeps and the hold current through the thyristor must be kept above 3 mA, a resistor with a value of 1.5–1.8 kΩ is connected in parallel with the buzzer. This may also be necessary with certain types of continuous-tone buzzers if the operating current is less than 3 mA.

The Zener diode must limit the operating voltage to 5.1 V, since the rated voltage of the GoldCap capacitor is 5.5 V.

R. Edlinger   (030332-1)

093
Triple Power Supply

Inexpensive miniature transformers normally provide one or two secondary voltages, which is sufficient for generating a set of positive and negative supply voltages, such as are needed for operational amplifier circuits. But what can you do if you need an additional voltage that is higher than either of the supply voltages (such as a tuning voltage for a receiver?). This circuit shows a simple solution to this problem, and it certainly can be extended to suit other applications. Using a 2×15 V transformer, it generates positive 24 V and 12 V supply voltages and a negative 12 V supply voltage. The little trick for generating the +24 V output consists of using IC1 to create a virtual ground. This is based on a well-known circuit with a voltage divider formed by two equal-valued resistors, which divide the voltage Ub across the rectifier from approximately 40 V down to 20 V. This Ub/2 potential is buffered by an opamp, which allows this virtual ground to drive a load. The present circuit uses the

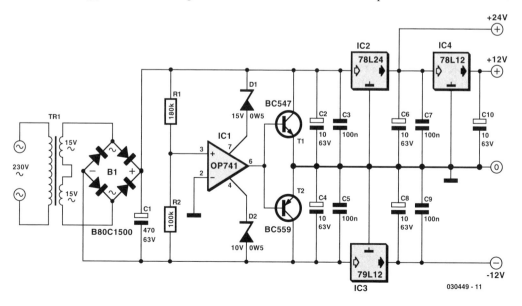

same principle, but instead of being divided by a factor of 2, the voltage across the rectifier (approximately 40 V) is divided unequally by R1 and R2. The resulting potential, which is buffered by the opamp and the subsequent transistor output stage, lies approximately 15 V above the lower potential, and thus around 25 V below the upper potential. The three voltages are stabilised using standard 100 mA voltage regulators, as shown in the schematic.

The supply voltages for the opamp are also asymmetric. Thanks to the low current consumption, this can be managed using two Zener diodes. You must bear in mind that the secondary voltage generated by an unloaded miniature transformer is significantly higher than its rated secondary voltage. The following results were obtained in a test circuit using a 1.6 VA transformer with two 15 V secondary windings: the positive and negative 12 V outputs could be loaded at around 10 mA each, and the 24 V output could be loaded with approximately 20 mA, all without any drop in any of the output voltages. For small circuits such as a 0(4)–20 mA instrumentation loop, this is fully adequate. For more complex circuits or switched loads, additional compensation may be necessary.

Bernd Schädler   (030449-1)

094
3.3 V or 5 V Direct from the Mains

The SR03x range of voltage regulator chips from Supertex (www.supertex.com) connects directly to the rectified mains supply and provides a low-current 3.3 V or 5.0 V output without the need for any step-down transformer or inductor. The circuit requires a full-wave rectified mains voltage input (waveform a). A built-in comparator controls a series-pass configured MOSFET. The MOSFET is only switched on whenever the input voltage is below an 18 V threshold. A 220 µF capacitor is used to smooth out  fluctuations so that the resultant voltage has a sawtooth waveform (waveform b)

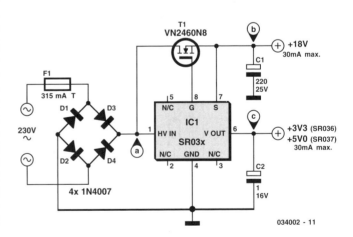

Warning: This circuit must only be used in a fully encapsulated enclosure with no direct connections to any external circuit. It is important to be aware that the circuit is connected to the mains and the chip has lethal voltages on its pins! All appropriate safety guidelines must be adhered to.

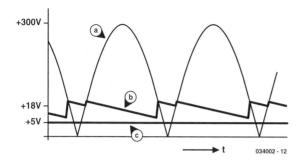

age regulator produces a regulated output (waveform c) of 3.3 V for the type SR036 or 5.0 V for the SR037. Normally you would expect to see a reservoir capacitor fitted across the output of a full wave rectifier in a power supply circuit but in this case it is important to note that one is not fitted. For correct operation it is necessary for the input voltage to fall close to zero during each half wave.

with a peak value of 18 V. This unregulated voltage is connected to the source input of the chip (pin 7) and an internal volt-

Gregor Kleine   (034002-1)

095
UV Torch Light

UV (ultra-violet) LEDs can produce eye-catching effects when their light is allowed to interfere with certain colours, particularly with reflected light under near-dark conditions. Also try shining some UV light on a diamond... Most UV LEDs require about 3.6 V (the 'blue' diode voltage) to light. Here, a MAX761 step-up switching IC is used to provide constant current to bias the UV diode. The IC employs PWM in high-current mode and automatically changes to PFM mode in low or medium power mode to save (battery) power. To allow it to be used with two AA cells, the MAX761 is configured in bootstrapped mode with voltage-adjustable feedback. Up to four cells may be used to power the circuit but they may add more weight than you would like for a torchlight. To prolong the switch life, R1 is connected to the IC's SHDN

(shutdown) pin. Less than 50 nA will be measured in shutdown mode. Electrolytic capacitor C1 is used to decouple the circuit supply voltage. Without it, ripple and noise may cause instability. The one inductor in the circuit, L1, may have any value between about 10 and 50 µH. It stores current in its magnetic field while the MOSFET inside the MAX761 is switched. A toroid inductor is preferred in this position as it will guarantee low stray radiation. D1 has to be a relatively fast diode so don't be tempted

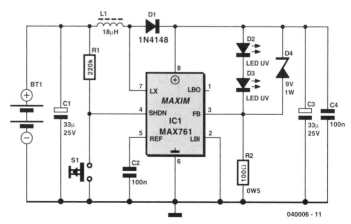

to use an 1N400x because it has a too slow recovery time. The circuit efficiency was measured at about 70%. R2, the resistor on the feed-back pin of the MAX761 effectively determines the amount of constant current, I, sent through the UV LEDs, as follows:

$$R2 = \frac{1.5}{I}$$

where I will be between 2 mA and 35 mA. Zener diode D4 clamps the output voltage when the load is disconnected, which may happen when one of the UV LEDs breaks down. Without a load, the MAX761 will switch L1 right up to the boost voltage and so destroy itself.

Myo Min   (040006-1)

096
Storage Battery Exerciser

A motorcycle or boat battery that is not needed over the winter is usually charged before being put away for the winter, after which it remains standing unused for

months on end. As a result, it accumulates deposits of lead sludge, which can result in reduced capacity or even complete failure of the battery. If you don't keep active, you

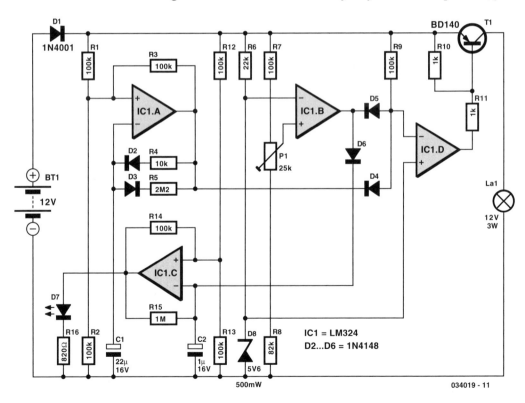

IC1 = LM324
D2...D6 = 1N4148

034019 - 11

rust! To avoid this, it's necessary to keep the battery active even during the winter. This circuit does such a good job of exercising the battery that it doesn't have to be recharged during the winter. It only has to be fully charged again in the spring before being used again. IC1.A is an astable multi-vibrator with an asymmetric duty cycle. The output is High for around 0.6 s and Low for around 40 s. IC1.B is wired as a comparator that constantly monitors the battery voltage. Its threshold voltage is set to 11.0 V using the trimpot. As soon as the battery voltage drops below this value, the comparator goes Low and D6 is cut off, allowing the second astable multi-vibrator IC1.C to oscillate at approximately 1.2 Hz. LED D7 then blinks to indicate that the battery must be charged. As long as the battery voltage is greater than 11 V, IC1.B is

High. IC1.A is Low most of the time, and in this state D4 conducts and the inverting input of IC1.D is Low. This means that IC1.D is High most of the time, with T1 cut off. T1 only conducts during the 0.6-s intervals when IC1.A is High. In this state it allows current to pass through the lamp (12 V / 3 W), which forms the actual load for the battery. After this, darkness prevails again for 40 s.

The average current consumption is approximately 5 mA. At this rate, a relatively new 40 Ah battery will take around one year to become fully discharged. However, this can vary depending on the condition of the battery, and it may be necessary to 'top up' the battery once during the winter.

Ludwig Libertin   (034019-1)

097
Simple Darkness Activated Alarm

Most darkness activated alarms employ opamps and some logic ICs. Here, a less expensive approach is shown based on the eternal 555, this time in monostable multivibrator mode. Components R2 and C1 represent a one-second network. When the LDR (light dependent resistor) is in the dark, its resistance is high, pulling pin 2 of the 555 to ground. This triggers the monostable and the (active!) 6 volt piezo buzzer will sound. Preset P1 is adjusted depending on ambient light levels.

The circuit may be fitted on a wall in your home. Assuming P1 has been set for the existing ambient light level, the shadow cast by anybody entering the room or hallway will trigger the alarm.

Myo Min   (030130-1)

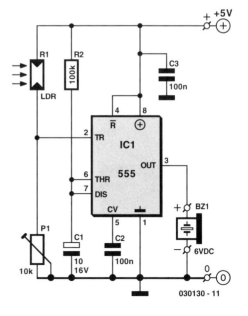

030130 - 11

098
Mains Failure Alarm

This circuit was designed to produce an audible alarm when the mains power is interrupted. Such an alarm is essential for anyone whose livelihood depends on keeping perishable foodstuffs in cold storage. The circuit is powered by a 12 V mains adapter. LED D5 will light when the mains voltage is present. When the mains voltage disappears, so does the +12 V supply voltage,

leaving the voltage regulator IC1 and relay driver T1-T2 without power. The relay driver, by the way, is an energy-saving type, reducing the coil current to about 50% after a few seconds. Its operation and circuit dimensioning are discussed in the article 'Relay Coil Energy Saver'. The value of the capacitor at the output of voltage regulator IC1 clearly points to a dif-

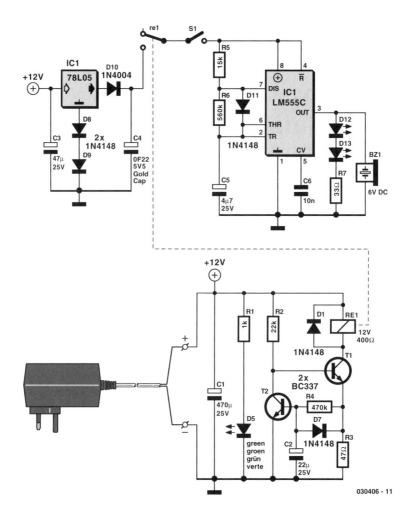

030406 - 11

ferent use than the usual noise suppression. When the mains power disappears, Re1 is de-energised and the 0.22 F Goldcap used in position C4 provides supply current to IC2. When the mains voltage is present, C4 is charged up to about 5.5 volts with IC1 acting as a 100 mA current limit and D10 preventing current flowing back into the regulator output when the mains voltage is gone. According to the Goldcap manufacturer, current limiting is not necessary during charging but it is included here for the security's sake. The CMOS 555 is config-

ured in astable multivibrator mode here to save power, and so enable the audible alarm to sound as long as possible. Resistors R5 and R6 R5 (at the expense of a higher average define a short 'on' time of just 10 ms. That current drawn from the Goldcap). is, however, sufficient to get a loud warning from the active buzzer. In case the pulses are too short, increase the value of R5 (at the expense of a higher average current drawn from the Goldcap).

Myo Min　　　　　　　　　　　　　　　(030406-1)

099
Simple NiCd Charger

A simple NiCd charger can be built using 'junk box' components and an inexpensive LM317 or 78xx voltage regulator. Using a current limiter composed of R3 and a transistor, it can charge as many cells as desired until a 'fully charged' voltage determined by the voltage regulator is reached, and it indicates whether it is charging or has reached the fully charged state. If the storage capacitor (C1) is omitted, pulsed charging takes place. In this mode, a higher charging current can be used, with all of the control characteristics remaining the same. The operation of the circuit is quite simple. If the cells are not fully charged, a charging current flows freely from the voltage regulator, although it is limited by resistor R3 and transistor T1. The limit is set by the formula:

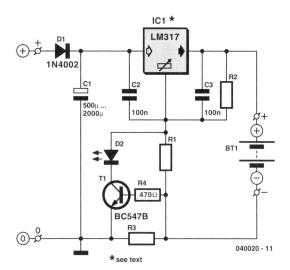

$$I_{max} \approx \frac{0.6\ V}{R3}$$

For I_{max} = 200 mA, this yields R3 = 3 Ω. The LED is on if current limiting is active,

which also means that the cells are not yet fully charged. The potential on the reference lead of the voltage regulator is raised by approximately 2.9 V due to the voltage across the LED. This leads to a requirement for a certain minimum number of cells.

For an LM317, the voltage between the reference lead and the output is 1.25 V, which means at least three cells must be charged (3 × 1.45 V > 2.9 V + 1.25 V). For a 78xx with a voltage drop of around 3 V (plus 2.9 V), the minimum number is four cells. When the cells are almost fully charged, the current gradually drops, so the current limiter becomes inactive and the LED goes out. In this state, the voltage on the reference lead of the regulator depends only on voltage divider R1/R2. For a 7805 regulator, the value of R2 is selected such that the current through it is 6 mA. Together with the current through the regulator (around 4 mA), this yields a current of around

10 mA through R1. If the voltage across R1 is 4 V (9 V – 5 V), this yields a value of 390 Ω.

The end-of-charge voltage can thus be set to approximately 8.9 V. As the current through the regulator depends on the device manufacturer and the load, the value of R1 must be adjusted as necessary. The value of the storage capacitor must be matched to the selected charging current. As already mentioned, it can also be omitted for pulse charging.

Wolfgang Schmidt   (040020-1)

100
The Eternal 555

You may not realise this, but the 555 timer IC has been in existence for over 30 years. The chip was originally manufactured by Signetics. In the first three months following its introduction (1972) over half a million of them were sold. Moreover, it has stayed successful: since that time the 555 has been the

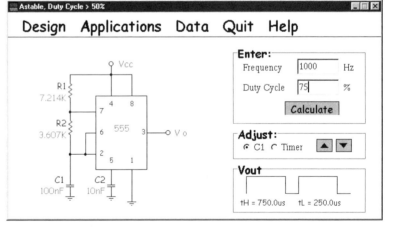

most popular IC sold every year! Nowadays it makes sense to use the CMOS version of this IC, since it consumes significantly less power. Virtually everything regarding the 555 can be found at www.schematica.com/555_Timer_design/555.htm. A program can be downloaded from this site, which easily calculates the values for the R-C components. The pro-

gram is suitable for both the astable and bistable modes. The 'adjust' buttons are used to switch is chosen for C1, the resistors change between the single 555 and the double version (the 556). When a different value is chosen for C1, the resistors change automatically.

Karel Walraven   (044020-1)

101
Reset from Multiple Power Supplies

Processor based systems usually require a voltage supervisor chip to produce a clean reset pulse to the processor whenever a 'brown-out' condition of the power supply is detected. More complex designs employing multiple power supplies can be unreliable if some of the supplies are not supervised. The circuit described here monitors all the supply rails in the system (here +12 V, –12 V and +5 V) and provides a reset pulse to the processor whenever it detects any are not within tolerance. IC1 (TL7705A) generates a processor reset if the 5 V rail falls below 4.55 V. The value of the capacitor fitted to pin 3 defines the reset pulse width td according to the formula:

$$t_d = 12 \cdot C_T \cdot 10^3$$

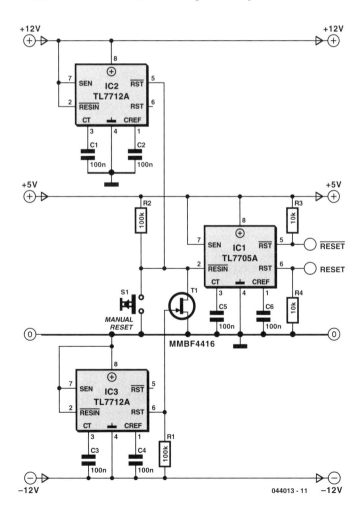

044013 - 11

With C_T in µF the value for td is given in µs. A capacitor of 100 nF for example, will produce a reset pulse of around 1.2 ms. Pin 6 (RESET) outputs an active-high pulse and Pin 5 ($\overline{\text{RESET}}$) an active-low pulse. The outputs are open collector types so an external pull-down and pull-up resistor (respectively) is required.

The $\overline{\text{RESIN}}$ input (Pin 2) of IC1 is driven from two TL7712A supervisors monitoring +12 V (IC2) and –12 V (IC3). The TL7712A generates a reset when the supply voltage falls below a threshold level of 10.8 V. The open collector output $\overline{\text{RST}}$ (Pin 5) of IC2 is connected to the $\overline{\text{RESIN}}$ pin of IC1 and pulled up to 5 V via a 100 kΩ resistor. The open collector out-put of IC2 can be directly connected to the reset input of IC1 but the output of IC3 must be connected via a level shifting device before it can be connected to the reset input of IC1 because the voltage level at the output of IC3 goes negative. JFET transistor T1 is

used to perform the necessary level shifting. The JFET turns off when the voltage at its gate-source junction is between –2.5 V and –6 V. When IC3 is issuing a reset signal the RST out-put (pin 6) will go up to ground potential and cause T1 to conduct and trigger a reset of IC1. At all other times the RST output of IC3 will be pulled to a minus voltage via the 100 kΩ resistor which then causes T1 to stop conducting and release the reset. A manual reset push button can also be connected to $\overline{\text{RESIN}}$ of IC1 if required.

The SENSE input (Pin 7) of the TL77xx chips is connected to the positive supply rail. The reference input (pin 1) is fitted with a 100 nF capacitor to reduce the effects of fast transients. The JFET type MMBF4416 is available from Conrad Electronic (www.conrad.de), order no. 14 28 08.

Gregor Kleine (044013-1)

102
Airflow Monitor

Fans are usually monitored by measuring their operating currents. If the current lies within a certain range, it is assumed that the fan is spinning properly and providing a stream of cooling air. If it falls below a lower threshold or exceeds an upper threshold, something is wrong with the fan: it is either defective or blocked by some sort of object. The cooling airflow generated by a fan can be directly monitored using the Analog Devices TMP12 sensor IC (www.ana-log.com). This IC contains a temperature sensor and a heater resistor, as well as two comparators and a reference-voltage source. Figure 1 shows the complete circuit diagram of an airflow monitor. The voltage divider formed by R1, R2 and R3 defines the temperature

thresholds and the hysteresis for the switching points (via the current I_{REF} flowing through the resistor chain). The internal heater resistor can be powered directly from the supply voltage via pin 5 (Heater), but an external resistor (R5) can also be connected in series between the supply voltage and pin 5 to reduce the internal power dissipation of the IC. The circuit output is provided here by two LEDs driven by the open-collector outputs $\overline{\text{UNDER}}$ (pin 6) and $\overline{\text{OVER}}$ (pin 7). The operating principle of the TMP12 IC is that it is warmed by the integrated heater resistor and cooled by the airs flow. If there is no airflow or the airflow is insufficient due to a defective fan or obstructed air inlet, the temperature increases until the amount of

1

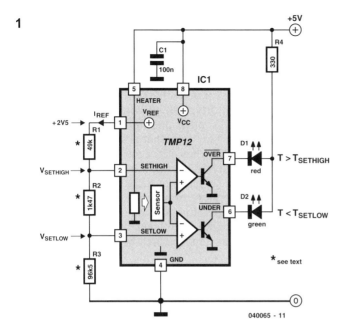

040065 - 11

2

040065 - 12

the temperature of the IC will rise can be read from the two curves, which show the situation with and without cooling airflow. As indicated, the temperature thresholds $T_{SETHIGH}$ and T_{SETLOW} are dimensioned such that with the amount of power converted into heat by the resistor (in this case, 250 mW), the temperature for the curve with cooling airflow lies between the two temperature thresholds. Here the threshold temperatures are +55 °C and +60 °C. The voltage divider R1/R2/R3 determines not only the absolute positions of the temperature thresholds, but also the hysteresis of the switching points. The hysteresis is determined by the current I_{REF} flowing through the resistor chain. The associated formulas are shown in Figure 3. Here ΔT is the hysteresis, which in this case is set to 2 °C and yields a value of 17 µA for I_{REF}. The node voltages for the voltage divider can now be determined from the threshold temperatures, which in this case yields $V_{SETHIGH} = 1.666$ V for an upper threshold of +60 °C and $V_{SETLOW} = 1.641$ V for a lower threshold of +55 °C As $V_{REF} = 2.5$ V, the values of R1, R2 and R3 can now be readily calculated from the current and the voltage drops across the resistors. The values calculated in this manner are shown in the schematic diagram, without taking into account whether such values are actually available. As the temperature thresholds used here are relatively close together, the actual values of the resistors must be quite

heat dissipated by the IC (by conduction to the circuit board or other means) balances the amount of heat generated inside the IC. Figure 2 shows this in the form of two curves. The power dissipation of the internal 100 Ω heater resistor is plotted on the X axis. This can be as much as 250 mW if pin 5 is connected directly to +5 V. If the heater resistor is not dissipating any power, the sensor will be at approximately ambient temperature, which is here taken to be +50 °C. If the power dissipated by the heater resistor increases, the level to which

3

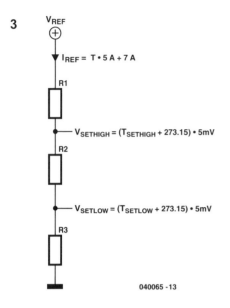

V_{REF}

$I_{REF} = T \cdot 5\,A + 7\,A$

R1

$V_{SETHIGH} = (T_{SETHIGH} + 273.15) \cdot 5mV$

R2

$V_{SETLOW} = (T_{SETLOW} + 273.15) \cdot 5mV$

R3

040065 -13

close to the calculated values. This can be achieved by connecting standard-value fixed resistors in series and/or parallel, or by using trimpots. The TMP12 can be used to generate digital monitoring signals for a processor or switch on a supplementary fan (via a driver stage connected to the outputs). Another possible application is controlling an oven that is switched off by the TMP12 when it reaches its set-point temperature. Such an oven can be used to operate a crystal oscillator at an elevated temperature in order to make it insensitive to temperature variations (a crystal oven). According to its data sheet, the IC can be used at temperatures between –40 °C und +125 °C.

Gregor Kleine   (040065-1)

103
Switchless NiCd/NiMH Charger

This circuit may be used to replace the single current limiting resistor often found in dirt cheap battery chargers. The alternative shown here will eventually pay off because you no longer have to throw away your NiCds after three months or so of maltreatment in the original charger. The circuit diagram shows an LM317 in constant-current configuration but without the usual fixed or variable resistor at the ADJ pin to determine the amount of output current. Also, there is no switch with an array of different resistors to select the charge currents for three cell or battery types we wish to charge: AAA, AA and PP3 (6F22). When, for example, an empty AAA cell is connected, the voltage developed across R1 causes T1 to be biased via voltage dropper D1. This results in about 50 µA flowing from the LM317's ADJ pin into the cell, activating the circuit into constant-current mode. D4 is included to prevent the battery

being discharged when the charger is switched off or without a supply voltage. The charging current I is determined by R1/R3/R3 as in

$$R(n) = \frac{1.25 + V_{sat}}{I}$$

where V_{sat} is 0.1 V.
The current should be one tenth of the nominal battery capacity – for example, 170 mA for a 1700 mAh NiCd AA cell. It should be noted that 'PP3' rechargeable batteries usually contain seven NiCd cells so their nominal voltage is 8.4 V and not 9 V as is often thought. If relatively high currents are needed, the power dissipation in R1/R2/R3 becomes an issue. As a rule of thumb, the input voltage required by the charger should be greater than three times the cell or battery (pack) voltage. This is necessary to cover the LM317's dropout voltage and the voltage across R(n). Two

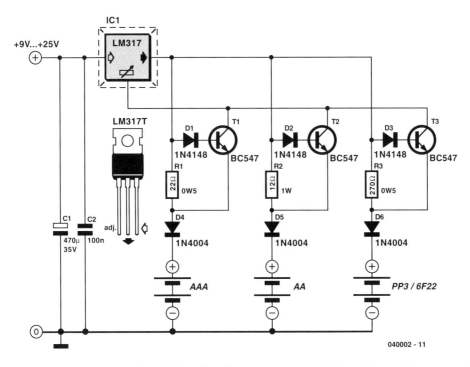

040002 - 11

final notes: the LM317 should be fitted with a small heat sink. With electrical safety in mind the use of a general-purpose mains adapter with DC output is preferred over a dedicated mains transformer/rectifier combination.

Myo Min

(040002-1)

104
Reset Sequencer

It is often necessary in complex designs to provide a sequence of reset pulses to different parts of a circuit to ensure the whole design functions reliably. The DS1830 from Maxim (www.maxim-ic.com) provides three sequenced open-drain reset outputs. This chip is designed for 5 V systems but a 3.3 V version (DS1830A) is also available. Both are offered in a range of package outlines including DIP, SO and µSOP. Two inputs give the chip some degree of programmability of its characteristics: The TOL input defines the chips toler-

ance to power supply fluctuations before a reset sequence is triggered. Jumper JP1 allows the TOL to be connected to Ub (Vcc), ground or left open circuit and will result in the following three reset thresholds:

TOL	5 V	3,3 V
+Ub	Ub × 0.95	Ub × 0.95
0 V	Ub × 0.90	Ub × 0.90
open	Ub × 0.85	Ub × 0.80

The TD input allows the length of the reset signal to be programmed and jumper JP2 gives the following three possibilities:

TD	TR1	TR2	TR3
0 V	10 ms	50 ms	100 ms
open	20 ms	100 ms	200 ms
+Ub	50 ms	250 ms	500 ms

The PBRST (pushbutton reset) allows a manual reset button to be connected to the chip.

This input has a built-in 40 kΩ pull up resistor and can also be driven by a digital output or used to cascade additional devices to provide more sequenced reset signals.

Gregor Kleine

(034006-1)

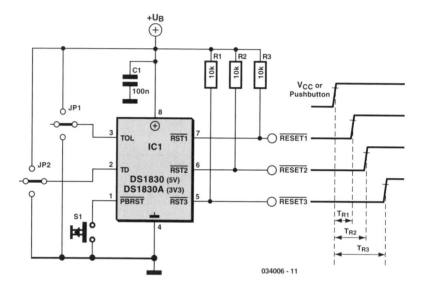

034006 - 11

105
Home Network for ADSL

The increased availability of fast ADSL Internet connections has made it more attractive to install a small RJ45 Ethernet network in the home. Not only can you exchange files between computers, you will also have fast Internet access for everybody! This does of course require an ADSL modem with a router. It's not possible to use a simple USB modem on its own. For laptops we recommend wireless Ethernet connections. If you find the lay- ing of cables too difficult or inconvenient you can also add wireless capabilities to 'ordinary' PCs. You should bear in mind that the range of wireless connections could sometimes be disappointing. When a network is set up round a router you should use a star configuration for the cabling. This means that only a single PC is connected to each router socket. The connecting cable may have a maximum length of 90 m and usually terminates at a connec-

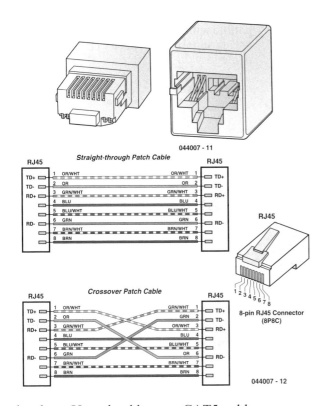

Straight-through Patch Cable

Crossover Patch Cable

044007 - 11

044007 - 12

8-pin RJ45 Connector
(8P8C)

plugs are attached to the cable using a special crimping tool. It is also possible without the tool, using just a screwdriver, but this isn't easy and we don't recommend that you try it. The wires in the cable have different colours and there are no official standards in Europe how you use them (EN50173). However, the colour code in the American T568B standard is often

used:	
1	orange/white
2	orange
3	green/white
4	blue
5	blue/white
6	green
7	brown/white
8	brown

tion box. You should use a CAT5 cable with 8 conductors for this, which is suitable for speeds up to 100 Mb/s. The 8 conductors are arranged in 4 pairs, with each pair twisted along the length of the cable. It is extremely important that the wires of each pair are kept together and that they are kept twisted as much as possible. At the connector ends you should therefore make sure that the non-twisted sections of the cable are kept as short as possible, at most a few centimetres. Should you fail to do this you may find that the network won't operate at the full rated speed or possibly cause interference. The wiring itself is very simple. Connect the plugs to the cables such that each pin connects to the corresponding pin at the other end. So pin 1 to 1, 2 tot 2 and so on. This also applies to all patch leads between the connection boxes and PCs (or if you prefer, the cable can go directly to the PC, without a connection box). It is only when two computers are connected directly together without a router that a crossover cable is required. The

The coloured/white wires and the solid coloured wires alternate nicely. For Ethernet cabling you only need connections 1, 2, 3 and 6. The central contacts on pins 4 and 5 are in the middle of the green pair and may be used for analogue telephones. You then have to make sure that 4 and 5 aren't connected to the Ethernet plugs because the voltages found on analogue telephone lines are high enough to damage an Ethernet card and/or router. Wires 4 and 5 should then be routed to an RJ11 telephone socket. We don't recommend it, but it is possible. It is also possible to pass ISDN signals through the same RJ45 plugs and cabling. In this case you can't use the same cable for both Ethernet and ISDN, since the latter uses pins 3/6 and 4/5. If you use patch cables it helps to keep things organised by using coloured cables. Blue for Ethernet (red for a crossover cable), yellow for analogue telephones and green for ISDN. Sticky labels or coloured cable markers can also be used for identification when you can't get hold of coloured cables. A new standard has recently been introduced, although you probably won't use it in the home for a while. Since around two years ago you can also use a GG45 connector, which is compatible with RJ45. This has 4 extra contacts and is suitable for speeds up to 600 Mb/s (Category 7/Class F).

Karel Walraven   (044007-1)

106
Master/Slave switch

In this age of enlightenment any sort of relationship that could be described as master/slave would be questionable but for the purposes of this circuit it gives a good idea of how it functions. The circuit senses mains current supplied to a 'master' device and switches 'slave' equipment on or off. This feature is useful in a typical hi-fi or home computer environment where several peripheral devices can all be switched on or off together. A solid-state relay from Sharp is an ideal switching element in this application; a built-in zero crossing detector ensures that switching only occurs when the mains voltage passes through zero and any resultant interference is kept to an absolute minimum. All of the triac drive circuitry (including optical coupling) is integrated on-chip so there are very few external components and no additional power supply necessary. This makes the finished design very compact. Diodes D1, D2, D3 and D4 perform the current sensing function and produce a voltage on C2 when

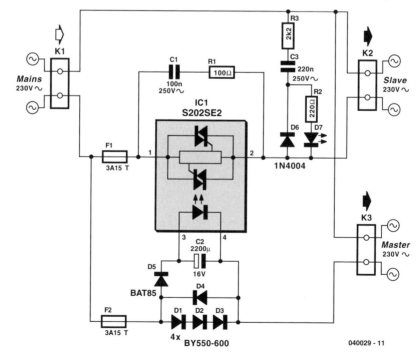

the master equipment is switched on. A Schottky diode is used for D5 to reduce forward voltage losses to a minimum. The circuit is quite sensitive and will successfully switch the slave even when the master equipment draws very little mains current. The RC network formed by R1 and C1 provides some protection for the solid-state relay against mains-borne voltage transients.

Warning: This circuit is connected to the mains. it is important to be aware that the chip has lethal voltages on its pins and all appropriate safety guidelines must be adhered to! This includes the LED, for safety it must be fitted behind a transparent plexiglass shield.

Karl Köckeis   (040029-1)

107
Reset IC with Selectable Voltages

Modern digital systems work with a supply voltage of +3.3 V, and sometimes they also need an additional, lower supply voltage, such as 1.8 V, 1.5 V or even 1.2 V. To generate a reset signal from these two voltages, it was previously necessary to use a separate reset IC for each voltage, and each IC had to be individually dimensioned for the voltage it monitored.

The Linear Technology LTC2904/5 (www.linear.com/pdf/29045f.pdf) can be programmed for two voltages. The voltages are selected using the S1, S2 and TOL inputs according to whether they are connected to V1, connected to ground or left open. The IC can be configured for the voltages shown in the table. The tolerance for the two voltages can be set using the TOL pin. The effect this has on the internally determined reset threshold is that the larger the tolerance, the lower the internal threshold is set. The $\overline{\text{RST}}$ output (pin 3) is an

open-drain output. It goes Low when at least one of the two voltages drops below the programmed threshold level. There is a time delay before the reset signal is deactivated after the voltages rise above the threshold level. With the LTC2904, this delay has a fixed value of 200 ms, while

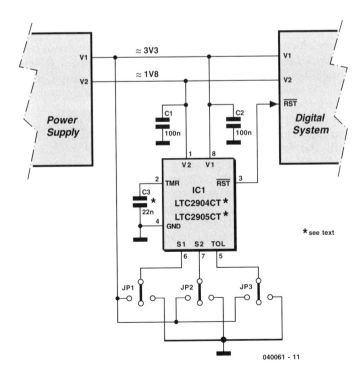

040061 - 11

S1	S2	V1	V2
V1	V1	5.0 V	3.3 V
open	ground	3.3 V	2.5 V
V1	open	3.3 V	1.8 V
open	V1	3.3 V	1.5 V
open	open	3.3 V	1.2 V
ground	ground	2.5 V	1.8 V
ground	open	2.5 V	1.5 V
ground	V1	2.5 V	1.2 V
V1	ground	2.5 V	1.0 V

TOL	Tolerance
V1	5 %
open	7.5 %
ground	10 %

$$t_{delay} = \frac{9 \text{ ms}}{\text{nF}}$$

This expression is valid for delay times between 1 ms and 10 s. In place of the TMR connection (pin 2),the LTC2904 has an open-drain RST output that is complementary to the $\overline{\text{RST}}$ output, which means it is active High.

with the LTC2905 it depends on the value of the capacitor connected to the TMR pin:

Gregor Kleine   (040061-1)

108
Mains Voltage Monitor

Many electronics hobbyists will have experienced the following: you try to finish a project late at night, and the mains supply fails. Whether that is caused by the electricity board or your carelessness isn't really important. In any case, at such times you may find yourself without a torch or with flat batteries. There is no need to panic, as this circuit provides an emergency light. When the mains fails, the mains voltage monitor turns on five super bright LEDs, which are fed from a 9 V battery (NiCd or NiMH) or 7 AA cells. A buzzer has also been included, which should wake you from your sleep when the mains fails. You obviously wouldn't want to oversleep because your clock radio had reset, would you? When the mains voltage is present, the battery is charged via relay Re1, diode D8 and resistor R10. D8 prevents the battery voltage from powering the relay, and makes sure that the relay

switches off when the mains voltage disappears. R10 is chosen such that the charging current of the battery is only a few milliamps. This current is small enough to

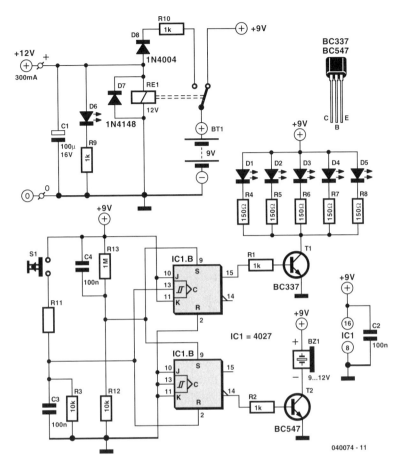

040074 - 11

prevent overcharging the battery. D6 acts as a mains indicator. When the relay turns off, IC1 receives power from the battery. The JK flip-flops are set via R12 and C4. This causes T1 and T2 to conduct, which turns on D1-D5 and the buzzer. When the push button is pressed, a clock pulse appears on the CLK input of flip-flop IC1b. The output then toggles and the LEDs turn off. At the same time IC1a is reset, which silences the buzzer. If you press the button again, the LEDs will turn on since IC1b receives another clock pulse. The buzzer remains off because IC1a stays in its reset state. R11, R3 and C3 help to debounce the push button signal. In this way the circuit can also be used as a torch, especially if a separate mains adapter is used as the power supply. As soon as the mains voltage is restored, the relay turns on, the LEDs turn off

and the battery starts charging. The function of R13 is to discharge C4, preparing the circuit for the next mishap. If mains failures are a regular occurrence, we recommend that you connect pairs of LEDs in series. The series resistors should then have a value of 100 Ω. This reduces the current consumption and therefore extends the battery life. This proves very useful when the battery hasn't recharged fully after the last time.

In any case, you should buy the brightest LEDs you can get hold of. If the LEDs you use have a maximum current of 20 mA, you should double the value of the series resistors! You could also consider using white LEDs.

Goswin Visschers   (040074-1)

109
Blinker Indicator

This circuit represents a somewhat unusual blinker indicator for use in a car or model. The running-light display progresses to-ward the left or the right depending on which directional signal is activated. That's pretty cool if you're fond of

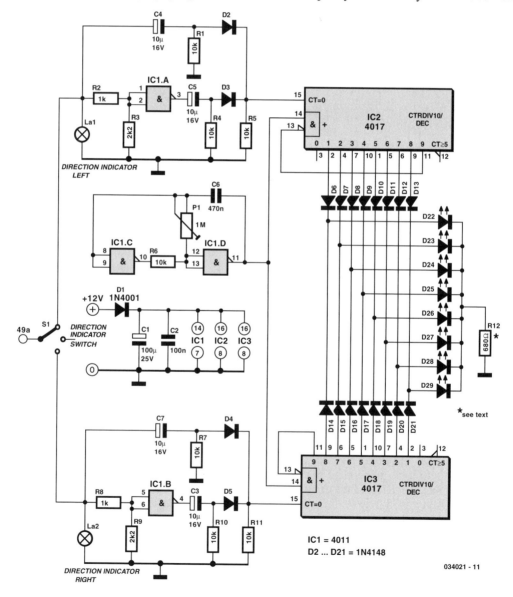

034021 - 11

light-show effects. The circuit consists of two counters (IC2 and IC3), which are reset to zero via C4 or C7 respectively whenever a blinker lamp (La) illuminates. The running-light display thus runs through once and then stops, since the highest counter output is connected to the Enable input. When the lamp goes out, a new reset pulse is issued to the relevant counter by NAND gate IC1.A or IC1.B respectively, and the counter counts all the way up again. The progression rate of the display can be adjusted to the right speed using P1. Only one LED is on at a time (except for the hazard blinker). This allows the brightness to be easily adjusted using R12. Incidentally, the circuit can also be modified by replacing the normal diodes with LEDs, with all of the cathodes connected to ground via R12.

Ludwig Libertin (034021-1)

110
Programmable-Gain Amplifier

The gain of an operational amplifier is usually set using two external resistors. If you wish to have adjustable gain, you can use a digitally controlled multiplexer to select several different gain-setting resistors. Such an arrangement using several ICs can now be replaced by the Linear Technology LTC 6910 single amplifier or LTC 6911 dual amplifier. These ICs incorporate all of the gain-setting components and can be programmed to eight different gain settings using three digital control inputs. The amplifier is always configured in the inverting mode and features rail-to-rail input and output. The input and output can be driven to within a few tens of millivolts of the sup-

1

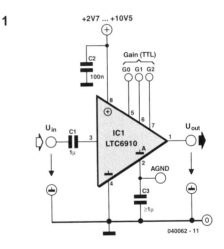

040062 - 11

G2	G1	G0	LTC6910-1	LTC6911-1	LTC6911-2	LTC6910-3
0	0	0	0	0	0	0
0	0	1	−1	−1	−1	−1
0	1	0	−2	−2	−2	−2
0	1	1	−5	−5	−4	−3
1	0	0	−10	−10	−8	−4
1	0	1	−20	−20	−16	−5
1	1	0	−50	−50	−32	−6
1	1	1	−100	−100		

2

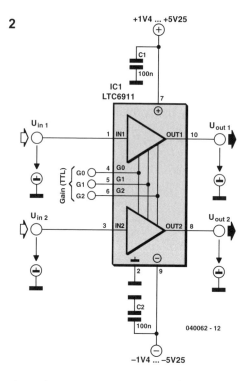

+1V4 ... +5V25

C1
100n

IC1
LTC6911

U_in 1 ... IN1 ... OUT1 ... U_out 1

Gain (TTL)
G0 ... 4 ... G0
G1 ... 5 ... G1
G2 ... 6 ... G2

U_in 2 ... 3 ... IN2 ... OUT2 ... 8 ... U_out 2

C2
100n
040062 - 12

–1V4 ... –5V25

ply voltages. At a gain of 100, the bandwidth still extends to approximately 100 kHz. With a unipolar supply, the supply voltage for the LTC 6910/6911 can range from +2.7 V to +10.5 V. With a bipo-

lar supply, the IC can be operated at ±1.4 V to ±5.25 V. There are several different versions of the IC, which are identified by the suffix -1, -2 or -3. The gains for the various combinations of the digital control signals are shown in the table. It should be noted that due to the internal arrangement of the resistors, the input resistance of the amplifier can range from 1 kΩ to 10 kΩ, depending on the gain setting. This means that a low-impedance signal source must be used to avoid affecting the configured gain setting. The AGND pin (pin 2) is the non-inverting input of the internal opamp. It is connected to an internal voltage divider consisting of two 5 kΩ resistors between V+ und V–. When a single supply voltage is used, a capacitor with a value of at least 1μF must be connected to this pin (Figure 1). With a bipolar supply, AGND can be connected directly to signal ground (Figure 2). Note also that with a unipolar supply, a coupling capacitor is required at the input, and possibly also at the output, since the input and output are internally biased to half the supply voltage. These coupling capacitors will determine the lower corner frequency of the amplifier.

Gregor Kleine   (040062-1)

111
Simulator for Bridge Measurements

The arrangement of resistors shown here can be used to test bridge amplifiers with differential inputs. Such amplifiers are used in conjunction with strain gauges, for example, or in measuring inductances or capacitances. Because of their symmetrical construction, bridges allow the tiniest variations in the quantity in question (for example, resistance) to be amplified without resorting to complex compensation techniques. The circuit here simulates small

variations in one of the resistors that forms the bridge, as happens in the example of strain gauges. Since only standard metal film resistors are used, it is suitable only for general testing of the connected amplifier. Strain gauges typically have a sensitivity of 2 mV/V. This means that if the supply voltage to the bridge is 10 V, the maximum variation in differential voltage taken from the bridge to the differential amplifier is 20 mV. The resistance of this

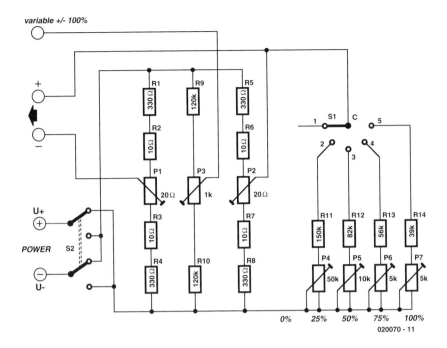

020070 - 11

circuit is rather greater than the 350 Ω typical of strain gauges, but in practice this is not important.

Calibrating the bridge simulator

Ensure that the power supply voltages connected to the positive and negative supply inputs have exactly equal magnitude (+5.000 V and –5.000 V). Now connect a voltmeter between the negative output and ground and adjust P1 so that the meter reads as close to 0 mV as possible.

Now connect the voltmeter across the two outputs and set S1 to '0 %'. Adjust P2 until the voltmeter reads as close to 0 mV as possible. Finally, switch S1 to the '25 %', '50 %', '75 %' and '100 %' positions in turn and adjust the corresponding trimmer potentiometers P4 to P7 to obtain voltages of 5mV, 10 mV, 15 mV and 20 mV respectively, with the voltmeter still connected between the positive and negative outputs. Once this calibration process has been carried out, you can now simulate a varying bridge resistance by switching

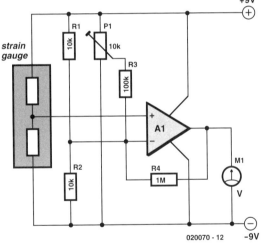

020070 - 12

S1 with a differential amplifier connected across the positive and negative outputs. If the variable output is used instead of the positive output, the voltage can be varied continuously over the full positive and negative range using potentiometer P3. Switch S2 allows the polarity of the supply voltage to be reversed.

Bernd Schaedler (020070-1)

112
1 MHz Frequency Counter

The XR 2206 function generator chip is a very popular device and forms the basis of many analogue function generator designs. A disadvantage of this chip is that in its most basic configuration the output frequency is adjusted manually by rotating an output frequency control knob and lining it up on a printed scale. Without a frequency counter its difficult to tell the value of the output frequency, especially at the high end of the scale. The frequency counter design presented here offers a low cost solution to this problem and achieves good accuracy. Power for the circuit is derived from the dc power supply of the existing function generator. The input signal to the counter is taken in parallel from the TTL output of the XR2206. The frequency counter circuit contains an Atmel AT90S2313 microcontroller together with an LCD display. The frequency counter program is written in Basic. The fundamental components needed to build a frequency counter are a timer, which is used to time an accurate measurement window and a counter to count the number of transitions that the input signal makes during the mea-

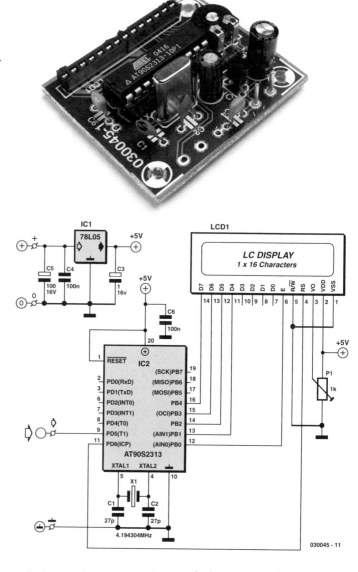

surement window period. The microcontroller contains two internal hardware counters of 8 and 16 bit length and

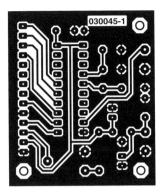

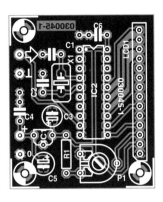

these can be configured as either a timer or counter. Operating with a crystal frequency of 8.388608 MHz the counter achieves an accuracy of 1 Hz at 1 MHz or with a more readily available 4.194304 MHz crystal the accuracy will be 1 Hz with a 500 kHz input signal. The 16-bit counter is configured as the frequency counter and the input signal is connected to pin 9 at TTL logic levels. A 16-bit counter can only count to a maximum value of 65536 so to extend its range the software keeps note of the number of counter timer overflow interrupts that occur during the measurement window. At the end of the measurement window it multiplies this value by 65536 and adds it to the current value of the counter:

$$F = \text{interrupts} \times 65536 + \text{Timer1}$$

The time base is generated by the 8-bit timer which together with a prescaler counts to 1024. Using the crystal frequency from above the timer 0 overflow occurs at a frequency of 32 Hz. The measurement timing window is closed when 32 interrupts have been counted and the frequency of the input signal is calculated and displayed.

The interrupt rate restricts the maximum input to 1.5 MHz and at this frequency the display will exhibit a jitter of 4 Hz i.e. the displayed frequency will 'wander' by 4 Hz. All input frequencies below 200 kHz should not display any jitter. The program is written in AVR Basic and is small enough to be compiled and downloaded to the microcontroller by the free demo version of the BASCOM AVR compiler (Download from: www.mcselec.com). The single line LCD display interface is designed for a Hitachi 16 × 1 or compatible display. Preset R1 allows adjustment of the display contrast. IC1 provides a regulated 5 V for the microcontroller and display. The frequency counter accuracy depends on the precision of the crystal and its frequency stability together with capacitors C1 and C2. PAL crystals generally are produced to close tolerances and exhibit low drift. During testing the crystal showed an offset of 60 ppm, which would produce a

Components list

Resistors:
R1 = 10 kΩ
P1 = 1 kΩ preset

Capacitors:
C1, C2 = 27 pF
C3 = 1 µF / 16V radial
C4, C6 = 100 nF

Semiconductors:
IC1 = 78 LO5
IC2 = AT90S2313-1OPI, programmed, order code 030045-41 *

Miscellaneous:
X1 – 4.194304MHz quartz crystal (see text)
LCD1 = LCD module, 1 line, 16 characters
PCB, ref. 030045-1 from The PCB Shop
Disk, project software, order code 030045-11 * or Free Download

* see Elektor SHOP pages or http://www.elektor-electronics.co.uk.

display error of 60 Hz on a 1 MHz signal. Adjustment of the capacitive loading (C1 and C2) on the crystal was able to 'pull' the crystal frequency and reduce the error to zero. At audio frequencies the counter measures to an accuracy of 1 Hz across the entire audio spectrum without any need for circuit alignment.

H. Breitzke

(030045-1)

113
12 V Dimmer

A dimmer is quite unusual in a caravan or on a boat. Here we describe how you can make one. So if you would like to be able to adjust the mood when you're entertaining friends and acquaintances, then this circuit enables you to do so. Designing a dimmer for 12 V is tricky business. The dimmers you find in your home are designed to operate from an AC voltage and use this AC voltage as a fundamental characteristic for their operation. Because we now have to start with 12 V DC, we have to generate the AC voltage ourselves. We also have to keep in mind that we're dealing with battery-powered equipment and have to be frugal with energy. The circuit that we finally arrived at can easily drive 6 lamps of 10 W each. Fewer are also possible, of course. In any case, the total current has to be smaller than 10 A. L1 and S1 can be adapted to suit a smaller current, if required. Note that the whole circuit will also work from 6 V. IC1 is a dual timer. You could also use the old faithful NE556, but it

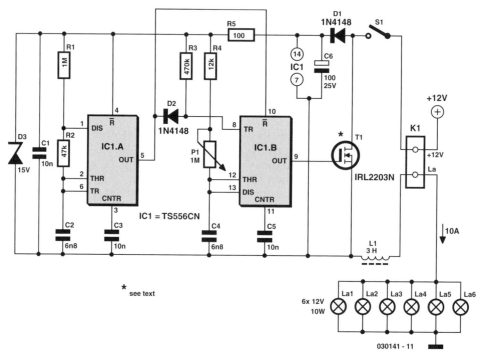

*
see text

030141 - 11

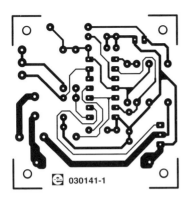

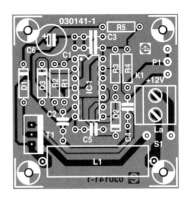

draws a little more current. IC1a is wired as an astable multivibrator with a frequency of 180 Hz. IC1b is configured as a monostable and is triggered, via D2, from the positive edge at the output of IC1a. The length of the pulse that now appears at the output of IC1b is dependant on the position of P1. IC1b will be reset whenever the output of IC1a goes low, independent of the pulse duration, set with P1, R4 and C4. This guarantees that the dimming is smooth, even when the brightness is set to maximum. The output of IC1b (pin 9) drives the gate of MOSFET T1. When the duration of the pulses at the gate increases, the average time that the MOSFET is in conduction will also increase. In this way the brightness of the lamps is controlled. T1 conducts about 96 % of the time when the brightness is set to maximum.

In this configuration, this can never be 100 %, because the 4 % of the time that the FET does not conduct is necessary to charge C6. If the FET were to conduct with 100 % duty cycle, the power supply for the circuit would be effectively short-circuited. C6 allows the circuit to ride through the conduction period of the FET. D1 ensures that the charge cannot leak away via the FET during the 'on' period. In the schematic, an IRL2203N is indicated for T1, but in principle you could use just about any power transistor (for example, BUK455, BUZ10, BUZ11 or BUZ100). The IRL2203 does however, have a very low 'on' resistance (RDS ON) of only 7 mΩ and can switch 12 V loads up to 5 A without a heatsink. If you choose a dif-

ferent MOSFET (with higher RDS ON) or use the circuit in a 6 V system then you will likely need a heatsink. Using the IRL2203N with six lamps rated 12 V / 10 W, T1 dissipates only 170 mW. At 6 V and 10 A this becomes 680 mW. The circuit itself consumes about 0.35 mA at maximum brightness and about 1.25 mA at minimum. The power supply is derived from the power system via D1 and C6.

Components list

Resistors:
R1 = 1MΩ
R2 = 47kΩ
R3 = 470kΩ
R4 = 12kΩ
R5 = 100Ω
P1 = 1MΩ preset

Capacitors:
C1, C3, C5 = 10 nF
C2, C4 = 6 nF8
C6 = 100 µF 25V radial

Semiconductors:
D1, D2 = 1N4148
D3 = zener diode 15V / 1.3W
T1 = IRL2203, BUZ10, Buz11, BUZ100 or BUK455 *
IC1 = TS556CN (CMOS) or NE556N (not CMOS)

Miscellaneous:
L1 = 3 µH 9A suppressor coil * (e.g., Farnell # 976-416)
K1 = 2-way PCB teminal block, lead pitch 5mm
S1 = on/off switch, 10A
PCB, ref. 030141-1 from The PCBShop
* see text

Zener diode D3 provides protection against voltage surges. The main purpose of R5 is to limit the current through D3, in the event it becomes active. L1 reduces interference that could be caused by the fast switch-ing of the transistor. We have designed a small printed circuit board for the

circuit. The construction is very simple. The PCB is quite compact in order to facilitate the replacement of existing switches. Keep in mind that the CMOS-parts, IC1 and T1, are sensitive to static electricity.

John Dakin   (030141-1)

114
Precision Headphone Amplifier

Designs for good-quality headphone amplifiers abound, but this one has a few special features that make it stand out from the crowd. We start with a reasonably conventional input stage in the form of a differential amplifier constructed from dual FET T2/T3. A particular point here is that in the drain of T3, where the amplified signal appears, we do not have a conventional current source or a simple resistor. T1 does in-

deed form a current source, but the signal is coupled out to the base of T5 not from the drain of T3 but from the source of T1. Notwithstanding the action of the current source this is a low impedance point for AC signals in the differential amplifier. Measurements show that this trick by itself results in a reduction in harmonic distortion to considerably less than −80 dB (much less than 0.01 %) at 1 kHz. T5 is

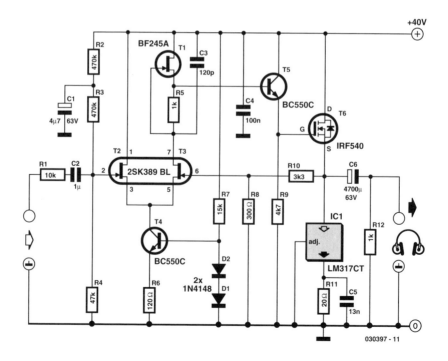

030397 - 11

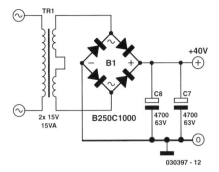

030397 - 12

connected as an emitter follower and provides a low impedance drive to the gate of T6: the gate capacitance of HEXFETs is far from negligible. IC1, a voltage regulator configured as a current sink, is in the load of T6. The quiescent current of 62 mA (determined by R11) is suitable for an output power of 60 mWeff into an impedance of 32 Ω, a value typical of high-quality headphones, which provides plenty of volume. Using higher-impedance headphones, say of 300 Ω, considerably more than 100 mW can be achieved. The gain is set to a useful 21 dB (a factor of 11) by the negative feedback circuit involving R10 and R8. It is not straightforward to change the gain because of the single-sided supply: this voltage divider also affects the operating point of the amplifier. The advantage is that excellent audio quality can be achieved even using a simple unregulated mains supply. Given the relatively low power output the power supply is considerably overspecified. Noise and hum thus remain more than 90 dB below the signal (less than 0.003 %), and the supply can also power two amplifiers for stereo operation. The bandwidth achievable with this design is from 5 Hz to 300 kHz into 300 Ω, with an output voltage of 10 Vpp. The damping factor is greater than 800 between 100 Hz and 10 kHz. A couple of further things to note: somewhat better DC stability can be achieved by replacing D1 and D2 by low-current red LEDs (connected with the right polarity!). R12 prevents a click from the discharge of C6 when headphones are plugged in after power is applied. T6 and IC1 dissipate about 1.2 W of power each as heat, and so cooling is needed. For low impedance headphones the current through IC1 should be increased. To deliver 100 mW into 8 Ω, around 160 mA is required, and R11 will need to be 7.8 Ω (use two 15 Ω resistors in parallel). To keep heat dissipation to a reasonable level, it is recommended to reduce the power supply voltage to around 18 V (using a transformer with two 6 V secondaries). This also means an adjustment to the operating point of the amplifier: we will need about 9V between the positive end of C6 and ground. R4 should be changed to 100 Ω, and R8 to 680 Ω. The gain will now be approximately 6 (15 dB). The final dot on the 'i' is to increase C7 by connecting another 4700 μF electrolytic in parallel with it, since an 8 Ω load will draw higher currents.

Hergen Breitzke   (030397-1)

115
Intelligent Flickering Light

Whether it is required to simulate an open fire in a nativity scene, a forest fire in a model railway landscape, a log fire in a doll's house or simply for an artificial candle, neither steady light nor the commercially-available regularly flickering lights are very realistic. The circuit described here imitates much better the irregular flickering of a fire. For maximum flexibility, and to reduce the component count to a

COMPONENTS LIST

Resistors:
R1,R4,R5 = 4kΩ7
R2 = 10kΩ
R3 = 1kΩ
R6 = 220Ω

Capacitors:
C1,C2 = 100nF
C3 = 10µF 16V

Semiconductors:
D1 = 5.1V zener diode, 400 mW
T1,T2 = BC547
IC1 = ATtiny11-6PI (programmed)

Miscellaneous:
L1,L2 = 6V / 80mA miniature lamp
PCB no. 040089-11, available from The
PCBShop Project software: file 040089-11,
Free Download

minimum, a microcontroller from the Atmel ATtiny range has been selected to generate the flickering pattern. Two miniature light bulbs, each driven by a transistor, are controlled using a PWM signal to produce eight different light levels. Potentiometer P1 in the RC network adjusts the speed of the clock to the microcontroller, and hence the speed of the flickering. Generating the light levels in software is straightforward in practice, but the underlying theory is far from simple: hence the 'intelligent' in the title. Using a digital pseudo-random number generator (an 8-bit shift register with feedback arranged according to the coefficients of a primitive polynomial) a sequence of period 255 can be produced. In order that the flickering is not too violent, the sequence is smoothed using a digital FIR low-pass filter which takes the average of the last two sample values. If

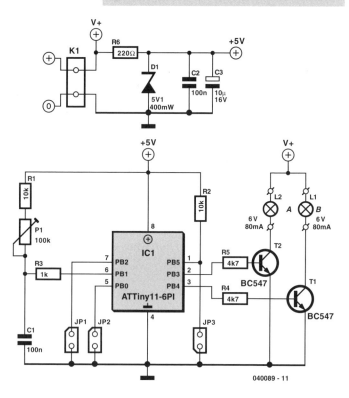

desired, a jumper can be fitted that compresses the dynamic range of the output by adding in a fixed basic intensity. The result is an irregular flickering which closely resembles that of a fire. A further option al-

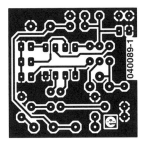

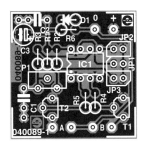

nect three miniature light bulbs in red, green and blue (or an RGB LED) and generate arbitrary colour patterns. The printed circuit board is just the size of a postage stamp and so should be easy to fit within small models or model landscapes. The board is single-sided, and

lows the brightness values to be read from a look-up table instead of using the sequence generator; this option obviously gives the greatest flexibility. A jumper gives the choice of two different tables. The look-up tables can be used to produce other decorative light effects, such as a light fader, or the continous mixing of two differently-coloured lights. It could even be used to imitate rotating flashing lights on a model. If the design were expanded to three channels, it would be possible to con-

making the board and populating it, should not present any difficulties, thanks to the absence of SMDs. The total component cost is very low, at around two or three pounds, not including the circuit board. Power can be obtained from any regulated 5 V supply. If only an unregulated supply is available, then this should be connected to V+. Current consumption is of course mostly dependent on the choice of lamp.

Andre Frank   (040089-1)

116
You Have Mail!

If your letterbox is some distance from your house, you will find a monitoring device useful to indicate when new post has arrived. This can take the form of a US-style visible flag; a more modern alternative uses a 433 MHz radio transceiver. The big advantage of the solution presented here is that is can use an existing two-core bell cable, without requiring any further power source. The arrival of post is signalled by a blinking LED; for added effect, a digital voice recorder can also be connected which will, at regular intervals, announce the fact that the letterbox needs emptying. The device is silenced by a reset button. The circuit uses one half-cycle of the AC supply to power the bell or buzzer, and the other half-cycle for the post indicator. Suitably-oriented diodes in the device

and in the letterbox ensure that the two signals are independent of one another (Figure 1). The bell current flows from K1.A through D3, bell-push S2, D1 and the relay back to K1.B. C1 provides adequate smoothing of the current pulses to ensure that the relay armature does not vibrate. The bell is operated by the normally-open relay contact. If the bell is actually a low-current piezo buzzer, then it can be connected directly and the relay dispensed with. During the half-cycle for the letterbox monitor current flows from connection K1.B on the bell transformer through current-limiting resistor R1, the LED in the optocoupler, reed contact S1 (a microswitch can also be used) and D2 and finally back to K1. If the reed contacts are closed, the LED in the optocoupler will

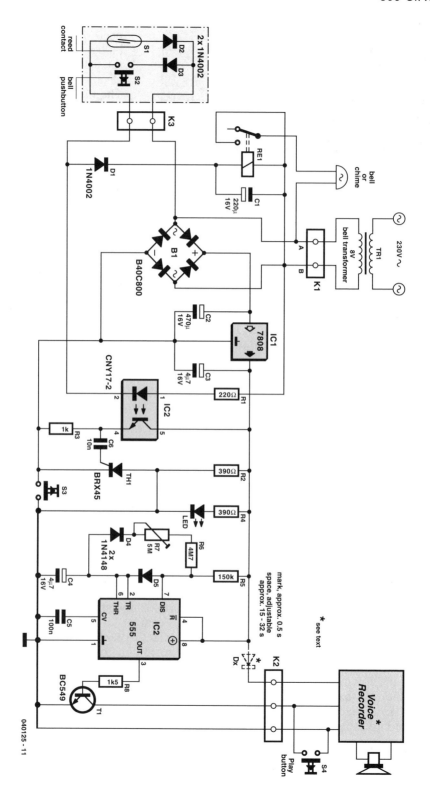

040125 - 11

light and switch on the phototransistor. A positive voltage will then appear across R3 which will turn the thyristor on via C6. The red LED will indicate that post has arrived. Pressing S3 shorts out the thyristor, reducing the current through it below the holding value. A small extra circuit can be added to provide continuous letterbox monitoring. This takes the form of a voice recorder whose 'play' button is operated by transistor T1. T1 in turn is driven by a 555 timer IC. In the usual 555 timer circuit, where the device is configured as an astable multi-vibrator, the mark-space ratio cannot be set with complete freedom. Here two diodes provide separate charge and discharge paths for capacitor C4. When capacitor C4 is charging, D5 conducts and D4 blocks: the charge rate is determined by R5. When discharging, D4 conducts and R6 and the potentiometer determine the rate. The values shown give a pulse length of approximately 0.5 s with a delay of between 15 s and 32 s. The short pulse is sufficient to trigger the voice recorder module via tran-

sistor T1 connected across its 'play' button. The voice recorder module (e.g. Conrad order code 115266) is designed to run from a 6 V supply. The maximum recording time is 20 s and the current consumption is 20 mA when recording and between 40 mA and 60 mA when playing back. Since our supply is at 8 V, the excess voltage must be dropped using between 1 and 3 series-connected 1N4148 diodes (shown as Dx in the circuit diagram). The final voltage should be checked using a multimeter. Alternatively, a 7806 can be used without suffering a significant loss in volume. If it is desired to use a piezo buzzer to provide an acoustic signal, the pulse length should be increased to at least 2 s. In this case, R5 should be increased to 560 kΩ or 680 kΩ: the pulse length, t_{on}, is $0.7 \cdot R5 \cdot C4$, and the interval between pulses, t_{off}, is $0.7 \cdot (R6+R7) \cdot C4$. Suitable buzzers are available with a wide range of rated voltages.

Robert Edlinger   (040125-1)

117
Two-LED Voltage Indicator

There are many applications where the accuracy of a digital or analogue (bar graph) is not required but something better than a simple low/high indicator is desirable. A battery charge level indicator in a car is a good example. This simple circuit requiring only two LEDs (preferably one with a green and red LED in a single package), a cheap CMOS IC type 4093 and a few resistors should fulfil many such applications. With a suitable sensor, the indicator will display the relevant quantity as a colour ranging from red through orange and yellow to green. IC1.A functions as an oscillator running at about 10 kHz with the component values given, although this is not

critical. Assuming for the moment that R1 is not commented, the output of IC1.A is a square wave with almost 50% duty cycle. The voltage at the junction of R2 and C1 will be a triangular wave (again, almost) with a level determined by the difference in the two threshold voltages of the NAND Schmitt trigger gate IC1.A. IC1.B, IC1.C and IC1.D form inverting and non-inverting buffers so that the outputs of IC1.C and IC1.D switch in complementary fashion. With a 50% duty cycle, the red and green LEDs will be driven on for equal periods of time so that both will light at approximately equal brightness resulting in an orange-yellow display. With R1 in circuit,

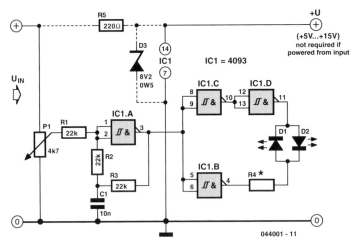

044001 - 11

in response to input voltages of 2.5 V and 5.6 V respectively. To monitor a car battery voltage, the battery itself could be used to power the circuit provided a zener diode and dropper resistor are added to stabilise the IC supply voltage. This is shown in dashed outlines in the circuit diagram. With an 8.2 V zener the dropper resistor should be around 220 Ω and R1 has to be reduced to 4.7 kΩ. The LED brightness is determined by R4. As a rule of thumb,

the actual input voltage to IC1.A will consist of the triangular waveform added to the dc input Vin. As the input voltage varies, so will the oscillator duty cycle causing either the red or the green LED to be on for longer periods and so changing the visible colour of the combi-LED. The actual range over which the effect will be achieved is determined by the relative values of R1 and R2, enabling the circuit to be matched to most supply voltages. With the component values given and a supply of 8 volts, the LED will vary from fully red to fully green

$$R4 = \frac{V_{supply} - 2}{3} [k\Omega]$$

and remember that the 4093 can only supply a few mA's of output current. Applications of this little circuit include 'non critical' ones such as go/non-go battery testers, simple temperature indicators, water tank level indicators, etc.

Bart Trepak   (044001-1)

118
Linear RF Power Meter

The National Semiconductor LMV225 is a linear RF power meter IC in an SMD package. It can be used over the frequency range of 450 MHz to 2000 MHz and requires only four external components. The input coupling capacitor isolates the DC voltage of the IC from the input signal. The 10 kΩ resistor enables or disables the IC according to the DC voltage present at the input pin. If it is higher than 1.8 V, the detector is enabled and draws a current of

around 5–8 mA. If the voltage on pin A1 is less than 0.8 V, the IC enters the shutdown mode and draws a current of only a few microampères. The LMV225 can be switched between the active and shutdown states using a logic-level signal if the signal is connected to the signal via the 10 kΩ resistor. The supply voltage, which can lie between +2.7 V und +5.5 V, is filtered by a 100 nF capacitor that diverts residual RF signals to ground. Finally, there is an out-

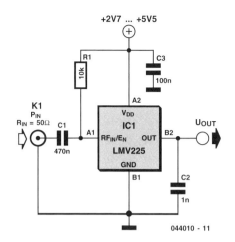

044010 - 11

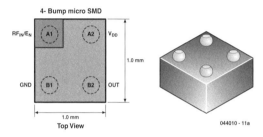

4- Bump micro SMD

Top View

044010 - 11a

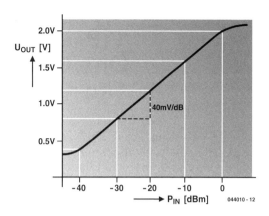

044010 - 12

put capacitor that forms a low-pass filter in combination with the internal circuitry of the LMV225. If this capacitor has a value of 1 nF, the corner frequency of this low-pass filter is approximately 8 kHz. The corner frequency can be calculated using the formula

$$f_c = \frac{1}{2 \cdot \pi \cdot C_{out} \cdot R_o}$$

where R_o is the internal output impedance (19.8 kΩ). The output low-pass filter determines which AM modulation components are passed by the detector. The output, which has a relatively high impedance, provides an output voltage that is proportional to the signal power, with a slope of 40 mV/dB. The output is 2.0 V at 9 dBm and 0.4 V at –40 dBm. A level of 0 dBm corresponds to a power of 1 mW in 50 Ω. For a sinusoidal wave-form, this is equivalent to an effective voltage of 224 mV. For modulated signals, the relationship between power

and voltage is generally different. The table shows several examples of power levels and voltages for sinusoidal signals. The input impedance of the LMV225 detector is around 50 Ω to provide a good match to the characteristic impedance commonly used in RF circuits. The data sheet for the LMV225 shows how the 40-dB measurement range can be shifted to a higher power level using a series input resistor. The LMV225 was originally designed for use in mobile telephones, so it comes in a tiny SMD package with dimensions of only around 1 × 1 mm with four solder bumps (similar to a ball-grid array package). The connections are labelled A1, A2, B1 and B1, like the elements of a matrix. The corner next to A1 is belevelled.

dBm	mW	Ueff (Sinusoid)
–40	0.0001	2.24 mV
–30	0.001	7.07 mV
–20	0.01	22.4 mV
–10	0.1	71 mV
0	1	224 mV
10	10	707 mV

Gregor Kleine   (044010-1)

163

119
One Component Oscillator for 1 to 10 MHz

Maxim (www.maxim-ic.com) has produced a completely self-contained TTL oscillator chip in a very small three-pin outline. The MAX7375 family of oscillators operates in the range of 1 MHz up to around 10 MHz, (depending on device suffix) and requires no external components. It may be necessary to fit a 100 nF decoupling capacitor across the supply pins if the chip is sited further than a few centimetres from any other decoupling capacitor. The specified supply voltage range is between 2.7 V and 5.5 V while current consumption is dependant on clock frequency; at 4 MHz the chip takes 4 mA while at 8 MHz this rises to 6.5 mA.

The device is available in an SOT23 package outline (MAX7375AUR) or in the even smaller SC70 outline (MAX7375AXR). Note that the pin-out definitions for these two outlines are not identical, the functions of pins 1 and pin 2 are swapped. The accuracy of the output frequency is guaranteed to be within ±2% of nominal with a supply voltage of 3 V. Over the entire temperature range this rises to a maximum of ±4 %.

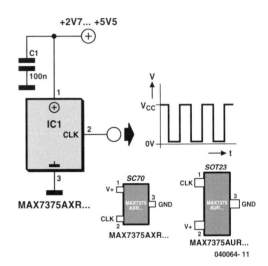

MAX7375AXR...

040064-11

order of 5 ns while the markspace ratio lies between 45 % and 57 %. The MAX7375 offers a smaller, more robust alternative to the conventional crystal or ceramic filter type of oscillator. The device has a wide operating temperature range of –40 °C to +125 °C and this makes it cost-effective and mechanically more particularly suitable for automotive applications.

Gregor Kleine   (040064-1)

This chip is currently available in a range of seven frequencies shown in the table below. The TTL push-pull output stage can sink and source up to 10 mA. The rise and fall times of the oscillator output are in the

SOT23 MAX 7375AUR...	SC70 MAX 7375AXR...	Nominal Output Frequency
...105		1 MHz
...185		1.8432 MHz
...365		3.579545 MHz
...375		3.6864 MHz
...405		4 MHz
...425		4.1943 MHz
...805		8 MHz

120
Meter Adapter with Symmetrical Input

In contrast to an ordinary voltmeter, the input of an oscilloscope generally has one side (GND) connected to ground via the mains lead. In certain situations this can be very problematic. When the measuring probe is connected to a circuit that is also connected to ground, there is a chance that a short is introduced in the circuit. That the circuit, and hence the measurement, is affected by this is the least of your problems. If you were taking measurements from high current or high voltage (valve equipment) circuits, the outcome could be extremely dangerous! Fortunately it is not too difficult to get round this problem. All

you have to do is make the input to the os-cilloscope float with respect to ground. The instrumentation amplifier shown here does that, and functions as an attenuator as well. The AD621 from Analog Devices amplifies the input by a factor of 10, and a switch at the input gives a choice of 3 ranges. A 'GND' position has also been included, to calibrate the zero setting of the oscilloscope. The maximum input voltage at any setting may never exceed 600 VAC. Make sure that R1 and R8 have a working voltage of at least 600 V. You could use two equal resistors connected in series for these, since 300 V types are

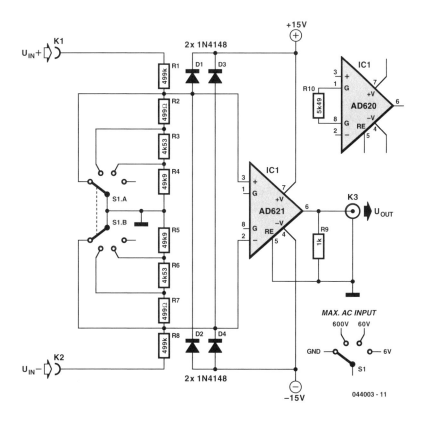

044003 - 11

165

more easily obtainable. You should also make sure that all resistors have a tolerance of 1% or better. Other specifications for the AD621 are: with an amplification of 10 times the CMRR is 110 dB and the bandwidth is 800 kHz. If you can't find the AD621 locally, the AD620 is a good alternative. However, the bandwidth is then limited to about 120 kHz. The circuit can be housed inside a metal case with a mains supply, but also works perfectly well when powered from two 9 V batteries. The current consumption is only a few milliamps. You could also increase R9 to 10 kΩ to reduce the power consumption a bit more.

Aart Rombout   (044003-1)

121
100 V Regulators

Standard three-legged voltage regulator chips like the LM317 can cope with an input voltage of up to about 30 V, a few high-voltage types can handle 60 V but if that is still not sufficient for your application the company Supertex (www.supertex.com) produce devices that can withstand much higher input voltages. The regulator type LR8 has a maximum input voltage of 450 V and can supply an output current of 20 mA. The LR 12 has a better output current of 50 mA but with a maximum input voltage of 100 V, and the output voltage can be adjusted up to 88 V. The output voltage is defined by a potential divider chain connected between the output and the ADJ (adjust) input pin. The regulator simply changes its output voltage until the divided voltage at the ADJ input is equal to 1.2 V. The output voltage can be more precisely expressed as a function of R1 and R2 in the formula:

$$V_{out} = 1.2 \quad V \cdot \left(1 + \frac{R2}{R1}\right)$$

$$R2 = R1 \cdot \left(\frac{V_{out}}{1.2\ V} - 1\right)$$

The current through R1 and R2 must be greater than 100 µA. Just like conventional voltage regulators the LR12 can also be configured as a constant current source. Again the regulator simply adjusts its output voltage until it measures 1.2 V at the ADJ input. For a constant current of 10 mA the value of the series resistor is equal to

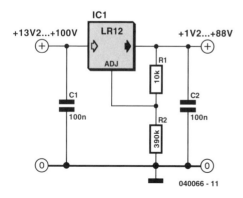

040066 - 11

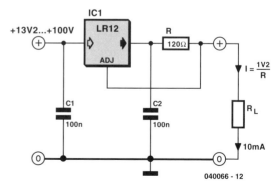

040066 - 12

the resistance that will produce a voltage drop of 1.2 V when 10 mA passes through it. As mentioned above the maximum output current is limited to 50 mA. A capacitor of 100 nF is necessary at the output to en-sure stable regulator operation. The LR12 is available in SO-8, TO-92 and TO-252/D-PAK outlines.

Gregor Kleine   (040066-1)

122
Gated Alarm

Sometimes the need arises for a simple, gated, pulsed alarm. The circuit shown here employs just four components and a piezo sounder and is unlikely to be outdone for simplicity. While it does not offer the most powerful output, it is likely to be adequate for many applications. A dual CMOS timer IC type 7556 is used for the purpose, with each of its two halves being wired as a simple astable oscillator (a standard 556 IC will not work in this circuit, nor will two standard 555's). Note that the CMOS7556 is supplied by many different manufacturers, each using their own type code prefix and suffix. The relevant Texas Instruments product, for instance, will be marked 'TLC556CN'. The circuit configuration used here is seldom seen, due probably to the inability of this oscillator to be more than lightly loaded without disturbing the timing. However, it is particularly useful for high impedance logic inputs, since it provides a simple means of obtaining a square wave with 1:1 mark-space ratio, which the 'orthodox' configuration does not so easily provide. IC1.A is a slow oscillator which is enabled when reset pin 4 is taken High, and inhibited when it is taken Low. Output pin 5 of IC1.A pulses audio oscillator IC1.B, which is similarly enabled when reset pin 10 is taken High, and inhibited when it is taken Low. In order to simplify oscillator IC1.B, piezo

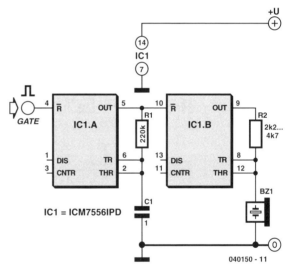

sounder X1 doubles as both timing capacitor and sounder. This is possible because a passive piezo sounder typically has a capacitance of a few tens of nanofarads, although this may vary greatly. As the capacitor-sounder charges and discharges, so a tone is emitted. The value of resistor R2 needs to be selected so as to find the resonant frequency of the piezo sounder, and with this its maximum volume.

The circuit will operate off any supply voltage between 2 V and 18 V. A satisfactory output will be obtained at relatively high supply voltages, but do not exceed 18 V.

Rev. Thomas Scarborough   (040150-1)

123
Steam Whistle

This circuit consists of six square wave oscillators. Square waves are made up of a large number of harmonics. If six square waves with different frequencies are added together, the result will be a signal with a very large number of frequencies. When you listen to the result you'll find that it is very similar to a steam whistle. The circuit should be useful in modelling or even in a sound studio. This circuit uses only two

ICs. The first IC, a 40106, contains six Schmitt triggers, which are all configured as oscillators. Different frequencies are generated by the use of different feedback resistors. The output signals from the Schmitt triggers are mixed via resistors. The resulting signal is amplified by IC2, an LM386. This IC can deliver about 1 W of audio power, which should be sufficient for most applications. If you leave out R13

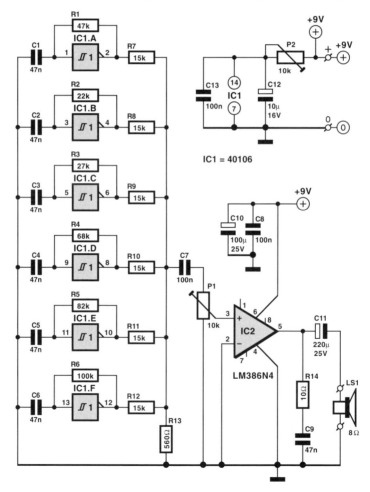

and all components after P1, the output can then be connected to a more powerful amplifier. In this way a truly deafening steam whistle can be created. The 'frequency' of the signal can be adjusted with P2, and P1 controls the volume.

Gert Baars   (044002-1)

124
Adjustable Duty Cycle

The circuit shown here can be used to convert a digital input signal having any desired duty cycle into a output signal having a duty cycle that can be adjusted between 10% and 80% in steps of 10%. The circuit is built around a 74HC4017 decade Johnson counter IC. Individual pulses appear on the ten outputs (Q0–Q9) of this IC at well-defined times, depending on the number of input pulses (see the timing diagram).

This characteristic is utilised in the circuit. The selected output is connected via a jumper to the Reset input (MR, pin 2) of a 74HC390 counter. A High level resets the output signals of the 74HC390 counter. Q9 of the 74HC4017 is permanently connected to the CP0 input of the counter to set the Q0 output of the 74HC390 (pin 3) High on its negative edge. As can be seen from the timing diagram, which shows the signals for a duty cycle of 30% as an example, this produces a signal with exactly the desired duty cycle.

The circuit cannot be used to produce a duty cycle of 10% (which would be equivalent to taking the signal directly from the Q0 output of the 74HC4017) or 90%. In both cases, the edges of the pulses used for the count input (CP0) and the asynchronous reset input (MR) of the 74HC390 would coincide, with the result that the output state of the 74HC390 would not be unambiguously defined. The input frequency must be ten times the desired output frequency. If the second half of the

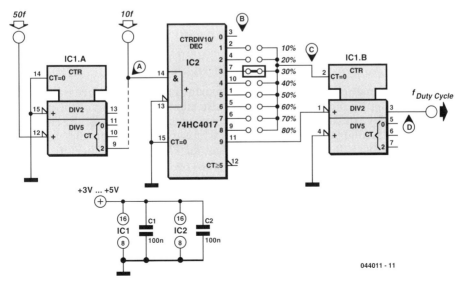

044011 - 11

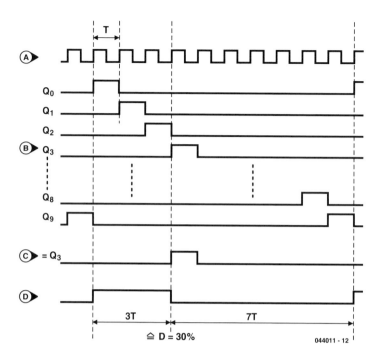

74HC390 is wired as a prescaler, a prescaling factor of 2, 5 or 10 can be achieved, thus allowing the ratio of the of input frequency to the output frequency to be 20, 50 or 100. If the circuit is built us- ing components from the 74HC family, it can be operated with supply voltages in the range of 3–5 V.

Gregor Kleine (044011-1)

125
Xport

The Gameboy Advance (GBA) already has its own power supply, processor, keypad and an LCD display. In addition, the system bus is made available externally. All this is ideal as the basis for your own embedded system. In the October 2000 issue of Elektor Electronics we published an expansion for the Gameboy: a digital oscilloscope. With the arrival of the Xport, made by Charmed Labs, the development of an embedded system based on the GBA has become a lot easier. The Xport is a complete development system. Apart from the expansion board, the necessary software is also supplied. The hart of the circuit on the expansion board is an FPGA made by Xilinx. Depending on the version you'll get an FPGA with either 50 K or 150 K gates on the board. Using the free development software from Xilinx, you can program your own designs into the FPGA. The board also has a 4 Mbyte flash memory. This memory stores the program for the GBA as well as the configuration for the FPGA. Since the FPGA loses its configuration when power is removed, it must reload the configuration every time that it is powered up. This takes place automatically thanks to a CPLD on the expansion board. Two version of the Xport come with an extra 16 Mbyte of SDRAM. This memory can be used by both the processor and the FPGA. Communication with the outside world is well provided for, with 64 I/O signals on board, in addition to the programming and debug connector! As mentioned earlier, the system consists not only of hardware. The PC software included is a C-compiler (GCC), complete with essential libraries,

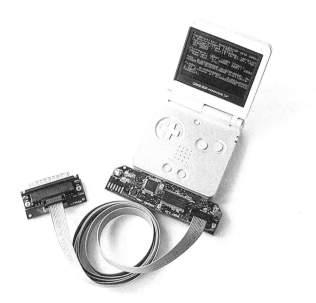

debugger and a programmer application. On top of this, there is an operating system (eCos) and its bootloader. There are also various examples included (which should be in every good development kit), so you can start using the Xport soon after get your hands on it.

Internet: www.charmedlabs.com

Paul Goossens　　　　　　　　　　(040153-1)

126
Long-Interval Pulse Generator

A rectangular-wave pulse generator with an extremely long period can be built using only two components: a National Semiconductor LM3710 supervisor IC and a 100 nF capacitor to eliminate noise spikes. This circuit utilises the watchdog and reset timers in the LM3710. The watchdog timer is reset when an edge appears on the WDI input (pin 4). If WDI is continuously held at ground level, there are not any edges and the watchdog times out. After an interval T_B, it triggers a reset pulse with a duration T_A and is reloaded with its initial value. The cycle then starts all over again. As a result, pulses with a period of $T_A + T_B$ are

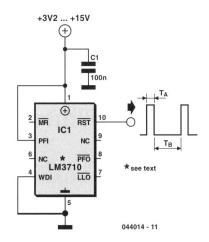

	$T_A = 1,4$ ms (1-2 ms)	T_A 28 ms (20-40 ms)	$T_A = 200$ ms (140-280 ms)	$T_A = 1,6$ s (1.12-2.24 s)
$T_B = 6,2$ ms (4.3-9.3 ms)	LM3710 XExxx	LM3710 XFxxx	LM3710 XGxxx	LM3710 XHxxx
$T_B = 102$ ms (71-153 ms)	LM3710 XJxxx	LM3710 XKxxx	LM3710 XLxxx	LM3710 XMxxx
$T_B = 1,6$ s (1.12-2.4 s)	LM3710 XNxxx	LM3710 XPxxx	LM3710 XQxxx	LM3710 XRxxx
$T_B = 25,6$ s (17.9-38.4 s)	LM3710 XSxxx	LM3710 XTxxx	LM3710 XUxxx	LM3710 XVxxx

T_A = Reset Timeout Period
T_B = Watchdog Timeout Period

LM3710 type designation, LM3710 a b cc ddd
a = output circuit: X = CMOS,
Y = open drain
b = timing (see table)
cc = package: MM = MSOP,
BP = micro SMD
ddd = reset threshold voltage (e.g. '450' for 4.50 V)

present at the RESET output (pin 10). As can be seen from the table, periods ranging up to around 30 seconds can be achieved in this manner. The two intervals T_A and T_B are determined by internal timers in the IC, which is available in various versions with four different ranges for each timer. To obtain the desired period, you must order the appropriate version of the LM3710. The type designation is decoded in the accompanying table. The reset threshold voltage is irrelevant for this particular application of the LM3710. The versions shown in bold face were available at the time of printing. Current information can be found on the manufacturer's home page (www.national.com). The numbers in brackets indicate the minimum and maximum values of intervals T_A and T_B for which the LM3710 is tested. The circuit operates with a supply voltage in the range of 3–5 V.

Gregor Kleine   (044014-1)

127
White LED Lamp

Did it ever occur to you that an array of white LEDs can be used as a small lamp for the living room? If not, read on. LED lamps are available ready-made, look exactly the same as standard halogen lamps and can be fitted in a standard 230 V light fitting. We opened one, and as expected, a capacitor has been used to drop the voltage from 230 V to the voltage suitable for the LEDs. This method is cheaper and smaller

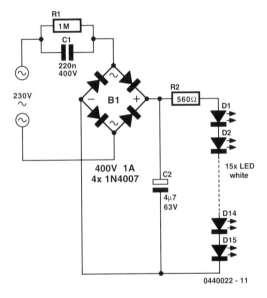

0440022 - 11

The circuit operates in the following manner: C1 behaves as a voltage dropping 'resistor' and ensures that the current is not too high (about 12 mA). The bridge rectifier turns the AC voltage into a DC voltage. LEDs can only operate from a DC voltage. They will even fail when the negative voltage is greater then 5 V. The electrolytic capacitor has a double function: it ensures that there is sufficient voltage to light the LEDs when the mains voltage is less than the forward voltage of the LEDs and it takes care of the inrush current peak that occurs when the mains is switched on. This current pulse could otherwise damage the LEDs. Then there is the 560-ohm resistor, it ensures that the current through the LED is more constant and therefore the light output is more uniform. There is a voltage drop of 6.7 V across the 560 Ω resistor, that is, 12 mA flows through the LEDs. This is a safe value. The total voltage drop across the LEDs is therefore 15 LEDs times 3 V or about 45 V. The voltage across the electrolytic capacitor is a little more than 52 V. To understand how C1 functions, we can calculate the impedance (that is, resistance to AC voltage) as follows:

$$\frac{1}{2 \cdot \pi \cdot f \cdot C} \text{ of}$$

$$\frac{1}{2 \cdot 3.14 \cdot 50 \cdot 220 \cdot 10^{-9}} = 14.4 \text{k}\Omega$$

When we multiply this with 12 mA, we get a voltage drop across the capacitor of 173 V.

This works quite well, since the 173 V capacitor voltage plus the 52 V LED voltage equals 225 V. Close enough to the mains voltage, which is officially 230 V. Moreover, the latter calculation is not very accurate because the mains voltage is in practice not quite sinusoidal. Furthermore, the mains voltage from which 50 V DC has

compared to using a transformer. The lamp uses only 1 watt and therefore also gives off less light than, say, a 20 W halogen lamp. The light is also somewhat bluer.

been removed is far from sinusoidal. Finally, if you need lots of white LEDs then it is worth considering buying one of these lamps and smashing the bulb with a hammer (with a cloth or bag around the bulb to prevent flying glass!) and salvaging the LEDs from it. This can be much cheaper than buying individual LEDs...

(044022-1)

128
Stepper Motor Generator

Stepper motors are a subject that keeps recurring. This little circuit changes a clock signal (from a square wave generator) into signals with a 90-degree phase difference, which are required to drive the stepper motor windings. The price we pay for the simplicity is that the frequency is reduced by a factor of four. This isn't really a problem, since we just have to increase the input frequency to compensate. The timing diagram clearly shows that the counter outputs of the 4017 are combined using inverting OR gates to produce two square waves with a phase difference. This creates the correct sequence for powering the windings: the first winding is negative and the second positive, both windings are negative, the first winding is positive and the second negative, and finally both windings are positive. Internally, the 4017 has a divide-by-10 counter followed by a decoder. Output '0' is active

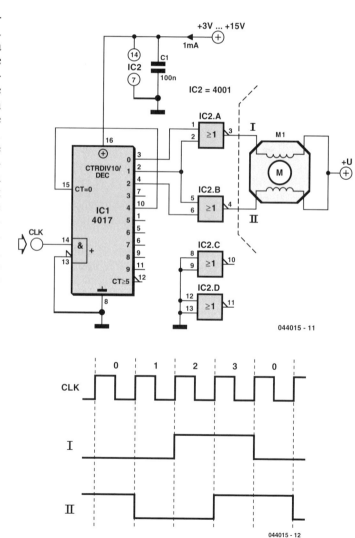

(logic one) as long as the internal counter is at zero. At the next positive edge of the clock signal the counter increments to 1 and output '1' becomes active. This continues until output '4' becomes a logic one. This signal is connected to the reset input, which immediately resets the counter to the 'zero' state. If you were to use an oscilloscope to look at this output, you would have to set it up very precisely before you would be able to see this pulse; that's how short it is. The output of an OR gate can only supply several mA, which is obviously much too little to drive a stepper motor directly. A suitable driver circuit, which goes between the generator and stepper motor, was published in the May 2004 issue of Elektor Electronics.

(044015-1)

129
Lifespan of Li-Ion Batteries

New technologies can introduce new problems. We haven't really had enough experience in the use of Lithium-Ion batteries to make a precise statement on their lifespan. Stories are floating around of a short lifespan of only a few years when used intensively in notebooks, whereas it should be possible for them to last anywhere between 500 and 1,000 cycles. Should you use the full capacity of the battery 200 days per year, it should in theory have a lifespan of about three years. But even when the battery has gone through only 100 cycles it appears to have lost some capacity. With nickel-cadmium and nickel-hydride cells it is recommended that they are never fully discharged, nor fully charged. The NiMH battery used by Toyota in the 'Prius' car operates between 40 % and 80 % of capacity and has an 8-year guarantee. If it was used between 0 and 100 % it wouldn't even survive one year of intensive use. Lithium-Ion batteries appear to behave differently. Discharging by 20 % and recharging often also seems to reduce the lifespan. With this type of battery it is therefore better to complete the discharge/charge cycle as much as possible, since half a cycle appears to count as a whole one.

A second aspect is the oxidation of the electrodes. They begin to deteriorate right from the moment of manufacture and that process is unavoidable. This causes a gradual reduction in the useable capacity. Although this process can't be stopped, it can be slowed down. The key words here are 'low temperature' and 'not fully charged'. It is ironic that this is the exact opposite to the conditions found in a typical notebook: the battery is kept fully charged and the temperature is often around 40 degrees Celcius.

There have been reports of batteries losing half their capacity after only three months when they've been kept fully charged at a temperature

of 60 degrees Celcius. Therefore, if you have a battery that won't be used for a while, you should charge it to 50% and keep it at a cool temperature (room temperature is fine). You can charge a battery to 50 % of its capacity by reducing the charging voltage to about 3.9 V. In any case, you should check the output voltage of the charger and take away a few tenths of a volt. Accidents can happen when the charging voltage is too high! Another cause of failure is when the battery is deeply discharged due to self-discharge. To avoid damage the battery voltage should never drop below 2 V. At room temperature this means that the battery should be checked once or twice a year, and recharged if necessary.

Karel Walraven

(044023-1)

130
Doorbell Cascade

Sometimes you have to do it the hard way, even if doing it the easy way is an option. That is the case here. The intention is to add a second doorbell in parallel with the existing bell. This does not, in principle, require any electronic components. You would simply connect the second bell to the first one. But if the existing bell transformer is not rated for the additional load then this is not a good idea! An option is to buy a new and larger transformer. But bigger also means more expensive! Moreover, replacing the existing transformer can be an awkward job, for example when it is built into the meter box. So we follow different approach.

This circuit is connected in parallel with the existing bell. This is possible because the current consumption is very small compared to the load of the bell. The bridge rectifier rectifies the bell voltage when the push-button is pressed. This will then close the relay contacts. These contacts are the which is powered from its own cheap bell 'electronic' button for the second bell, transformer.

René Bosch   (044024-1)

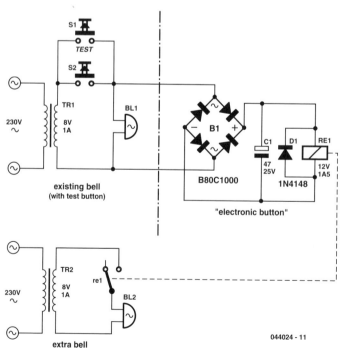

044024 - 11

131
Motor Turn/Stall Detector

In single phase AC induction motors, often used in fridges and washing machines, a start winding is used during the starting phase. When the motor has reached a certain speed, this winding is turned off again. The start winding is slightly out of phase to the run winding. The motor will only start turning when the current through this winding is out of phase to that of the run winding. The phase difference is normally provided by placing a capacitor of several µF in series with the start winding. When the motor reaches a minimum speed, a centrifugal switch turns off the start winding. The circuit diagram doesn't show a centrifugal switch; instead it has a triac that is turned on during the staring phase. For clarity, the series capacitor isn't shown in the diagram. Once the motor turns it will continue to do so as long as it isn't loaded too much. When it has to drive too heavy a load it will almost certainly stall. A large current starts to flow (as the motor no longer generates a back EMF), which is limited only by the resistance of the winding. This causes the motor to overheat after a certain time and causes permanent damage. It is therefore important to find a way to detect when the motor turns, which happens to be surprisingly easy. When the motor is turning and the start winding is not used, the rotation induces a voltage in this winding. This voltage will be out of phase since the winding is in a different position to the run winding. When the motor stops turning

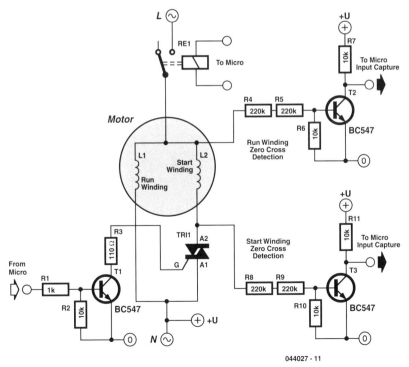

044027 - 11

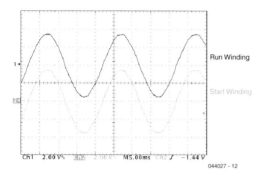

Run Winding

Start Winding

Ch1 2.00 V   2.00 V   M5.00ms Ch2 ∫ −1.44 V

044027 - 12

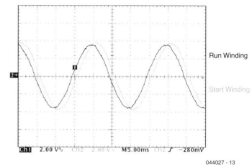

Run Winding

Start Winding

Ch1 2.00 V   2.00 V   M5.00ms Ch2 ∫ −280mV

044027 - 13

this voltage is no longer affected and will be in phase with the mains voltage. The graph shows some of the relevant waveforms. More information can be found in the application note for the AN2149 made by Motorola, which can be downloaded from their website at www.motorola.com. We think this contains some useful ideas,

but keep in mind that the circuit shown is only partially completed. As it stands, it certainly can't be put straight to use. We should also draw your attention to the fact that mains voltages can be lethal, so take great care when the mains is connected!

Karel Walraven   (044027-1)

132
Stable Zener Reference

Nowadays some first-rate voltage references are available. Take the LM385 for example: this is available for different voltages and even comes in an adjustable version. What is more, the current consumption may be kept very small (10 µA). But as often happens, you may not have one to hand when you need one for an experimental circuit. In that case, you could use an ordinary zener diode for the reference. Unfortunately, they have a somewhat higher internal resistance (about 5 Ω), which means they won't be very stable when the supply voltage varies.

The solution is right in front of us: use the stabilised zener voltage as the supply voltage! This is obviously only possible if the stabilised voltage is higher than the zener voltage. It therefore has to be amplified a little. This is exactly what this circuit

does: it amplifies it by a factor of two. The current limiting resister should be chosen such that a current of 1 to 3 mA flows through the zener diode. Manufacturers usually state the zener voltage at a current

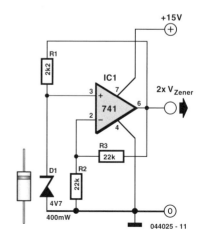

+15V

R1
2k2

IC1

3 +   7
   741
2 −   6   2x V$_{Zener}$
   4

R3
22k

D1   R2
         22k
4V7
400mW

044025 - 11

between 3 to 5 mA. The zener diode is fed from a stabilised voltage and hence has a very stable operating point, which is independent from the supply voltage. The graph speaks for itself. It is clear that the output voltage is much more stable. The graphs have been plotted to different scales to make the comparison easier. In reality the opamp output is twice the zener voltage. Zener diodes also have a temperature coefficient, which is smallest for types with a zener voltage around 5 volts. Virtually any type

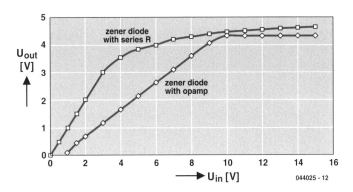

of opamp should be suitable; even our old friend the 741 works well enough.

Karel Walraven   (044025-1)

133
Save Energy

Despite our best efforts, a lot of energy is still wasted imperceptibly. We insulate our homes, install high efficiency boilers and buy low energy light bulbs. But it doesn't end there as far as electrical consumption is concerned. Many other items in the home consume electrical power, but here we concentrate on mains adapters (also called 'wall cubes' or 'battery eliminators'). Take a good look around the house to see how many you have, and you could soon find about ten of them: phone charger(s), battery chargers, minivac, telephone, answering machine, the radio in the kitchen, modem, and so on. The disadvantage of these adapters is that they easily consume from 1 to 2.5 W under no load, without you getting anything in return (apart from some heat, of course). When five mains adapters are in use, each consuming 2 W, they'll take one kilowatt-hour every 100 hours, at a cost of 7 p. And 100 hours amounts to only 4 days! In a year, this is 87.6 times as much, or a bit over £6 per year. And if ten

adapters are in use this amounts to over £12. Something can be done about this, of course. The simplest way is to switch off all adapters when they are not in use. Most of you do this already, surely. There are probably a few adapters that have to remain switched on at all times though. There is an alternative for these as well:

take a look at those modern switched-mode adapters! They no longer have a bulky transformer, just a switched-mode supply. They are (unfortunately) a bit more expensive, but tend to be smaller and give a better regulated output voltage. The quiescent power consumption of these adapters really is very small.

(044028-1)

134
Servo Tester using a 4538

There are times when a small servo tester for modelling comes in very useful. Everybody who regularly works with servos will know several instances when such a servo tester will come in handy. The function of a servo tester is to generate a pulsing signal where the width of the positive pulse can be varied between 1 and 2 ms. This pulse-width determines the position the servo should move to. The signal has to repeat itself continuously, with a frequency of about 40 to 60 Hz. We have already published other servo testers in the past. These circuits often use an NE555 or one of its derivatives to generate the pulses. This time we have used a 4538 for variety. This IC contains two astable multi-vibrators.

You can see from the circuit diagram that not many other components are required besides the 4538. The astable multi-vibrator in a 4538 can be started in two ways. When input I0 (pin 5 or 11) is high, a rising edge on input I1 (pin 4 or 12) is the start signal to generate a pulse.

The pulse-width at the output of IC1a is equal to $(R1+P1)\times C1$. This means that when potentiometer P1 is turned to its minimum resistance, the pulse-width will be $10\,\text{k} \times 100\,\text{n} = 1\,\text{ms}$. When P1 is set to maximum (10 k), the pulse-width becomes $20\,\text{k} \times 100\,\text{n} = 2\,\text{ms}$. At the end of this pulse inverting output Q generates a rising edge. This edge triggers IC1.B, which then generates a pulse. The pulse-width here is

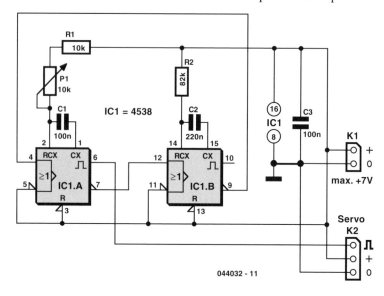

044032 - 11

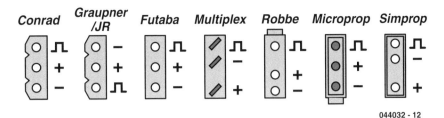

044032 - 12

$82\ k \times 220\ n \approx 18$ ms. At the end of this pulse the Q output will also generate a rising edge.

This in turn makes IC1.A generate a pulse again. This completes the circle. Depending on P1, the total period is between 19 and 20 ms. This corresponds to a frequency of about 50 to 53 Hz and is therefore well within the permitted frequency range.

Paul Goossens (044032)

135
Universal Mains Filter

There are plenty of mains filters available, with or without sockets, with or without a switch, etc. In any case, this filter is very versatile as far as the maximum working current is concerned. It so happens that the footprint of the coil remains the same for different currents. Depending on the required specification, you can use a specific coil (which the manufacturer calls a 'Current-Compensated Ring Core Double Choke'). The table below lists a suitable range of coils made by Epcos. When we designed the PCB we

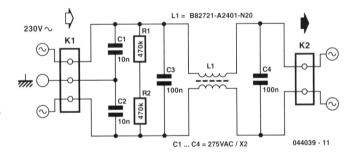

chose the smallest size that still gave a wide selection of maximum currents. Be careful that you don't use a vertical version of the coil. These appear to be pin-compatible, but they have different connections.

COMPONENTS LIST

Resistors:
R1,R2 = 470kΩ

Capacitors:
C1,C2 = 10nF 275 VAC X2, lead pitch 10 or 15mm
C3,C4 = 100 n/275 VAC X2, lead pitch 15mm

Inductors:
L1 = B82721-A2401-N20, Epcos (Farnell 976-477)

Miscellaneous:
K1 = 3-way PCB terminal block, lead pitch 7.5mm
K2 = 2-way PCB terminal block, lead pitch 7.5mm

I_R (A)	L_R (mH)	$L_{S,typ.}$ (µH)	$R_{S,typ.}$ (mΩ)	Ordering code horizontal version
0.4	39	450	2000	B82721-A2401-N20
0.4	27	270	1700	B82721-A2401-N21
0.5	18	260	1500	B82721-A2501-N1
0.7	10	90	600	B82721-A2701-N20
1.2	6.8	70	280	B82721-A2122-N20
1.5	3.3	37	190	B82721-A2152-N1
2.0	1.0	13	90	B82721-A2202-N1
2.6	0.4	6	60	B82721-A2262-N1
3.6	0.4	6	35	B82721-A2362-N1

They would cause a severe short-circuit on this PCB! For this reason we haven't included them in the table (they have a K instead of an A in their order code). This filter is suitable for use up to 3.6 A (a good 800 VA). In practice there should be no need to use different values for the capacitors. You should make sure that you obtain the correct lead spacing and that they're X2 types. For the 100 nF capacitors provision is only made for a lead spacing of 15 mm (as these are the most common). For the 10 nF capacitors you can use either 10 or 15 mm types. A final warning: always take great care when working with mains voltages!

Ton Giesberts   (044039-1)

136
Xilinx JTAG Interface

In September 2002 we published a JTAG interface that was compatible with the programming software from Altera. Unfortunately, the software from Xilinx didn't work in combination with this interface. The interface published here is compatible with the software from Xilinx, so you can use it to program their range of CPLDs and FPGAs. The circuit is very simple and consists of just two ICs and a handful of discrete components. Connector K1 is connected to the PC using a 1:1 printer cable with a 25-way sub-D connector at each end. Connector K2 is connected to the JTAG interface of the device being programmed. The pin-out for con-

Table 1. Pinout for K2

	FPGA	CPLD
1	V_{dd}	V_{dd}
2	GND	GND
3	CCLK	TCK
4	D/P	TDO
5	DIN	TDI
6	$\overline{PROG}$	TMS

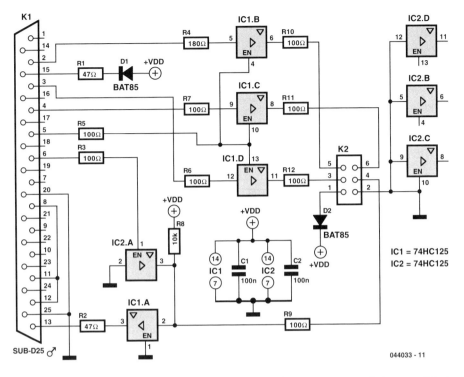

044033 - 11

nector K2 is shown in Table 1. If the device uses a different type of programming connector, K2 will have to be adapted accordingly. The circuit can be easily built on a piece of prototyping board. Since there isn't any real standard for the pro-gramming connector, it is very likely that the connections to K2 have to be modified; in that case, a ready-made PCB isn't very useful.

Paul Goossens   (044033-1)

137
PWM Modulator

If you ever thought of experimenting with pulse-width modulation, this circuit should get you started nicely. We've kept simplicity in mind and used a dual 555 timer, making the circuit a piece of cake. We have even designed a small PCB for this, so building it shouldn't be a problem at all. This certainly isn't an original circuit, and is here mainly as an addition to the 'Dimmer with MOSFET' article elsewhere in this issue. The design has therefore been tailored to this use. A frequency of 500 Hz was chosen, splitting each half-period of the dimmer into five (a low frequency generates less interference). The first timer is configured as a standard astable frequency generator. There is no need to explain its operation here, since this can easily be found on the Internet in the datasheet and application notes. All we need to mention is that the frequency equals $1.49 / ((R1+2R2) \times C1)$ [Hz]. R2 has been kept small so that the frequency can be varied easily by adjusting the values of R1 and/or C1. The second timer works as a monostable multi-vibrator and is triggered by the differentiator constructed using R3 and C3. The trigger input reacts to a rising edge. A low level at the trigger input forces the output of the timer low. R3 and C3 have therefore been added, to make the control range as large as possible. The pulse-width of the monostable timer is given by $1.1 \cdot R4 \cdot C4$ and in this case equals just over a millisecond. This is roughly half the period of IC1a. The pulse-width is varied using P1 to change the voltage on the CNTR input. This changes the voltage to the internal comparators of the timer and hence varies the time required to charge up C4. The control range is also affected by the supply

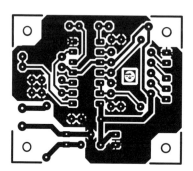

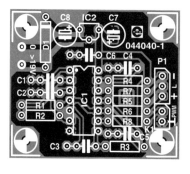

COMPONENTS LIST

Resistors:
R1 = 270kΩ
R2,R3 = 10kΩ
R4 = 100kΩ
R5,R8 = 1kΩ
R6,R7 = 220Ω
P1 = 2kΩ2, linear, mono

Capacitors:
C1,C4 = 10nF
C2,C5,C6 = 100nF
C3 = 1nF
C7 = 2µF2 63V radial
C8 = 100µF 25V radial

Semiconductors:
D1 = 1N4002
IC1 = NE556
IC2 = 78L15

Miscellaneous:
P1 = 3-way pinheader
K1 = 2-way pinheader

voltage; hence we've chosen 15 V for this. The voltage range of P1 is limited by R6, R7 and R5. In this design the control voltage varies between 3.32 V and 12.55 V (the supply voltage of the prototype was 14.8 V). Only when the voltage reaches 3.51 V does the output become active, with a duty-cycle of 13.5 %. The advantage of this initial 'quiet' range is that the lamp will be off. R8 protects the output against short circuits. With the optocoupler of the dimmer as load, the maximum current consumption of the circuit is about 30 mA.

Ton Giesberts   (044040-1)

138
USB Converter Controlled via HTML

In this issue we publish an ActiveX component, which can be used to control the USB analogue converter (Elektor Electronics, November 2003). In this way, programmers can use C/C++, Delphi, VB etc. to include the converter in their own application. It is maybe less well known that these ActiveX components can also be used from web browsers that support scripts and ActiveX. For this reason we've created an example HTML file, which uses JavaScript and ActiveX to control the USB converter. This file is available as a Free Download from http://www.elektor-electronics.co.uk/ (044034-11). To place an ActiveX component onto a web page you have to make use of the <OBJECT> tag. In this a name and a CLASSID have to be specified. This CLASSID is a number that indicates which type of ActiveX component should be used. Since it is impractical to remember all these number by heart, Microsoft have made a program available called ActiveX Control Pad. With this program it becomes easy to place an ActiveX component onto a web page and adapt its

properties to your own liking. Now that we've placed the ActiveX component onto the page, we can use JavaScript to send commands to this component or get it to return information. Here the JavaScript part sets up a communication channel with the USB converter when the page is opened. It also starts a timer that calls the function ShowInput() every half a second. The functions in JavaScript are very similar to those found in C.

The three functions used in this example are simple enough for anyone with a bit of programming experience to follow. An important detail that should be mentioned is that every ActiveX component on the page is given a name during the initialisation. In this case, we have given the meaningful name 'USB' to the component that takes care of the communication with the USB module. The two labels on this page have been creatively named as 'Label1' and 'Label2'. The previous tale sounds good, but does it work in practice? Everybody who has a USB analogue converter from the November 2003 issue of Elektor Elec-

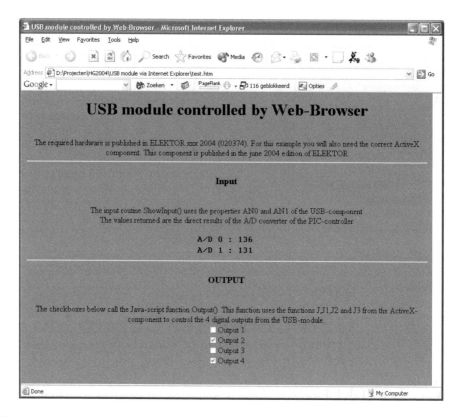

tronics and who has installed this month's ActiveX component, can try it out very quickly. The USB converter first has to be connected via a USB cable. Then you should open the file 'test.htm'. If you have a web browser that supports ActiveX and JavaScript (such as Internet Explorer), you should see the web page as shown here. The values of the A/D converter are refreshed on the screen twice per second and the tick-boxes at the bottom of the screen let you change the state of the four digital outputs. JavaScript is not very powerful when com- pared with other programming languages such as C, C++, Delphi, etc, but has the advantage that it is relatively easy to understand. Furthermore, everybody who has a standard Windows operating sys- tem installed on their PC can get started straight away.

Paul Goossens   (044034-1)

139
Codec Complete

Digital audio equipment usually contains an A/D and a D/A converter. In practice a codec is used for this. This is a chip where both converters are built-in and which often includes standard inputs and outputs for digital audio, such as I2S. Apart from

FUNCTIONAL BLOCK DIAGRAM

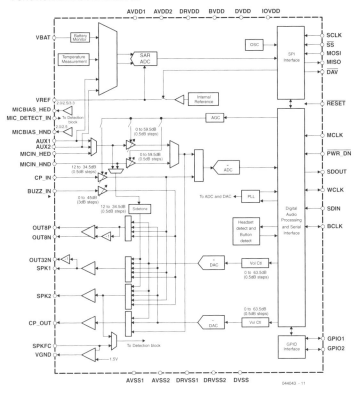

this chip, such as 2 I/O pins for use in push-button control, microphone detector, etc. It is therefore extremely suitable for use in combination with headsets. The chip can be controlled via an SPI interface, which means that most microcontrollers can communicate easily with this codec. As we said earlier, the audio interface can take an I2S signal, but the audio interface is very flexible, as with many other codecs, and can cope with several other audio formats. Should you be on the lookout for a codec and you intend to use a microphone input and headphone output, then this one makes an excellent choice. More

this codec there is often a requirement for a microphone input and headphone output as well. Texas Instruments have made a new codec, the TLV320AIC28, which has an integrated microphone pre-amplifier and a 400 mW headphone amplifier. A few other practical functions have also been added to

information for this codec can be found in its datasheet at the website of Texas Instruments:

http://focus.ti.com/docs/prod/folders/print/tlv320aic28.html

Paul Goossens (044043-1)

140
Telephone Line Indicator

Many 'busy' indicators for use in tele-phone systems present undesirable loading of the telephone line. Some circuits are very simple indeed to the extent of only loading the line when it is not in use. The downside is that a (usually green) LED

lights when the line is not occupied. The author feels that a LED should flash when the line is actually in use by another extension and that the circuit should present a minimal load of the line. The circuit shown here fulfills both requirements. We should,

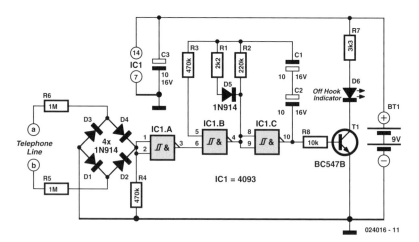

024016 - 11

however, not forget to mention its only drawback: its needs to be powered from a battery or an energy-friendly battery eliminator (a.k.a. wall cube or mains adapter). If a high-efficiency LED is used then the current drain from the 9 volt supply will be so small that a standard (170 mAh) 9 V PP3 battery will last for months. Considering that the LED is powered at current of 2 mA by T1, theoretically some 85 hours of 'LED on' time can be obtained. If, for some reason, you wish to

change the flash frequency or on/off ratio (duty cycle) then do feel free to experiment with the values and ratio of R1 and R2. The effect will also depend on the brand of the 4093 IC, its exact logic High/Low switching thresholds and hysteresis.
The circuit is not approved to BABT standards for connection to the public switched telephone network (PSTN). Please check local/national regulations.

Flemming Jensen   (024016-1)

141
An Electronic Watering Can

Summertime is holiday time but who will be looking after your delicate houseplants while you are away? Caring for plants is very often a hit or miss affair, sometimes you under-water and other times you over-water. This design seeks to remove the doubt from plant care and keep them optimally watered. The principle of the circuit is simple: first the soil dampness is measured by passing a signal through two electrodes placed in the soil. The moisture content is inversely proportional to the measured resistance. When this measure-

ment indicates it is too dry, the plants are given a predefined dose of water. This last part is important for the correct function of the automatic watering can because it takes a little while for the soil to absorb the water dose and for its resistance to fall. If the water were allowed to flow until the soil resistance drops then the plant would soon be flooded. The circuit shows two 555 timer chips IC1 and IC2. IC1 is an astable multi-vibrator producing an ac coupled square wave at around 500 Hz for the measurement electrodes F and F1. An ac signal

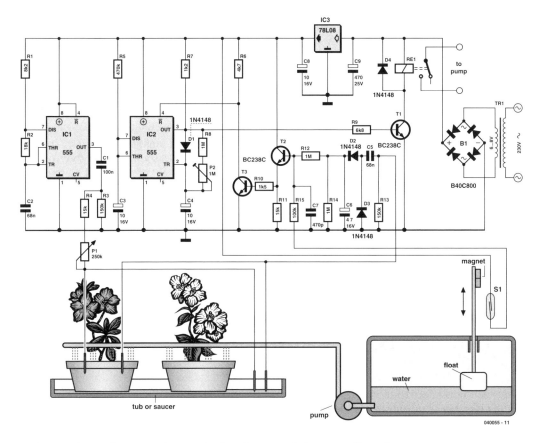

040055 - 11

reduces electrode corrosion and also has less reaction with the growth-promoting chemistry of the plant. Current flowing between the electrodes produces a signal on resistor R13. The signal level is boosted and rectified by the voltage doubler produced by D2 and D3. When the voltage level on R13 is greater than round 1.5 V to 2.0 V transistor T2 will conduct and switch T3. Current flow through the soil is in the order of 10 μA. T2 and T3 remain conducting providing the soil is moist enough. The voltage level on pin 4 of IC2 will be zero and IC2 will be disabled. As the soil dries out the signal across R13 gets smaller until eventually T2 stops conducting and T3 is switched off. The voltage on pin 4 of IC2 rises to a '1' and the chip is enabled. IC2 oscillates with an 'on' time of around 5 s and an 'off' time (adjustable via P2) of 10 to 20 s. This signal switches the water pump via T1. P1 allows adjustment of the minimum soil moisture content necessary before watering is triggered. The electrodes can be made from lengths of 1.5 mm^2 solid copper wire with the insulation stripped off the last 1 cm. The electrodes should be pushed into the earth so that the tips are at roughly the same depth as the plant root ball. The distant between the electrodes is not critical; a few centimetres should be sufficient. The electrode tips can be tinned with solder to reduce any biological reaction with the copper surface. Stainless steel wire is a better alternative to copper, heat shrink sleeving can used to insulate the wire with the last 1 cm of the electrode left bare.

Two additional electrodes (F1) are connected in parallel to the soil probe electrodes (F). The F1 electrodes are for safety to ensure that the pump is turned off if for some reason water collects in the plant pot saucer. A second safety measure is a float switch fitted to the water reservoir tank. When the water level falls too low a floating magnet activates a reed switch and turns off the pump so that it is not damaged by running with a dry tank. Water to the plants can be routed through closed end plastic tubing (with an internal diameter of around 4 to 5 mm) to the plant pots. The number of 1 mm to 1.5 mm outlet holes in the pipe will control the dose of water supplied to each plant. The soil probes can only be inserted into one flowerpot so choose a plant with around average water consumption amongst your collection. Increasing or decreasing the number of holes in the water supply pipe will adjust water supply to the other plants depending on their needs. A 12 V water pump is a good choice for this application but if you use a mains driven pump it is essential to observe all the necessary safety precautions. Last but not least the electronic watering can is too good to be used just for holiday periods, it will ensure that your plants never suffer from the blight of over or under-watering again; provided of course you remember to keep the water reservoir topped up…

Robert Edlinger   (040055-1)

142
Simple Microphone Preamplifier

The technical demands on microphones used with radio equipment are not stringent in terms of sound quality: a frequency response from around 50 Hz to 5 kHz is entirely adequate for speech. For fixed CB use or for radio amateurs sensitivity is a more important criterion, so that good intelligibility can be achieved without always having to hold the microphone directly under your nose. Good microphones with extra built-in amplifiers can be bought, but, with the addition of a small preamplifier, an existing microphone will do just as well. The project described here uses only a few discrete components and is very undemanding. With a supply voltage of between 1.5 V and 2 V it draws a current of only about 0.8 mA. If you prefer not to use batteries, the adaptor circuit shown, which uses a 10 kΩ resistor, three series-connected diodes and two 10 μF electrolytic smoothing capacitors, will readily generate the required voltage from the 13.8 V supply that is usually available. There is little that need be said about the amplifier itself. Either an ordinary dynamic microphone or a cheaper

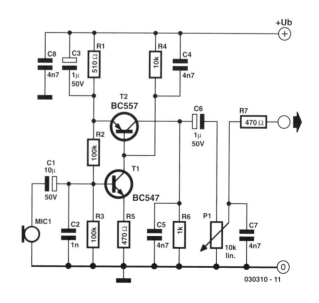

030310 - 11

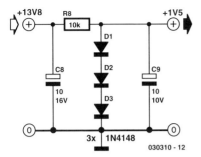

electret capsule type can be connected to the input. In the latter case a 1 kΩ resistor needs to be connected between the 1.5 V supply voltage and the positive input connection. The impedance of the microphone and of the following stage in the radio apparatus are not of any great importance

since the available gain of 32 dB (a factor of 40) is so great that only in rare cases does P1 have to be set to its maximum position. With a frequency range from 70 Hz to about 7 kHz, low distortion, and small physical size, the preamplifier is ideal for retrofitting into the enclosure of the radio equipment or into the base of a micro-phone stand. In case you are concerned about our somewhat cavalier attitude towards distortion: for speech radio the 'fi' does not need to be 'hi'. Quite the reverse, in fact: the harmonics involved in a few percent of distortion can actually improve intelligibility – it's not a bug, it's a feature!

Ludwig Libertin   (030310-1)

143
Long-Delay Stop Switch

Pre-settable times for train stops in stations are indispensable if you want to operate your model railway more or less realistically according to a timetable. This circuit shows how a 555 timer can be used with a relatively small timing capacitor to generate very long delay times as necessary by using a little trick (scarcely known among model railway electronic technicians): pulsed charging of the timing network. Such long delays can be used in hidden yards with through tracks, for instance. As the timer is designed for half-wave operation, it requires only a single lead to the transformer and one to the switching track or reed contact when used with a Märklin AC system (H0 or H1). The other lead can be connected to any desired grounding point for the common ground of the track and lighting circuits. As seen from the outside, the timer acts as a monostable flip-flop. The output (pin 3) is low in the quiescent state. If a negative signal is ap-

plied to the trigger input (pin 2), the output goes high and C4 starts charging via R3 and R4. When the voltage on C4 reaches 2/3 of the supply voltage, it discharges via an internal transistor connected to pin 7 to 1/3 of the supply voltage and the output (pin 3) goes low. The two threshold values (1/3 and 2/3) are directly proportional to the supply voltage. The duration of the output signal is independent of the supply voltage:

$$t = 1.1(R4 + R5) \times C4$$

if the potentiometer is connected directly to the supply line (A and B joined). The maximum delay time that can be generated using the component values shown in the schematic diagram is 4.8 minutes. However, it can be increased by a factor of approximately 10 if the timing network is charged using positive half-waves of the AC supply voltage (reduced to the 10–16 V

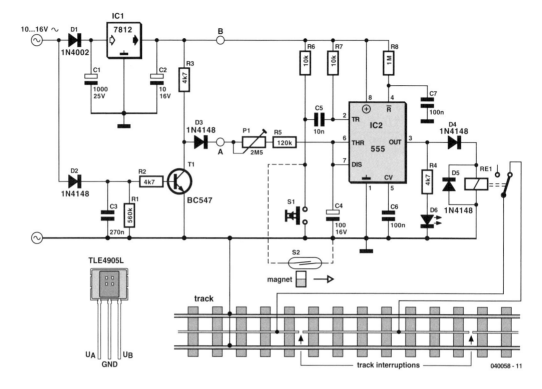

level) instead of a constant DC voltage. The positive half-waves of the AC voltage reach the timing network via D2, the transistor, and D3. Diode D3 prevents C4 from being discharged between the pulses. The total resistance of R4 and R5 should not be too high (no more than 10 MΩ if possible), since electrolytic capacitors (such as are needed for C4) have significant leakage currents. Incidentally, the leakage current of aluminium electrolytic capacitors can be considerably reduced by using a supply voltage well below the rated voltage. Capacitor C6 is intended to suppress noise. It forms a filter network in combination with an internal voltage-divider resistor. If a vehicle happens to remain standing over the reed switch so the magnet holds the contacts constantly closed, the timer will automatically be retriggered when the preset delay times out. In this case the relay armature will not release and the locomotive will come to the 'end of the line' in violation of the timetable. This problem can be reliably eliminated using R6, R7 and C5.

This trigger circuit ensures that only one trigger pulse is generated, regardless of how long the reed switch remains closed. RC network R8/C7 on the reset pin ensures that the timer behaves properly on switch-on (which is far from being something to be taken for granted with many versions of the 555 or 556 dual timer). Reed switches have several special characteristics that must be kept in mind when fitting them. The contact blades, which are made from a ferromagnetic material, assume opposite magnetic polarities under the influence of a magnetic field and attract each other. Here the position and orientation of the magnet, the distance between the magnet and the reed switch, and the direction of motion of the magnet relative to the switch are important factors. The fragility of the glass housing and the thermal stress from soldering (stay at least 3 mm away from the glass housing) require a heat sink to be used between the soldering point and the glass/metal seal. A suitable tweezers or flat-jawed pliers can be used for this purpose. If you need to bend the leads, use flat-jawed pliers to protect the glass/metal seal against mechanical stresses. Matching magnets in various sizes are available from toy merchants and electronics mail-order firms. They should preferably be fitted underneath the locomotive or carriage. However, the magnet can also be fitted on the side of a vehicle with a plastic body. In this case the reed switch can be hidden in a mast, bridge column or similar structure or placed in a tunnel, since the distance must be kept to less than around 10 mm, even with a strong magnet. If fitting the circuit still presents problems (especially with Märklin Z-gauge Mini-Club), one remedy is to generate the trigger using a unipolar digital Hall switch, such as the Siemens TLE4905L or Allegro UGN3120. To avoid coupled-in interference, the stop timer should be fitted relatively close to the Hall sensor (use screened cable if necessary). Pay attention to the polarity of the magnet when fitting it to the bottom of the vehicle.

With both types of sensors, the South pole must point toward the front face of the Hall IC (the face with the type marking). The North pole is sometimes marked by a dab of paint. Generally speaking, the polarity must be determined experimentally. Fitting the circuit is not a problem with Z-gauge and 1-gauge tracks, since the dis-tance between the iron parts (rails) and the Hall switch is sufficiently large. In an HO system, some modifications must be made to the track bed of the Märklin metal track. Cut a suitably sized 'window' between one wheel rail and the centre rail in order to prevent secondary magnetic circuits from interfering with the operation of the sensor. Keep the distance between the magnet and the case of the Hall switch between 5 and 10 mm, depending on the strength of the magnet, to ensure reliable actuation.

Robert Edlinger   (040058-1)

144
Slave Flash with Red-Eye Delay

Digital cameras are becoming more and more affordable. At the economy end of the market cameras are usually equipped with a small built-in flash unit that is ideal for close-ups and simple portraiture. The power rating of the built-in flash unit is quite low so that any subject further away than about 2 to 3 metres (maybe 4 m if you are lucky) tends to disappear into the gloom. You soon become aware of the limitations if you need to photograph a larger group of people say at a function under artificial light in a large hall or outdoors. The majority of these cameras are not fitted with an accessory socket so it is not possible to simply connect a second flash unit to increase the amount of light. Single lens re-

flex cameras also need additional lighting (e.g. fill-in flash) to reduce the harsh contrast produced by a single light source. For

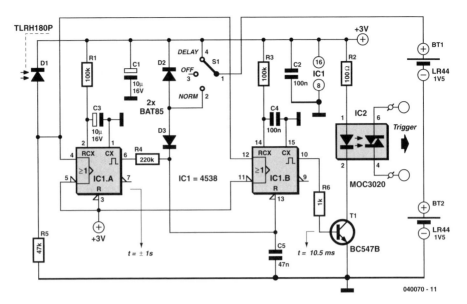

040070 - 11

all these cases an additional slave flashgun is a useful addition to the equipment bag. Rather than shelling out lots of cash on a professional slave flashgun, the circuit here converts any add-on flashgun into a slave flash unit triggered by light from the camera flash. Simple slave flash circuits can have problems because most modern cameras use a red eye reduction pre-flash sequence. This pre-flash is useful for portraiture. It is designed to allow time for the subjects pupils to contract so that the red inner surcus. A simple slave flash circuit will be triggered by the pre-flash sequence

Components list

Resistors:
R1, R3 = 100kΩ
R2 = 100Ω
R4, R5 = 220kΩ
R6 = 1kΩ

Capacitors:
C1, C3 = 10µF 16V radial
C2, C4 = 100nF
C5 = 47 nF

Semiconductors:
D1 = TLRH180P (Farnell # 352-5451)
D2, D3 = BAT85
IC1 = 4538P
IC2 = MOC3020
T1 = BC547B

Miscellaneous:
Bt1 = two 1.5V batteries (LR44) with PCB mount holder
S1 = 3-position slide switch
PCB, ref. 040070-1 from The PCBShop Cable or adaptor for external flasher

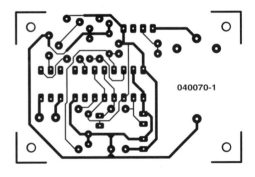

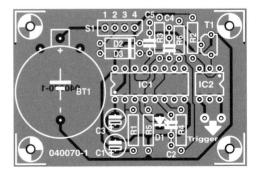

and will therefore not provide any additional lighting when the main flash occurs and the picture is actually taken. The circuit shown here is quite simple but neatly solves the pre-flash problem. With switch S1 set to 'Normal', the pulse produced by D1 when it detects the camera flash will trigger both monoflops IC1a and IC1b. The output of IC1.A does not perform any useful action in this mode because the logic level on the other side of resistor R4 is pulled high by D3. The output of IC1.B will go high for approximately 10 ms switching T1 on and causing the triac to conduct and trigger the slave flash. The use of a triac optocoupler here has the advantage that the circuit can be used on older types of flashgun triggered by switching a voltage of around 100 V as well as newer types that require only a few volts to be switched. With switch S1 in the delay position the first flash will trigger IC1.A and its output will enable IC1.B but the low pass

characteristics of the filter formed by R4 and C5 slow the rising edge of this waveform so that IC1.B will only be enabled 10 ms after the first flash is detected. IC1.B is now enabled for a period of about 1 s (governed by R1 and C3). When the main flash occurs in this time window it will immediately trigger IC1.B and the triac will be switched as described above. The circuit requires a supply of 3 V and draws very little current from the two 1.5 V button cells. It will run continuously for quite a few days, should it be accidentally left on. Switch S1 can be either a three-position toggle or slider type. Circuit construction is greatly simplified and the finished unit looks much neater if it is built on the available PCB. Space is also provided to fit the PCB mounted battery holders. A suitable flash extension cable or adapter can be found in most photo shops.

Paul Goossens   (040070-1)

145
Filter-based 50 Hz Sinewave Generator

When it comes to designing a reliable 50-Hz oscillator, the disadvantage of the good old Wien Bridge oscillator is the difficulty to adjust its own gain. If the gain is higher or lower than the optimum value, the Wien Bridge often fails to work properly. The circuit shown here combines the functions of a low-pass filter and an integrator, presenting a novel approach to creating a precision 50-Hz source with relatively low distortion. The circuit is free from any kind of gain setting network. Opamp IC.1B and R3/C4/C5 act as an inverting integrator effectively converting the incoming sine wave (from IC.1A) into a square wave with a good amount of harmonics. R4, D1 and D2 divide the square wave to the desired level. For optimum

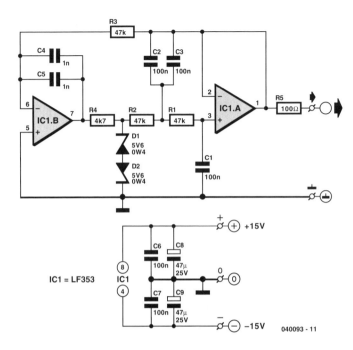

switching speed, a matching number of series-connected low-power switching diodes like the ubiquitous 1N4148 may be used instead of the zener diodes. The output voltage is directly proportional to the zener diode values. The second opamp, IC.1A and its surrounding components R1, C1, acts as a two-pole low-pass filter supplying the 50-Hz output signal via R5. Theoretically, the filter roll-off is at 24 Hz, which means the base frequency of 50 Hz is also attenuated to some degree. That is not too serious as long as higher harmonics are properly attenuated. The design with

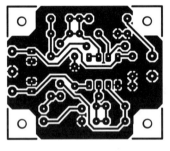

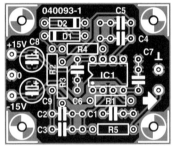

the component values shown here will supply an output voltage of 1.24 V at a frequency of 49.6 Hz. Current consumption was measured at less than 5 mA, while distortion was 3.7% when using an LF353 opamp. We performed an FFT (Fast Fourier Transform) analysis on the generator's output signal. Even-numbered harmonics were found clearly less in level than their odd-numbered counterparts. This is caused by a slight asymmetry in the generator's internal square wave. In our prototype, the FFT graph showed the 3rd harmonic at a relatively high level of -29.2 dB; the 2nd harmonic did much better at just –67.7 dB.

Myo Min   (040093-1)

Components list

Resistors:
R1, R2, R3 = 47kΩ
R4 = 4kΩ
R5 = 100Ω

Capacitors:
C1, C2, C3, C6, C7 = 100nF 16V radial
C4, C5 = 1nF
C8, C9 = 47µF 25V radial

Semiconductors:
D1, D2 = zener diode 5.6V 400mW
IC1 = LF353DP

Miscellaneous:
PCB, ref. 040093-1 from The PCBShop

146
PDM Pulse Generator

This pulse generator, which offers a mark-space ratio that is adjustable in exponentially increasing steps, is a novel use for a familiar IC. It can come in handy in a range of situations: for example, when testing the dynamic regulation characteristics of a power supply. In this test a load is applied to a power supply that draws a square-wave, rather than constant, current. An oscilloscope can then be used to check how quickly the regulator in the power supply responds and to observe any undesirable artefacts in the output such as overshoots or oscillation. A disadvantage of such an arrangement is that the dynamic characteristics of the power supply are only measured at one av-erage current, determined by the mark-space ratio of the current and by its maximum value. Modern computers, for example, draw varying amounts of current depending on what they are doing, and the

1

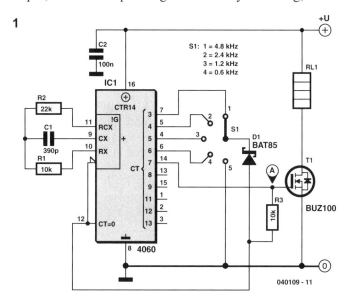

040109 - 11

Table 1. Pulse width in exponentially increasing steps						
Reset pulse		**Q 7 (junction A)**				
Switch position	**CLR pin connected to**	**Duty cycle**		**Frequency***	**Period***	**Pulse width***
		as percentage	as a fraction**	Hz	μs	μs
1	Q 3	5.9 % approx.	1/17	566	1765	103
2	Q 4	11 1/9 %	1/9	535	1868	207
3	Q 5	20.0 %	1/5	480	2082	415
4	Q 6	33 1/3 %	1/3	401	2495	830
5	GND	50.0 %	1/2	302	3320	1660

197

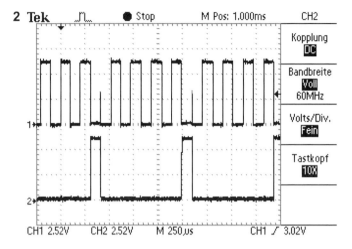

pulses would only be of the order of the ten nanoseconds. A low level via D1 and S1, however, delays the resetting of the IC for exactly as many clock periods as it takes for the output selected by S1 to go high. For example, if Q3 is selected, this happens after a further eight periods. The cycle time is then 128+8 clock periods, of which 128 clock periods are space and just eight are mark. This gives a duty cycle, as shown in the table, of 8/136 or 1/17. If Q6 were used the pulse would be $0.5 \times 27 = 64$ clock periods wide with the same space period of 128 clocks.

The duty cycle in this case is 1/3. If the cathode of D1 is tied to ground, the IC does not receive a reset signal and so a symmetrical square wave with a 128-clock mark and 128-clock space is produced. T1 functions as a power switch. Load R4 is chosen so that the maximum rated current of the power supply flows when T1 conducts. The average load on the power supply can be calculated using the duty cycle values given in the table. It should go without saying that the current drawn must be within the specification of the power supply, and that R4 must be capable of the necessary power dissipation. At higher currents a heatsink is recommended for T1. It can pass a maximum of 60 A and although it has a very low resistance in the conducting state (just 18 mΩ) it is no superconductor! At currents above about 10 A it starts to get warm.

current can vary over very short timescales. A modern PC power supply must therefore deliver accurate and stable voltages not only under high and intermediate load, but also at low load (in stand-by mode). The PDM pulse generator is ideal for this situation. As can be seen from the table, S1 allows the mark-space ratio to be adjusted from 1:17 to 1:2. The frequency varies from 566 Hz to 302 Hz, making it suitable for testing PC power supplies.

The variation in frequency with mark-space ratio is a consequence of how the circuit works. At the heart of the circuit is our old friend the CMOS 4060. This includes an RC oscillator circuit and a 14 bit binary counter. The basic clock frequency, which can be calculated using the formula shown, is around 77 kHz for the given values of R1 and C1. The space time is constant at around 1660 μs. When the IC is reset, all outputs are initially low. Consider now the voltage at the gate of T1. If it were not for D1, when Q7 goes high after the 128th clock pulse (the divide-by-28 output spends 128 clock periods low and then 128 periods high) IC1 would be reset immediately via R3. T1, and therefore the load, would thus receive extremely short pulses with long pauses between. Because of the switching speed of the IC, the width of the

The oscilloscope traces show the circuit operating at one of the possible mark-space ratios. The upper trace is the basic clock on pin 11 of IC1, and the lower trace shows the signal at test point A, which has a duty cycle of 1/5.

Klaus-Juergen Thiesler   (040109-1)

147
Garage Timer

The circuit described here is a testament to the ingenuity of two young designers from a specialist technical secondary school. The 'garage timer' began as a school electronics project and has now made it all the way to publication in our Summer Circuits special issue of Elektor Electronics. The circuit demonstrates that the application possibilities for the 555 and 556 timer ICs are by no means exhausted. So what exactly is a 'garage timer'?

When the light switch in the garage is pressed, the light in the garage comes on for two minutes. Also, one minute and forty-five seconds after the switch is pressed, the outside light also comes on for a period of one minute. The timer circuit is thus really two separate timers. Although the circuit for the interior light timer is relatively straightforward, the exterior light timer has to deal with two time intervals. First the 105 second period must expire; then the exterior light is switched on, and after a further 60 seconds the light is turned off. To realise this sequence of events, a type 556 dual timer device, a derivative of the 555, is used. The first of the two timers triggers the second after

a period of 105 seconds. The second timer is then active for 60 seconds, and it is this timer that controls the exterior light. The interior light timer is triggered at the same moment as the dual timer. In this case a simple 555 suffices, with an output active for just two minutes from the time when the switch is pressed. Push-button S1 takes over the role of the wall-mounted light switch, while S2 is provided to allow power to be removed from the whole circuit if necessary. The circuit could be used in any application where a process must be run for a set period after a certain delay has

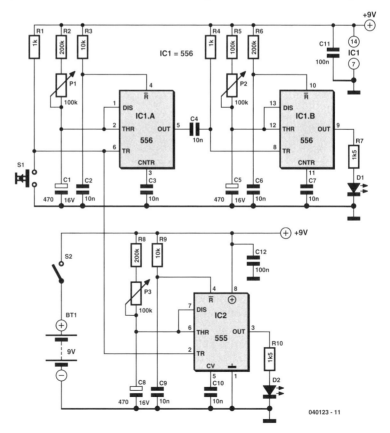

040123 - 11

expired. For the school project the two garage lights are simulated using two LEDs. This will present no obstacle to experienced hobbyists, who will be able to extend the circuit, for example using relays, to control proper lightbulbs. The principles of operation of type 555 and 556 timers have been described in detail previously in Elektor Electronics, but we shall say a few words about the functions of IC1a, IC1b and IC2.

When S1 is pressed (assuming S2 is closed!) the trigger inputs of both IC1a and IC2 are shorted to ground, and so the voltage at these inputs (pins 6 and 2 respectively) falls to 0 V. The outputs of IC1a and IC2 then go to logic 1, and D2 (the interior light) illuminates. Capacitors C1 and C8 now start to charge via P1 and R2, and R8 and P3 respectively. When the voltage on C8 reaches two thirds of the supply voltage, which happens after 120 seconds, the output of IC2, which is connected as a monostable multi-vibrator, goes low. D2 then goes out. This accounts for the interior

light function. Likewise, 105 seconds after S1 is closed, the voltage on C1 reaches two thirds of the supply voltage and the output of IC1a goes low. Thanks to C4, the trigger input of IC1b now receives a brief pulse to ground, exactly as IC1a was triggered by S1.

The second monostable, formed by IC1b, is thus triggered. Its pulse duration is set at one minute, determined by C5, R5 and P2. D1 thus lights for one minute. Potentiometers P1, P2 and P3 allow the various time intervals to be adjusted to a certain extent. If considerably shorter or longer times are wanted, suitable changes should be made to the values of C1, C5 and C8. The period of the mono-stable is given by the formula

$$T = 1.1\ RC$$

where T is the period in seconds, R the total resistance in ohms, and C the capacitance in farads.

148
Navigator Assistant

These days it is quite common that a handheld computer (PDA) is used in a car as a navigation system. The author uses, for example, a Dell Axim X5 with TomTom navigator. The PDA is installed in a holder in such a way that the screen is clearly visible to the driver. Such holders do not normally have power and audio connections for the PDA. This circuit offers a solution to both problems: a regulated power supply and an audio power amplifier. The author originally used an LM350 IC, but it proved to be barely capable of coping with the dissipation when the battery was nearly completely discharged. That is why, in the end,

a switching power supply was used. The power supply is designed around an LT1074, a bipolar switching regulator that contains just about all the components required for a so-called buck-regulator. The IC operates at a switching frequency of 100 kHz and can deliver up to 5 A. Coil L1 and diode D3 form part of the flyback circuit. Make sure that these are capable of dealing with the maximum desired output current. Potentiometer P1 is used to adjust the output voltage to the value that is optimum for the PDA that is being used. LED D2 indicates the presence of the input power supply. The second part, the audio

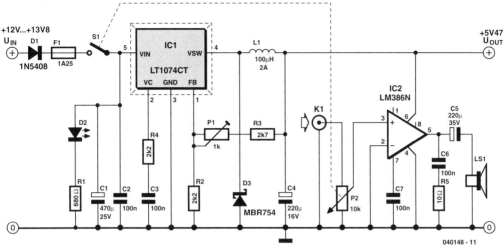

040148 - 11

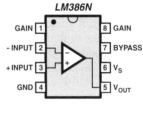

LM386N

GAIN	1		8	GAIN
- INPUT	2		7	BYPASS
+INPUT	3		6	V_S
GND	4		5	V_{OUT}

040148 - 12

amplifier, is an equally simple design. The circuit really speaks for itself. This is an LM386 in a standard configuration. P2 adjusts the volume. The amount of power that the LM386 can deliver is about 300 mW. The best results are obtained of you use a special communications loudspeaker. The navigation messages are then very easy to understand. The gain of IC2 is set to 20. This will usually be sufficient, but you can also increase the amplification. At a gain of 50, it is necessary to connect a series network consisting of a 1k2 resistor and 10 µF electrolytic between pins 1 and 8 of the LM386 (the negative of the capacitor connects to pin 8). Finding the correct power supply connector for the PDA can be a bit of a problem. So look for a suitable connector before you start to build this circuit. Also be very careful about the correct polarity of the connector. The circuit can be wired directly to the car's power circuit or connected to the cigarette-lighter socket. In either case a fuse (F1) can be connected in the power supply lead or the plug.

René Bosch   (040148-1)

149
Energy-saving Switch

Light do not always need to be on at full power. Often it would be useful to be able to turn off the more powerful lights to achieve softer illumination, but requires an installation with two separately switchable circuits, which is not always available. If the effort of chasing out channels and re-plastering for a complete new circuit is too much, the this circuit might help. Normal operation of the light switch gives gentle il-

Components list

Resistors:
R1 = 100Ω
R2 = 680Ω

Capacitors:
C1 = 4700µF 25 V

Semiconductors:
D1, D2 = 1N4001

Miscellaneous:
K1,K2,K3 = 2-way PCB terminal block, lead pitch 7.5 mm
F1 = fuse, 4AT (time lag) with PCB mount holder
TR1 = mains transformer, 12V @ 1.5 VA, short-circuit proof, PCB mount
B1 = B80C1400, round case (80V piv, 1.4A)
RE1 = power relay, 12V, 2 x c/o, PCB mount
RE2 = miniature relay, 12V, 2 x c/o, PCB mount

lumination (LA1). For more light, simply turn the switch off and then immediately (within 1 s) on again. The circuit then returns to the gentle light setting when switched off for more than 3 s. There is no need to replace the light switch with a dual version: simply insert this circuit between switch and lamp. How does it work? Almost immediately after switch-on, fast-acting miniature relay RE2 pulls in, since it is connected directly after the bridge rectifier. Its normally-closed contact then isolates RE1 from the supply, and thus current flows to LA1 via RE1's normally-closed contact. RE1 does not have time to pull in as it is a power relay and thus relatively slow. Its response is also slowed down by

the time constant of R1 and C1. If the current through the light switch is briefly interrupted, RE2 drops out immediately. There is enough energy stored in C1 to activate RE1, which then holds itself pulled in via a second, normally-open, contact. If current starts to flow again through the light switch within 1 s, LA2 will light. To switch LA1 back on it is necessary to turn

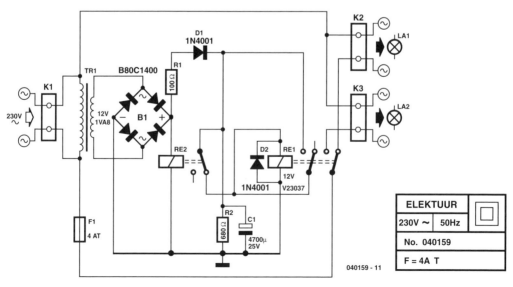

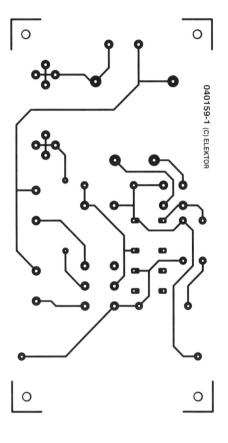

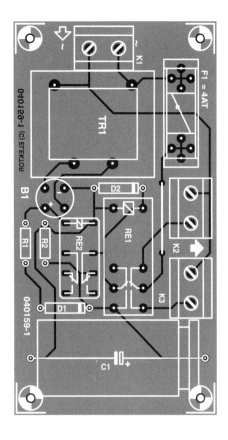

the light switch off for more than 3 s, so that C1 can discharge via R2 and RE1. The printed circuit board can be built into a well insulating plastic enclosure or be incorporated into a light fitting if there is sufficient space.

Helmut Kraus   (040159-1)

Caution:

The printed circuit board is connected directly to the mains-powered lighting circuit. Every precaution must be taken to prevent touching any component or tracks, which carry dangerous voltages. The circuit must be built into a well insulated ABS plastic enclosure.

150
Efficient Current Source for High-power LEDs

To get the maximum brightness and working life out of a high-power LED, it needs to be driven at the optimum specified cur-

rent. Allowing the current to exceed the permitted value is to be avoided at all costs, since it will severely affect the life of the device. A power supply or a battery with a small current-limiting resistor is not really an ideal solution, since not only is energy wasted in heating the resistor, but, if a small value is chosen to minimise this wastage, small changes in the applied voltage will lead to large changes in the current that flows. It is well known that LEDs have a small dynamic resistance in the neighbourhood of their optimal operating point. We will therefore need more in the way of electronics than a simple series resistor to meet our requirements.

The most direct way to provide a highly constant current in the face of relatively small changes in supply voltage is to use a conventional regulated current source. Unfortunately this type of circuit unnecessarily wastes energy in its series transistor, which rather detracts from the charm of using a semiconductor-based circuit. The inefficiency can be mitigated by using a modern device such as a power MOSFET as the series component. Power loss is then limited to that in any current sense resistor that might be used and the dissipation in the relatively small 'on' resistance of the switching transistor. The circuit suggested here drives a commercially-available Luxeon LED using a BUZ71. The 5 W version of this LED draws 0.7 A. This means that 0.175 V is dropped across R9, making

for a power dissipation of 122 mW. T1 has a typical resistance in the 'on' state of 85 mΩ. In the ideal scenario this means that about 60 mV is dropped across it, for a dissipation of at least 42 mW. The supply voltage therefore needs to be about 230 mV higher than the nominal voltage of the LED (6.85 V). To have something in reserve, 7.2 V, allowing 0.35 V for T1 and R9, is a good compromise. Serendipitously, a series of six NiCd or NiMH cells will give almost exactly this value under load! A further happy coincidence is that an unregulated mains power supply with a 6V transformer with bridge rectifier and smoothing capacitor will also give us almost exactly our target voltage when loaded. A 7.5 VA transformer is suitable, along with a 2200 µF/16 V electrolytic smoothing capacitor. Now to how it works. D1 acts as a reference, with a voltage of 2.5 V being dropped across it. IC1b, together with T1, form a current source, whose current can be set between 360 mA and 750 mA using P2. The otherwise unused opamp IC1a is connected to form an under-voltage cutout switch which prevents a connected battery from being discharged too deeply. The threshold point is set using P1. IC1a is configured as a comparator with a small hysteresis. If its output is high, IC1b is fooled into thinking that the

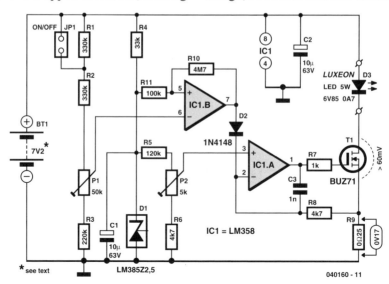

040160 - 11

current through R9 is too high, whereupon it switches off the LED. The same happens if R1 is not shorted by a switch. For the purposes of this circuit only opamps with input stages constructed using PNP transistors should be used. One last look at the energy budget: if six cells are used, the average voltage during discharge will be around 7.4 V. Subtract the nominal voltage of the LED and 0.55 V is left to be converted into heat. About 0.4 W will be dissipated by T1, which therefore will not require cooling. Efficiency is very good, at over 90 %.

Juergen Heidbreder   (040160-1)

151
Gentle Battery Regulator

This small but very effective circuit protects a lead-acid battery (12 V solar battery or car battery) against overcharging by a solar module when the incident light is too bright or lasts too long. It does so by energising a fan, starting at a low speed when the voltage is approximately 13.8 V and rising to full speed when the voltage exceeds 14.4 V (full-charge voltage). The threshold voltage (13.8 V) is the sum of the Zener diode voltage (12 V), the voltage across the IR diode (1.1 V), and the base-emitter voltage of the 2N3055 (0.7 V). In contrast to circuits using relays or IC amplifiers, the circuit has a gradual switching characteristic, which avoids relay chatter and the constant switching on and off near the switching point produced by a 'hard' switching point. The circuit does not draw any current at all (auto power-off) below 13 V. Pay attention to the polarisation of the Zener and IR diodes when building the circuit. The transistor must be fitted to a heat sink, since it becomes hot when the fan is not fully energised (at voltages just below 14 V). A galvanised bracket from a DIY shop forms an adequate heat sink. The indicated component values are for a 10 W solar module. If a higher-power module is used, a motor with higher rated power must also be used. The circuit takes advantage of the positive temperature coefficient of the lamp filament. The filament resistance is low at low voltages and increases as the voltage rises. This reduces the speed of the fan to avoid generating an annoying noise level. The lamp also provides a form of finger protection. If you stick your finger into the fan blade, the lamp immediately takes over the majority of the power dissipation and lights brightly. This considerably reduces

040192 - 11

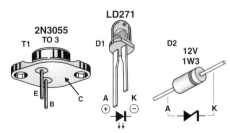

the torque of the fan. An ordinary 10 W or 20 W car headlight (or two 25 W headlights in parallel) can be used for the lamp. Don't try to replace the LED by two 1N4001 diodes or the like, replace the ZPY12 by a ZPY13, or fit a series resistor for the LED. That would make the 'on' region too large.

Wolfgang Zeiller   (040192-1)

152
Telephone Line Watchdog

Circuits have been published on earlier occasions that keep an eye on the telephone line. This simple circuit does it with very few components and is completely passive. The operating principle is simplicity itself. The circuit is connected in series with one of the two signal lines. It does not matter which one of these two is used. When the telephone receiver is lifted off the hook, or the modem makes a connection, a voltage will appear across the four diodes. This voltage is used to drive the duoLED. Depending on the direction of the current, either the red or the green part of the duoLED will light up. In some countries, the polarity of the telephone line voltage is reversed after a few seconds.

This does not matter with this circuit since a duoLED has been used. Depending on the polarity of the line, the current will flow through either one branch or the other. The 22 Ω resistor is used as a current limiter, so that both colours are about the same brightness. The duoLED can be ordered from, among others, Conrad Electronics (part number 183652). You can, of course, also use another, similar LED. For the diodes use the ubiquitous 1N4148.

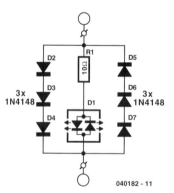

040182 - 11

Dick Sleeman   (040182-1)

153
Resistor Colour Band Decoder

Despite claims to the contrary by the non-initiated, electronics is still very much an exact science, so unless your memory is rock-solid you can not afford to make a mistake in reading a resistor value from the colour bands found on the device. So why not use the computer for the job? The program supplied by the author comes as an

Excel spreadsheet that does all the colour-to-value converting for you in response to a few mouse clicks. The program is extremely simple to use. Just click on the various colours to put them on the virtual resistor. Check the colour band structure against the real resistor on a board, on the floor or in the 'spares allsorts' drawer. The window below the colour bands will indicate the resistor's E series, nominal, high/low values and tolerance. The program supports the E6 and E12 through E192 series. The program may be obtained free of charge from www.elektor-electronics.co.uk as archive file 040203-11.zip (July/August 2005).

Carlos Alberto Gonzales　　　　(040203-1)

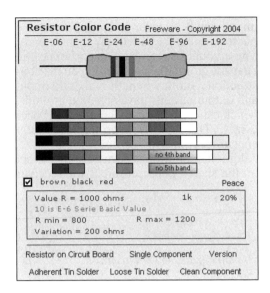

154
Optical Mixer

Mixing signals at different frequencies is common practice in many areas of electronics. Audio systems, communications systems and radio systems are typical application areas. With conventional frequency mixers, feedback capacitance can cause the signal sources to be affected by the output signal, thus making supplementary filter circuits necessary. The signals from the individual signal sources can also affect each other. In an optical mixer, LEDs or laser diodes are used to first convert the signals to be mixed into optical signals. The light beams are then aimed at a shared photosensor (a light-sensitive resistor, photo-diode, phototransistor, or photovoltaic cell). The current in the output circuit is thus controlled by the

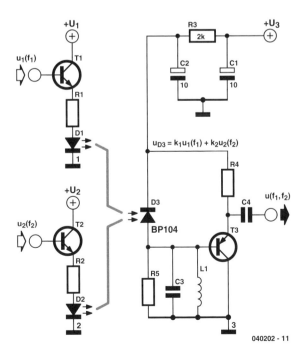

$u_{D3} = k_1 u_1(f_1) + k_2 u_2(f_2)$

040202 - 11

mixed input signals, so signal from the photosensor is the sum of the input signals. The amount of feed-back capacitance can be made quite small, depending on the construction. Another benefit is that the input and output circuits have separate grounds, which can be electrically connected if necessary. This operating principle directly encourages experimentation.

Additional input stages can be added to act on the shared photosensor. If the receiver

signal is applied to a component with a curved characteristic, such as a diode, this produces amplitude modulation, which can be used in a heterodyne receiver. If the difference between the frequencies of the two input signals is small, a beat effect occurs. The components must be selected according to the frequency range that is used.

Peter Lay   (040202-1)

Formulas

Input signal voltages:

$$u_1(f_1) = \hat{u}_1\sin(\omega_1 t + \varphi_1)$$
$$u_2(f_2) = \hat{u}_2\sin(\omega_2 t + \varphi_2)$$

Voltage across the photosensor:

$$u_{D3} = k_1\hat{u}_1\sin(\omega_1 t + \varphi_1) + k_2\hat{u}_2\sin(\omega_2 t + \varphi_2)$$

Key:
$u_1(f_1)$	first input signal at frequency f1
$u_2(f_2)$	second input signal at frequency f2
u_{D3}	receiver signal generated by superposition of the two input signals
k_1, k_2	optoelectronic coupling factors (empirically determined)

155
Solar-powered SLA Battery Maintenance

This circuit was designed to 'baby-sit' SLA (sealed lead-acid or 'gel') batteries using freely available solar power. SLA batteries suffer from relatively high internal energy loss which is not normally a problem until you go on holidays and disconnect them from their trickle current charger. In some cases, the absence of trickle charging current may cause SLA batteries to go completely flat within a few weeks. The circuit shown here is intended to prevent this from happening.

Two 3 volt solar panels, each shunted by a diode to bypass them when no electricity is generated, power a MAX762 step-up voltage converter IC. The '762 is the 15 volt-out version of the perhaps more familiar MAX761 (12 V out) and is used here

to boost 6 V to 15 V. C1 and C2 are decoupling capacitors that suppress high and low frequency spurious components produced by the switch-mode regulator IC. Using Schottky diode D3, energy is stored in inductor L1 in the form of a magnetic field. When pin 7 of IC1 is open-circuited by the internal switching signal, the stored energy is diverted to the 15 volt output of the circuit. The V+ (sense) input of the MAX762, pin 8, is used to maintain the output voltage at 15 V. C4 and C5 serve to keep the ripple on the output voltage as small as possible. R1, LED D4 and pushbutton S1 allow you to check the presence of the 15 V output voltage. D5 and D6 reduce the 15 volts to about 13.6 V which is a frequently quoted nominal standby

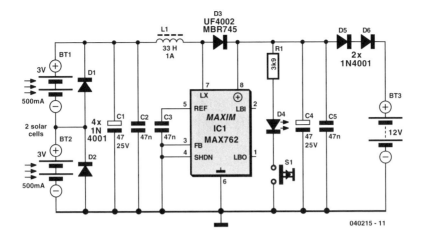

040215 - 11

trickle charging voltage for SLA batteries. This corresponds well with the IC's maximum, internally limited, output current of about 120 mA. The value of inductor L1 is not critical – 22 µH or 47 µH will also work fine. The coil has to be rated at 1 A though in view of the peak current through it. The switching frequency is about 300 kHz. A suggestion for a practical coil is type M from the WEPD series supplied by Würth

(www.we-online.com). Remarkably, Würth supply one-off inductors to individual customers. At the time of writing, it was possible, under certain conditions, to obtain samples, or order small quantities, of the MAX762 IC through the Maxim website at www.maxim-ic.com.

Myo Min (040215-1)

156
Proximity Switch

This circuit is for an unusually sensitive and stable proximity alarm which may be built at very low cost. If the negative terminal is grounded, it will detect the presence of a hand at more than 200mm. If it is not grounded, this range is reduced to about one-third. The Proximity Switch emits a loud, falling siren when a body is detected within its range. A wide range of metal objects may be used for the sensor, including a metal plate, a doorknob, tin foil, a set of burglar bars – even a complete bicycle. Not only this, but any metal object which comes within range of the sensor, itself becomes a sensor. For example, if a tin foil

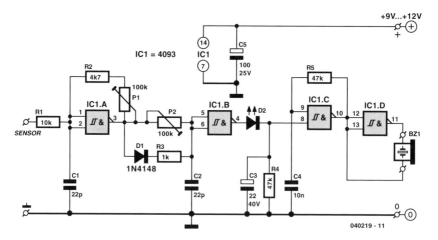

040219 - 11

sensor is mounted underneath a table, metal items on top of the table, such as cutlery, or a dinner service, become sensors themselves. The touch plate connected to the free end of R1 detects the electric field surrounding the human body, and this is of a relatively constant value and can therefore be reliably picked up. R1 is not strictly necessary, but serves as some measure of protection against static charge on the body if the sensor should be touched directly. As

a body approaches the sensor, the value of C1 effectively increases, causing the frequency of oscillator IC1.A to drop. Consequently capacitor C2 has more time to discharge through P2, with the result that the inputs at IC1.B go Low, and the output goes High. As the output goes High, so C3 is charged through LED D2. D2 serves a dual purpose – namely as a visual indication of detection, and to lower the maximum charge on C3, thus facilitating a sharper distinction between High and Low states of capacitor C3. The value of R4 is

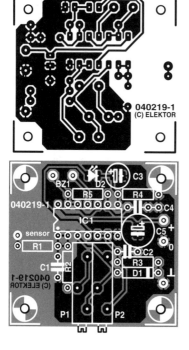

040219-1
(C) ELEKTOR

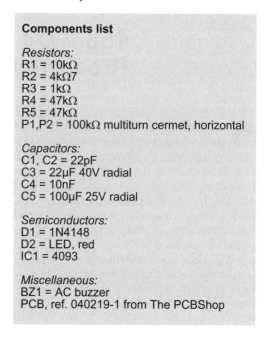

Components list

Resistors:
R1 = 10kΩ
R2 = 4kΩ7
R3 = 1kΩ
R4 = 47kΩ
R5 = 47kΩ
P1,P2 = 100kΩ multiturn cermet, horizontal

Capacitors:
C1, C2 = 22pF
C3 = 22µF 40V radial
C4 = 10nF
C5 = 100µF 25V radial

Semiconductors:
D1 = 1N4148
D2 = LED, red
IC1 = 4093

Miscellaneous:
BZ1 = AC buzzer
PCB, ref. 040219-1 from The PCBShop

chosen to enable C3 to discharge relatively quickly as pulses through D2 are no longer sufficient to maintain its charge. The value of C3 may be increased for a longer sounding of the siren, with a slight reduction in responsiveness at the sensor. When C3 goes High, this triggers siren IC1.C and IC1.D. The two NAND gates drive piezo sounder X1 in push-pull fashion, thereby greatly increasing its volume. If a piezo tweeter is used here, the volume will be sufficient to make one's ears sing. The current consumption of the circuit is so low a small 9 V alkaline PP3 battery would last for about one month. As battery voltage falls, so sensitivity drops off slightly, with the result that P1 may require occasional readjustment to maintain maximum sensitivity. On the down side of low cost, the hysteresis properties of the 4093 used in the circuit are critical to operation, adjustment and stability of the detector. In some cases, particularly with extremely high sensitivity settings, it will be found that the

circuit is best powered from a regulated voltage source. The PCB has an extra ground terminal to enable it to be easily connected to a large earthing system. Current consumption was measured at 3.5 mA stand-by or 7 mA with the buzzer activated. Usually, only P1 will require adjustment. P2 is used in place of a standard resistor in order to match temperature coefficients, and thus to enhance stability. P2 should be adjusted to around 50 k, and left that that setting. The circuit is ideally adjusted so that D2 ceases to light when no body is near the sensor. Multiturn presets must be used for P1 and P2. Since the piezo sounder is the part of the circuit which is least affected by body presence, a switch may be inserted in one of its leads to switch the alarm on and off after D2 has been used to check adjustment. Make sure that there is a secure connection between the circuit and any metal sensor which is used.

Rev. Thomas Scarborough   (040219-1)

157
1:800 Oscillator

Oscillators are ten a penny, but this one has something special. Its frequency can be adjusted over a range of 800:1, it is voltage controlled, and it switches off automatically if the control voltage is less than approximately 0.6 V. As can be seen from the chart, the characteristic curve $f = f(U_e)$ is approximately logarithmic. If the input voltage is less than 0.7 V, T1 and T3 are cut off. The capacitor then charges via the 10 kΩ

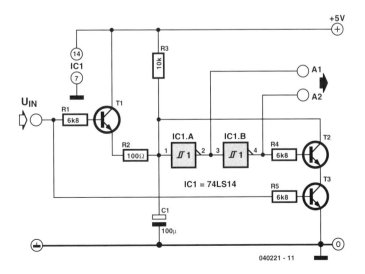

040221 - 11

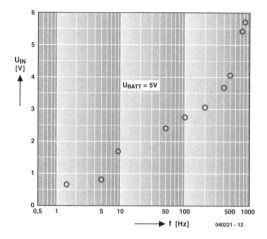

resistor. The combination of the capacitor, the two Schmitt triggers and T2 form the actual oscillator circuit. However, T2 can-

not discharge the capacitor, because T3 is cut off. In this state, a low level is present at A1 and a high level is present at A2. If the input voltage is increased, T3 starts conducting. This allows the capacitor to be discharged via T2, and the circuit starts to oscillate. If U_e is further increased, the capacitor receives an additional charging current via T1 and the 100 Ω resistor. That causes the oscillator frequency to increase. In situations where the duty cycle of the output signal is not important (such as when the circuit is used as a clock generator), this circuit can be used as a voltage-controlled oscillator (VCO) with a large frequency range and shutdown capability.

Bernd Oehlerking   (040221-1)

158
Two-Cell LED Torch

It sometimes comes as a bit of a shock the first time you need to replace the batteries in an LED torch and find that they are not the usual supermarket grade alkaline batteries but in fact expensive Lithium cells. The torch may have been a give-away at an advertising promo but now you discover that the cost of a replacement battery is more than the torch is worth. Before you consign the torch to the waste bin take a look at this circuit. It uses a classic two-transistor astable multivibrator configuration to drive the LEDs via a transformer from two standard 1.5 V alkaline batteries. The operating principle of the multivibrator has been well documented and with

the components specified here it produces a square wave output with a frequency of around 800 Hz. This signal is used to drive a small transformer with its output across two LEDs connected in series. Conrad Electronics supplied the transformer used

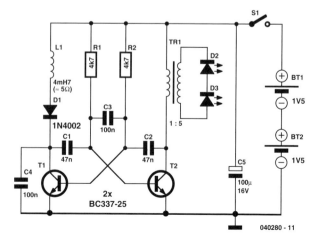

in the original circuit. The windings have a 1:5 ratio. The complete specification is available on the (German) company website at www.conrad.de part no. 516236. It isn't essential to use the same transformer so any similar model with the same specification will be acceptable. The LEDs are driven by an alternating voltage and they will only conduct in the half of the waveform when they are forward biased. Try reversing both LEDs to see if they light more brightly. Make sure that the transformer is fitted correctly; use an ohm-meter to check the resistance of the primary and secondary windings if you are unsure which is which.

The load impedance for the left hand transistor is formed by L in series with the 1N4002 diode. The inductance of L isn't critical and can be reduced to 3.3 mH if necessary. The impedance of the transformer secondary winding ensures that a resistor is not required in series with the LEDs. Unlike filament type light sources, white LEDs are manufactured with a built-in reflector that directs the light forward so an additional external reflector or lens glass is not required.

The LEDs can be mounted so that both beams point at the same spot or they can be angled to give a wider area of illumination depending on your needs. Current consumption of the circuit is approximately 50 mA and the design is even capable of producing a useful light output when the battery voltage has fallen to 1 V. The circuit can be powered either by two AAA or AA size alkaline cells connected in series or alternatively with two rechargeable NiMH cells.

Wolfgang Zeiller   (040280-1)

159
Simple Oscillator / Pipe Locator

Sometimes the need arises to construct a really simple oscillator. This could hardly be simpler than the circuit shown here, which uses just three components, and offers five separate octaves, beginning around Middle C (Stage 14). Octave # 5 is missing, due to the famous (or infamous) missing Stage 11 of the 4060B IC. We might call this a Colpitts 'L' oscillator, without the 'C'.

Due to the reactance of the 100 µH inductor and the propagation delay of the internal oscillator, oscillation is set up around 5 MHz.

When this is divided down, Stage 14 approaches the frequency of Middle C (Middle C = 261.626 Hz). Stages 13, 12, 10, and 9 provide higher octaves, with Stages 8 to 4 being in the region of ultrasound. If the oscillator's output is taken to the aerial of a

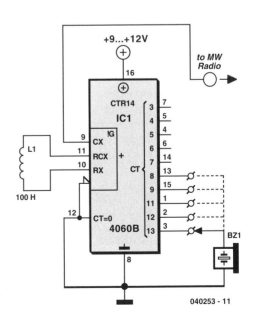

Medium Wave Radio, L1 may serve as the search coil of a Pipe Locator, with a range of about 50 mm. This is tuned by finding a suitable heterodyne (beat note) on the medium wave band. In that case, piezo sounder Bz1 is omitted. The Simple Oscillator / Pipe Locator draws around 7 mA from a 9-12 V DC source.

Rev. Thomas Scarborough   (040253-1)

160
Minimalist Microcontroller

They say that things were always better in the old days, although perhaps they were not thinking of microcontrollers and their complex support circuitry. In the ATmega8, microcontroller specialists Atmel have introduced a device that lets you construct a prototype circuit using just two resistors and one potentiometer in addition to the microcontroller chip. Not even a crystal is required: an internal 8 MHz oscillator provides the clock. We thus have a four-component circuit that is a powerful and practical development kit; not only that, it can be programmed directly from the parallel port of any PC without additional hardware. Incredible!

The circuit shown offers a number of I/O pins and an A/D converter input; not only that, it is ready to be connected to a commer-cially-available liquid crystal display. The whole thing can be built on a simple prototyping board, and no heroic soldering skills are required. Software, in the form of a C compiler (AVR-GCC under Linux or WinAVR under Windows) is available for free on the Inter-net. Example applications,

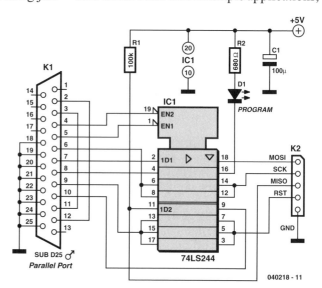

040218 - 11

Links:

Introduction to development tools:
for Linux etc.: www.nongnu.org/avr-libc/ user-manual/install_tools.html
for Windows: www.avrfreaks.net/Tools/ showtools.php?ToolID=376

Procyon AVRlib JTAG-Hardware:
(examples and libraries): http://avr.openchip.org/bootice http://hubbard.engr.scu.edu/ WinAVR:
embedded/avr/avrlib/index.html http://winavr.sourceforge.net

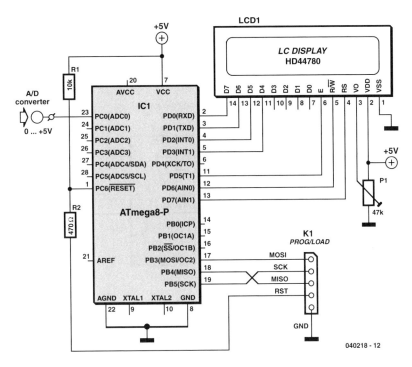

040218 - 12

expansion ideas, programming tools and code collections are also widely available. And, since the circuit is so simple, it can easily be modified to use other types of microcontroller from Atmel: just take a look at the relevant data sheets and determine which pins are used for the various functions.

Christoph Fritz   (040218-1)

161
Cheap Dot-mode Bargraph Display

The five-stage linear dot-mode bargraph display shown in here has a number of distinct advantages, which may be summarized as:

- the resistor chain is fully customizable;
- IC1's high input impedance results in minimal loading on circuits;
- IC1 and IC2 have a wide supply voltage range, between 3 V and 15 V;
- if a higher supply voltage is used, all colour LEDs may be employed with a single ballast resistor R3.

Four opamp comparators are used to provide a five-stage output, as follows: as the voltage at the signal input rises, so the outputs of comparators IC1.A to IC1.D go High in succession. This creates the sequence at IC2's binary channel select and output channels shown in the Table. The last binary number in this sequence is created as IC1.D goes High, thus pulling IC2 binary input 1 Low through T1. In this way, five of the (IC2) 8-to-1 analogue multiplexer/demultiplexer output channels

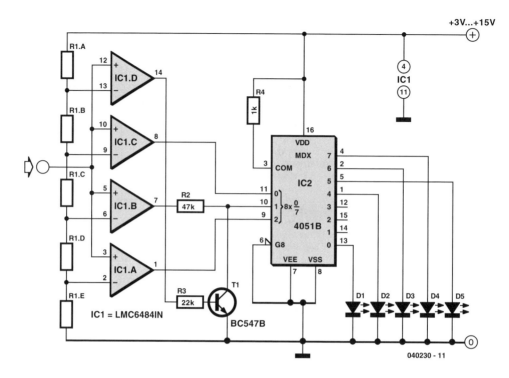

040230 - 11

are utilised to produce a five-stage linear dot-mode bargraph display. The signal input voltage required to switch any one of IC1's four outputs may be calculated by dividing the supply voltage by the values in the resistor chain. For example, if the supply voltage is 12 V, and all of the resistors in the chain are 10 kΩ, then IC1.C's output pin 8 (and IC2's output channel 6) will swing High as the voltage of the signal input exceeds:

$$12 \ V \cdot \frac{R3+R4+R5}{R1+R2+R3+R4+R5} = 7.2 \ V$$

The only aspect of the circuit which might require special clarification is T1. T1 pulls IC2's binary channel select input 1 Low by shorting it to 0 V, thus overriding IC1.B's output pin 7. This creates one additional binary input number (binary 101) for IC2, and adds a fifth stage to the bargraph. A 'logic' MOSFET such as a BUZ11 or IRF510 would also work in this position, and in this case R1 may be omitted. The LMC6484IN opamp indicated in the cir-

cuit diagram is a rail-to-rail type which may be difficult to source. If you use an 'old fashioned' opamp like the LM324, do remember that the input voltage may not exceed the supply voltage minus about 1.5 volts. In the case of the TL074 (TL084), the reverse applies: the input voltage must be comprised between 1.5 volts and the supply rail level. This is called the common-mode range of the opamp – check the datasheets! Resistor R3 determines the LED current and may need

IC2 Channel Select			IC2 Output Channel
2	1	0	
0	0	0	0
1	0	0	4
1	1	0	6
1	1	1	7
1	0	1	5

to be adapted to match whatever LEDs you have available for the readout. Since the output current of 4000 series CMOS ICs needs to be observed, it is a good idea to use high-efficiency LEDs of the 2 mA class. The value of R3 is calculated from:

$$R3 = \frac{V_b - 2}{3}$$

where the result is in kilo-ohms.

Rev. Thomas Scarborough   (040230-1)

162
CCO Metal Detector

To the best of the author's knowledge, the metal detector shown here represents another new genre. It is presented here merely as an experimental idea, and operates in conjunction with a Medium Wave radio. If a suitable heterodyne is tuned in on the medium waves, its performance is excellent. An old Victorian penny, at 180 mm, should induce a shift in frequency of one tone through the radio speaker. This suggests that the concept will match the performance of induction balance (IB) detector types, while employing a fraction of the components. In principle, the circuit is loosely based on a transformer coupled oscillator (TCO), a well known oscillator type. This essentially consists of an amplifier which, by means of a transformer, feeds the output back to the input, thus sustaining oscillation. On this basis, the author has named the detector a Coil Coupled Operation (CCO) Metal Detector. In fact the circuit would oscillate even without L2 and C1. However, in this case one would have nothing more than a beat frequency operation (BFO) detector. Coil L2 is added to bring the induction balance principle into operation, thus modifying the signal which is returned to the output, and greatly boosting performance. This does not mean, however, that we are dealing strictly with an IB detector type, since the design requires a beat frequency oscillator for detection. Also, unlike IB, its Rx section (L2) is active rather than passive, being an integral

part of a TCO. Nor is this strictly a BFO type, since its performance far outstrips that of BFO, and of course it uses two coils. Search oscillator IC1 oscillates at around 480 kHz, depending on the positioning of the coils on the search head. The presence of metal induces changes both in the inductance and coupling of the two coils, thereby inducing a shift in oscillator IC1's frequency. The output (pin 6) is taken via a screened cable to a Medium Wave radio aerial. A crocodile clip termination would make a convenient connection. The two coils are each made of 50 turns 30swg (0.315mm) enamelled copper wire, wound on a 120mm diameter former. Each has a Faraday shield, which is connected to 0V as shown. A sketch of the coil is shown in the separate drawing. The coils are positioned on the search head to partly overlap one another, in such a way as to find a low tone on the best heterodyne, which should

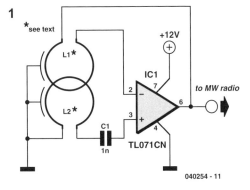

040254 - 11

2

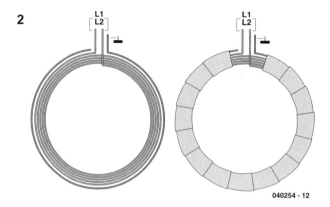

040254 - 12

match the performance mentioned above. Oscillator IC1 will sustain oscillation no matter which way the coils are orientated – however, orientation significantly affects performance. The correct orientation may be determined experimentally by flipping one of the coils on the search head. Ideally, the coils will finally be potted in polyester resin. The CCO Metal Detector's search head offers a wide area of sensitivity, so that it is better suited to sweeping an area than pinpointing a find. As with both BFO and IB, it offers discrimination between ferrous and non-ferrous metals, making it well suited to 'treasure hunting'. And if you get fed up with searching, there's always the radio to listen to.

Rev. Thomas Scarborough   (040254-1)

163
Reflection Light Barrier with Delay

This circuit can be used to check, for example, whether the door of a refrigerator has been properly closed. An LED sends out a beam of light, which, if the door is closed, is reflected. An optical sensor (CNY70) then detects the amount of light. If the sensor does not receive the right amount of light, the buzzer will sound after about a minute. When the door is closed (and the CNY70 receives enough light again), the buzzer turns off. The power supply for the circuit requires about 12 mA at 12 V. Potentiometer P1 adjusts

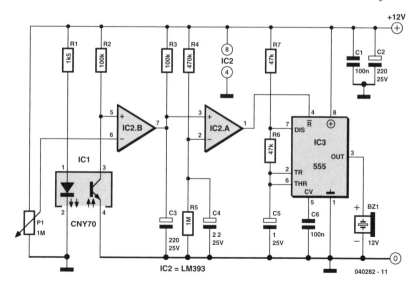

040282 - 11

the sensitivity of the sensor. The sensor works reliably from a distance of one centimetre. If the current through the LED is increased, the distance can be increased a little. The delay can be adjusted with C3. C4 provides extra filtering for the reference voltage. The buzzer would otherwise switch on with a 'chirping' sound. The well-known NE555 is used to drive the buzzer. The buzzer is driven with a duty cycle of 2:1, which improves the audibility.

Goswin Visschers   (040282-1)

164
Simple Cable Tester

This cable tester allows you to quickly check audio cables for broken wires. Because of the low power supply voltage, batteries can be used which makes the circuit portable, and therefore can be used on location. The design is very simple and well organised: using the rotary switch, you select which conductor in the cable to test. The corresponding LED will light up as indication of the selected conductor.

This is also an indication that the power supply voltage is present. If there is a break in the cable, or a loose connection, a second LED will light up, corresponding to the selected conductor. You can also see immediately if there is an internal short circuit when other than the corresponding LEDs light up as well.

You can also test adapter and splitter cables because of the presence of the different connectors. Two standard AA- or AAA-batteries are sufficient for the power supply. It is recommended to use good, low-current type LEDs. It is also a good idea not to use the cheapest brand of connectors, otherwise there can be doubt as to the location of the fault. Is it the cable or the connector?

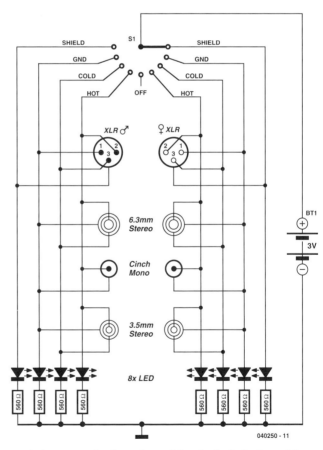

Bert Vink   (040250-1)

165
Simple Overcurrent Indicator

This circuit eventually surfaced while pondering over the design of a current indicator for a small power supply. Fortunately, it proved possible to employ the supply voltage as a reference by dividing it down with the aid of R1 and R2. C1 is an essential capacitor to suppress noise and surges. The half supply voltage level is applied to the non-inverting pin of opamp IC1. The value of the R3 determines the trip level of the indicator, according to

$$R3 = 0.4 \cdot \frac{\text{desired voltage drop}}{I_{trip}}$$

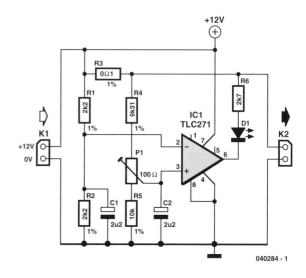

040284 - 1

Actually this is high side sensing but the method can be used as low side sensing, too! The desired voltage or sense voltage can be any value between 0.35 V and 0.47 V. If currents greater than about 1 A are envisaged, you should not forget to calculate R3's dissipation on penalty of smoke & smells. Another voltage divider network, R4, R5 and P1 divide the voltage between supply voltage and desired voltage. This divided voltage, filtered by C2, is fed to the inverting input of IC1 to compare levels. The result causes D1 to light or remain off. Turn P1 to the end of R4 to hold off D1. Then connect a load causing over current and adjust P1 towards the end of R5 until D1 lights. The accuracy of the circuit depends entirely on the tolerances of the resistors used – high stability types are recommended.

Myo Min   (040284-1)

166
DIL/SOIC/TSSOP
Experimenting Boards

Whenever you want to put a circuit together quickly, it will take too long to design a PCB for it first. In that case, the experimenting PCBs shown here are eminently suitable. Three variations are available: 20-pin DIL package, 20-pin SOIC package and a 20-pin TSSOP package. If you use an IC with fewer pins, then

you simply don't use some of the solder pads. The remaining part of each PCB is filled with solder pads, which are arranged in groups of four. In between is a continuous copper pattern. By connecting this copper grid to ground, the boards are also suitable for RF designs.

The layouts for the various PCBs are shown here at a reduced scale. You can download the true scale layouts from the Elektor Electronics website (in pdf format). The file number is 040289-1.zip. They are also available ready-made from Elektor Electronics (refer to SHOP pages or website), the part numbers are 040289-1, -2 and -3.

Wolfgang Steimle   (040289-1)

167
Transformerless 5 volt Power Supply

An increasing number of appliances draw a very small current from the power supply. If you need to design a mains-powered device, you could generally choose between a linear and a switch-mode power supply. However, what if the appliance's total power consumption is very small? Transformer-based power supplies are bulky, while the switchers are generally made to provide greater current output, with a significant increase in complexity, problems involving PCB layout and, inherently, reduced reliability. Is it possible to create a simple, minimum-part-count mains (230 VAC primary) power supply, without transformers or coils, capable of delivering about 100 mA at, say, 5 V? A general approach could be to employ a highly inefficient stabilizer that would rectify AC and, utilizing a zener diode to provide a 5.1 V output, dissipate all the excess from 5.1 V to $(230 \times \sqrt{2})$ volts in a resistor. Even if the load would require only about 10 mA, the loss would be approximately 3 watts, so a significant heat dissipation would occur even for such a small power consumption.

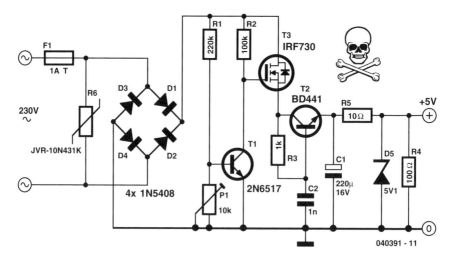

040391 - 11

At 100 mA, the useless dissipation would go over 30 W, making this scheme completely unacceptable. Power conversion efficiency is not a major consideration here; instead, the basic problem is how to reduce heavy dissipation and protect the components from burning out. The circuit shown here is one of the simplest ways to achieve the above goals in practice. A JVR varistor is used for over-voltage/surge protection. Voltage divider R1-R2 follows the rectified 230 V and, when it is high enough, T1 turns on and T3 cannot conduct. When the rectified voltage drops, T1 turns off and T3 starts to conduct current into the reservoir capacitor C1. The interception point (the moment when T1 turns off) is set by P1 (usually set to about 3k3), which controls the total output current capacity of the power supply: reducing P1 makes T1 react later, stopping T3 later, so more current is supplied, but with increased heat dissipation. Components T2, R3 and C2 form a typical 'soft start' circuit to reduce current spikes – this is necessary in order to limit

C1's charging current when the power supply is initially turned on. At a given setting of P1, the output current through R5 is constant. Thus, load R4 takes as much current as it requires, while the rest goes through a zener diode, D5. Knowing the maximum current drawn by the load allows adjusting P1 to such a value as to provide a total current through R5 just 5 to 6 mA over the maximum required by the load. In this way, unnecessary dissipation is much reduced, with zener stabilization function preserved. Zener diode D5 also protects C1 from overvoltages, thus enabling te use of low-cost 16 V electrolytics. The current flow through R5 and D5, even when the load is disconnected, prevents T3's gate-source voltage from rising too much and causing damage to device. In addition, T1 need not be a high-voltage transistor, but its current gain should exceed 120 (e.g. BC546B, or even BC547C can be used).

Srdjan Jankovic and
Branko Milovanovic   (040391-1)

CAUTION

168
Audio Click/Pop Suppressor

Audio amplifier circuits with a single supply voltage have output coupling capacitors that produce audible clicking or popping sounds when the supply voltage is switched on, since they must be initially charged to half the supply voltage. Similarly, a clicking or popping noise can be produced by the discharge current when the supply is switched off. The capacitance (Cout) of the output capacitors cannot be reduced, since it determines the lower limit of the frequency range. The process of establishing the DC operating point in upstream amplifier stages also generates switch-on and switch-off noises. For headphone outputs in particular, this can be remedied using an 8-pin IC from Maxim (www.maxim-ic.com), the MAX9890, which can be connected between the output stage and the output capacitors to suppress irritating clicks and pops. The secret of the MAX9890 is that it changes the shape of the charging current for the output capacitors from an abrupt (and thus audible) step to an optimised S-shaped curve that has such a low frequency that it does not produce any audible sound. After the capacitor has been charged, two integrated switches are enabled to connect the audio amplifier outputs to the already charged coupling capacitors. When the supply voltage is switched off, these switches open immediately and the coupling ca-

pacitors discharge slowly via internal 220 kΩ resistors. There is also an undervoltage detector that opens the switches if the supply voltage is less than +2.5 V. A shutdown input ($\overline{\text{SHDN}}$, pin 2) allows the headphone output to be selectively disabled. Inside the IC, the Startup and Shutdown Control section controls the switches and the Ramp Up and Ramp Down Control section controls capacitor charging and discharging. Capacitor CEXT generates a switching time delay after the supply voltage is applied. During the switch-off process, it powers the internal circuitry responsible for discharging the coupling capacitors. A 100 nF capacitor is adequate for this purpose.

The switch-on delay is 200 ms (MAX9890A) or 330 ms (MAX9890B). The A version is adequate for coupling capacitors up to 100 μF, with the B version being preferred for capacitors up to 220 μF. With coupling capacitor values larger than this, switch-on noises may still

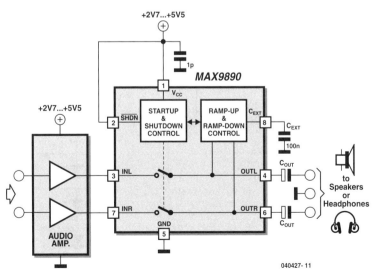

040427- 11

be audible under certain conditions. The MAX9890 operates over a supply voltage range of +2.7–5.5 V, draws only around 20 µA of current, and is specifically protected against electrostatic discharges up to ±8kV. The input voltage on INL and INR must lie between 0 V and the supply voltage level. Click and pop suppression is 36 dB. The additional distortion factor is specified by the manufacturer as 0.003 %

(THD+N) for a 32 Ω headphone load. The power supply rejection ratio is typically 100 dB. The IC is available in two different SMD packages; the pinout shown here is for the TDFN package.

http://pdfserv.maxim-ic.com/   en/ds/
MAX9890.pdf

Gregor Kleine   (040427-1)

169
Cable Tester

Microcontroller-based circuits for testing cables, sometimes in conjunction with a PC, are easy to use and very flexible. For the hobbyist, however, the complication of such devices is not justified. The circuit described here is an economical, but nevertheless easy-to-understand tester for cables with up to ten conductors. The basic idea for the cable tester is to apply a different voltage to each conductor in the cable at one end. The voltage seen at the other end of the cable is indicated by light-emitting diodes. The eight reference voltages are generated using a row of nine LEDs connected in series (D1 to D9). The first and and the tenth conductors are connected to the posi-

tive and negative terminals of the power supply respectively. The LEDs are pow-

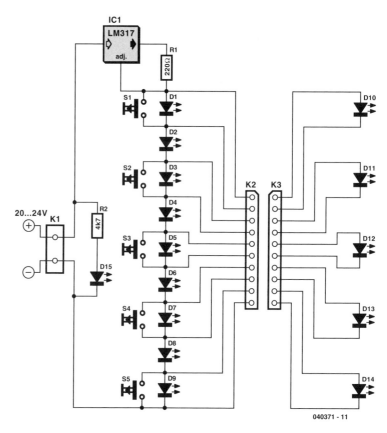

040371 - 11

ered from a constant current source, which allows us to dispense with the current-limiting series resistor that would otherwise be necessary. For the constant current source we use a type LM317 voltage regulator. R1 is selected using the formula

$$I_{const} = 1.25\ V\ /\ R1$$

to produce a current of 5 mA. This part of the circuit forms the transmitter end of the cable tester. The conductors of the cable under test can be connected to the transmitter in any order. The receiver consists of five LEDs whose connections are taken directly from terminal block X3. If the corresponding points in the two parts of the circuit are wired to one another using a working cable, all the LEDs on both receiver and transmitter sides will light. If there is a fault in the cable, the following situations are possible. Two LEDs opposite one another fail to light: two conductors are crossed or shorted. Only the LED on the transmitter side lights: one or both of the conductors in the pair is broken. One of the even-numbered LEDs on the transmitter side (D2, D4, D6 or D8) fails to light: there is a short between the outer conductors of the neighbouring pairs. Several neighbouring LEDs fail to light: the conductor corresponding to the first unlit LED is crossed with the one corre-sponding to the last unlit LED, or they are shorted. If all LEDs light on both sides, there is still a chance that two pairs might be interchanged. Buttons S1 to S5 can be used to test this: the same LED should extinguish on each side when the button is pressed. If the wrong LED goes out on the receiver side, a pair must be swapped over. More complicated effects can result from combinations of these five faults. Different colours of LED have different forward voltage drops, and so the same type of LED should be used throughout. The required current can be put into the formula to calculate R1, which can then be altered if necessary. Of course, these remarks do not apply to the power indicator LED (D15). The LM317 used for the constant current source can only deliver the calculated current if its input voltage is at least about 3 V higher than the voltage required at its output. The load voltage depends on the number of LEDs in the transmitter and on their forward voltage drop. For nine red LEDs at least 20 V is required.

Uwe Reiser (040371-1)

170
Negative-Output Switching Regulator

There are only a limited number of switching regulators designed to generate negative output voltages. In many cases, it's thus necessary to use a switching regulator that was actually designed for a positive voltage in a modified circuit configuration that makes it suitable for generating a negative output voltage. The circuit shown in Figure 1 uses the familiar LM2575 step-down regulator from National Semiconductor (www.national.com). This cir-cuit converts a positive-voltage step-down regulator into a negative-voltage step-up regulator. It converts an input voltage between –5 V and –12 V into a regulated –12 V output voltage. Note that the output capacitor must be larger than in the standard circuit for a positive output voltage. The switched current through the storage choke is also somewhat higher. Some examples of suitable storage chokes for this circuit are the PE-53113 from Pulse

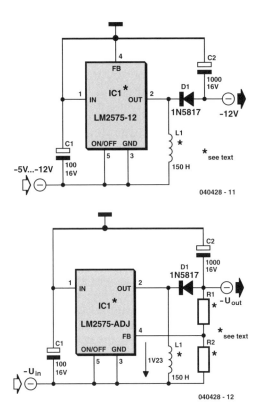

040428 - 11

040428 - 12

input voltage is more negative than –12 V (i.e., Vin < –12 V), the output voltage will not be regulated and will be lower than the desired –12 V. The LM2575 IC will not be damaged by such operating conditions as long as its maximum rated input voltage of 40 V is not exceeded. High-voltage (HV) types that can withstand up to 60 V are also available. Although the standard LM2575 application circuit includes circuit limiting, in this circuit the output current flows via the diode and choke if the output is shorted, so the circuit is not short-circuit proof. This can be remedied by using a Multifuse (PTC) or a normal fuse. There is also an adjustable version of the regulator with the type designation LM2575-ADJ (Figure 2). This version lacks the internal voltage divider of the fixed-voltage versions, so an external voltage divider must be connected to the feedback (FB) pin. The voltage divider must be dimensioned to produce a voltage of 1.23 V at the FB pin with the desired output voltage. The formula for calculating the output voltage is:

$$V_{out} = 1.23 \text{ V} \cdot \left(1 + \frac{R1}{R2}\right)$$

The electrolytic capacitors at the input and output must be rated for the voltages present at these locations.

Gregor Kleine   (040428-1)

(www.pulseeng.com) and the DO3308P-153 from Coilcraft (www.coilcraft.com). The LM2575-xx is available in versions for output voltages of 3.3V, 5 V, 12 V and 15 V, so various negative output voltages are also possible. However, you must pay attention to the input voltage of the regulator circuit. If the

171
Low-cost Step-down Converter

The circuit described here is mostly aimed at development engineers who are looking for an economical step-down converter which offers a wide input voltage range. As a rule this type of circuit employs a step-down converter with integrated

switching element. However, by using a more discrete solution it is possible to reduce the total cost of the step-down converter, especially when manufacturing in quantity. The TL5001A is a low-cost PWM controller which is ideal for this pro-

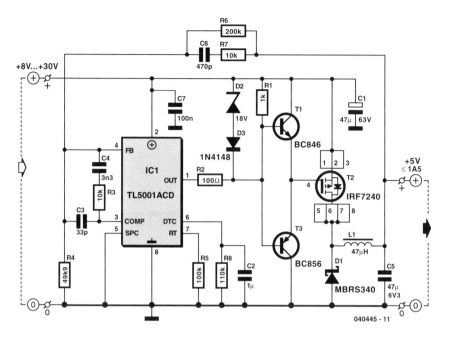

ject. The input voltage range for the step-down converter described here is from 8 V to 30 V, with an output voltage of 5 V and a maximum output current of 1.5 A. When the input voltage is applied the PWM output of IC1 is enabled, taking one end of the voltage divider formed by R1 and R2 to ground potential. The current through the voltage divider will then be at most 25 mA: this value is obtained by dividing the maximum input voltage (30 V) minus the saturation voltage of the output driver (2 V) by the total resistance of the voltage divider (1.1 kΩ). T1 and T3 together form an NPN/PNP driver stage to charge the gate capacitance of P-channel MOSFET T2 as quickly as possible, and then, at the turn-off point, discharge it again. The base-emitter junction of T3 goes into a conducting state when the PWM output is active and a voltage is dropped across R2. T3 will then also conduct from collector to emitter and the gate capacitance of T2 will be discharged down to about 800 mV. The P-channel MOSFET will then conduct from drain to source. If the open-collector output of the controller is deactivated, a negligibly small current flows through resistor R2 and the base of

T1 will be raised to the input voltage level. The base-emitter junction of T1 will then conduct and the gate capacitance of T2 will be charged up to the input voltage level through the collector and emitter of T1. The P-channel MOSFET will then no longer conduct from drain to source. This driver circuit constructed from discrete components is very fast, giving very quick switch-over times. Diodes D2 and D3 provide voltage limiting for the P-channel MOSFET, whose maximum gate-source voltage is 20 V. If the Zener voltage of diode D2 is exceeded it starts to conduct; when the forward voltage of diode D3 is also exceeded, the two diodes together clamp the gate-source voltage to approximately 19 V. The switching frequency is set at approximately 100 kHz, which gives a good compromise between efficiency and component size. Finally, a few notes on component selection. All resistors are 1/16 W, 1 %. Apart from electrolytic C1 all the capacitors are ceramic types. For the two larger values (C2 and C5) the following are used:
- C2 is a Murata type GRM21BR71C105KA01 ceramic capacitor, 1 μF, 16 V, X7R, 10 %;

227

- C5 is a Murata type GRM32ER60J476ME20 ceramic capacitor, 47 µF, 6.3 V, X5R, 10 %;

D1 (Fairchild type MBRS340T3) is a 40 V/3 A Schottky diode;
Coil L1 is a Würth WE-PD power choke type 744771147, 47 µH, 2.21 A, 75 mΩ;

T1 (BC846) and T3 (BC856) are 60 V, 200 mA, 310 mW complementary bipolar transistors from Vishay.
The TL5001AID (IC1) is a low-cost PWM controller with an open-collector output from Texas Instruments.

Dirk Gehrke   (040445-1)

172
Step-up Converter for 20 LEDs

The circuit described here is a step-up converter to drive 20 LEDs, designed to be used as a home-made ceiling night light for a child's bedroom. This kind of night light generally consists of a chain of Christmas tree lights with 20 bulbs each consuming 1 W, for a total power of 20 W. Here, in the interests of saving power and extending operating life, we update the idea with this simple circuit using LEDs. Power can be obtained from an unregulated 12 V mains

adaptor, as long as it can deliver at least about 330 mA. The circuit uses a low-cost current-mode controller type UCC3800N, reconfigured into voltage mode to create a step-up converter with simple compensation. By changing the external components the circuit can easily be modified for other applications. To use a current-mode controller as a voltage-mode controller it is necessary to couple a sawtooth ramp (rising from 0 V to 0.9 V) to the CS (current

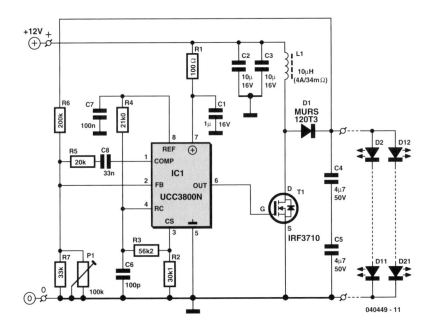

040449 - 11

sense) pin, since this pin is also an input to the internal PWM comparator. The required ramp is present on the RC pin of the IC and is reduced to the correct voltage range by the voltage divider formed by R3 and R2. The RC network formed by R4 and C6 is dimensioned to set the switching frequency at approximately 525 kHz. The comparator compares the ramp with the divided-down version of the output voltage produced by the potential divider formed by R6 and R7. Trimmer P1 allows the output voltage to be adjusted. This enables the current through the LEDs to be set to a suitable value for the devices used. The UCC3800N starts up with an input voltage of 7.2 V and switches off again if the input voltage falls below 6.9 V. The circuit is designed so that output voltages of between 20 V and 60 V can be set using P1. This should be adequate for most cases, since

the minimum and maximum specified forward voltages for white LEDs are generally between 3 V and 4.5 V. For the two parallel chains of ten LEDs in series shown here a voltage of between 30 V and 45 V will be required.

The power components D1, T1 and L1 are considerably overspecified here, since the circuit was originally designed for a different application that required higher power. To adjust the circuit, the potentiometer should first be set to maximum resistance and a multimeter set to a 200 mA DC current range should be inserted in series with the output to the LEDs. Power can now be applied and P1 gradually turned until a constant current of 40 mA flows. The step-up converter is now adjusted correctly and ready for use.

Dirk Gehrke   (040449-1)

173
Simple Window Comparator

A window comparator that monitors whether its input voltage Uin lies within a defined voltage window can be built using only a few components. The circuit shown in Figure 1 is a version that operates with complementary supply voltages. Two diode pairs are connected across the inputs of the comparator IC and supplied with bias currents via R1 and R2. If the input of the circuit is open, the current flowing through diode pair D2a and D2b (whose common terminal is connected to ground) causes the inverting input of the comparator to be at +0.7 V and the non-inverting input to be at –0.7 V. The comparator output is thus at the –U_b level, since the differential voltage at the comparator input is negative. The differential voltage is equal to the voltage on the non-inverting input minus the voltage on the inverting input. If a positive in-

put voltage is applied, diode D1b conducts and the voltage on the non-inverting input rises. This voltage is always 0.7 V lower than the input voltage. Diode D2b is cut off if the input voltage is positive. If the voltage at the circuit input rises to somewhat more than +1.4 V, the voltage on the non-inverting input will be slightly higher than the voltage on the inverting input, which is held to 0.7 V. The differential voltage between the comparator inputs will thus be positive, and the comparator output will switch to the +U_b level. The behaviour of the circuit is similar with a negative input voltage, with the difference being that the roles of the two diodes are swapped. In this case, the non-inverting input of the comparator remains at –0.7 V. If the input voltage drops below –1.4 V, the differential voltage between the comparator inputs

1

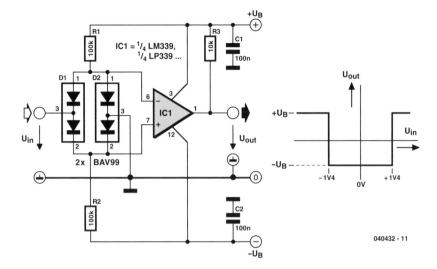

040432 - 11

again becomes positive, and the compara-tor output again switches to the +U$_b$ level. Note that the 'Low' output level of the win-dow comparator is –U$_b$ instead of ground, due to the complementary supply voltages. A version of the window comparator that works with a single supply voltage can be implemented by tying the common termi-nal of the diodes to a freely selectable inter-mediate voltage.

The schematic diagram for such an ar-rangement is shown in Figure 2, where opamp IC2 acts as a low-impedance source of reference voltage U$_{center}$. The voltage at

the centre of the window can be adjusted to the desired value using the trimpot. The width of the voltage window is determined by the diodes alone and amounts to twice the forward voltage drop (0.7 V) of a single diode. The width of the window can easily be enlarged by connecting additional di-odes in series with D1 and D2.

That can be done symmetrically, but it can also be done asymmetrically, such as by only wiring an additional diode in series with each of D1b and D2b. In the latter case, the window will not be symmetric about voltage U$_{center}$ (U$_{center}$ ± 0.7 V), but will instead extend from U$_{center}$ – 0.7 V to

2

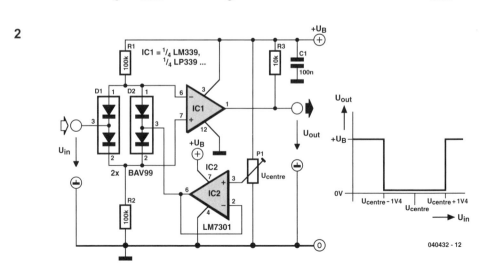

040432 - 12

$U_{center} + 1.4$ V. The circuit works with practically any type of comparator, such as the familiar LM399 or LP399 (low-power) quad comparator IC with supply voltages between ±2V and ±18 V. IC2 in Figure 2

must be a rail-to-rail type (such as the LM7301) that is rated for the same supply voltage range as the comparator.

Gregor Kleine   (040432-1)

174
IR Testing with a Digital Camera

If a device fails to respond to an IR remote control unit, the problem is often in the remote control, and it usually means that the batteries are dead. If the remotely controlled device still doesn't respond to the IR remote control after the batteries have been replaced, you're faced with the question of whether the remote control is not sending a signal or the device isn't receiving it properly. After checking for trivial errors, such as incorrectly fitted or defective batteries, the next thing you should check is whether the remote control transmits a signal. In the past, you would have needed an IR tester or a special IR detector card (as shown in the photo) for this. Nowadays you can use a digital camera (still or video), which is commonly available in most households. That's because the CCD chip is sensitive to infrared as well as visible light, which allows pictures to be taken at night to a certain extent. If

you switch on the camera and the display, aim the remote control unit toward the camera, and press one of the buttons on the remote control, you should see a blinking light coming from the IR unit is defective. If the LED remains dark, you can safely assume that the remote control unit is defective.

Dirk Gehrke   (040446-1)

175
Discharge Circuit

The author encountered a problem with a microcontroller system in which the +5 V supply voltage did not decay to 0 V sufficiently quickly after being switched off. A certain residual voltage remained, and it declined only very slowly. As a result, cer-

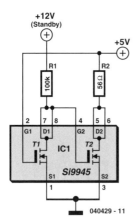

040429 - 11

tain system components could not perform a clean reset if the power was quickly switched on again. To remedy this problem, a very simple circuit was used to discharge the +5 V supply. It consists of two resistors and a type Si9945 dual MOSFET from Vishay Siliconix: (www.vishay.com/mosfets). These MOSFETs switch fully on at a threshold gate voltage between +1 V and +3 V. MOSFET

T2 connects discharge resistor R2 for the +5 V supply line to ground if the voltage on its gate exceeds the threshold voltage. When the +5 V supply is switched off, the first MOSFET (T1), whose gate is connected to the +5 V supply voltage, no longer connects pull-up resistor R1 to ground, so the standby voltage is applied to the gate of T2 via R1. This requires the standby voltage to remain available for at least as long as it takes to discharge the +5 V supply, even when the system is switched off. R2 is dimensioned to avoid exceeding the 0.25 W continuous power rating of a type 1206 SMD resistor. It may be necessary to change the component value for use in other applications. The circuit can be constructed very compactly, since the dual MOSFET is housed in an SO8 SMD package, but it can also be built using 'ordinary' individual FETs, such as the BS170.

http://www.vishay.com/doc?70758

Gregor Kleine   (040429-1)

176
USB Power Booster

Power shortage problems arise when too many USB devices connected to PC are working simultaneously. All USB devices, such as scanners, modems, thermal printers, mice, USB hubs, external storage devices and other digital devices obtain their power from PC. Since a PC can only supply limited power to USB devices, external power may have to be added to keep all these power hungry devices happy. This circuit is designed to add more power to a USB cable line.

A sealed 12 V 750 mA unregulated wall cube is cheap and safe. To convert 12 V to 5 V, two types of regulators, switching and linear are available with their own advan-

tages and drawbacks. The switching regulator is more suitable to this circuit because of high efficiency and compactness and now most digital circuits are immune to voltage ripple developed during switching. The simple switcher type LM2575-5 is chosen to provide a stable 5V output voltage.

This switcher is so simple it just needs three components: an inductor, a capacitor and a high-speed or fast-recovery diode. Its principle is that internal power transistor switch on and off according to a feedback signal. This chopped or switched voltage is converted to DC with a small amount of ripple by D1, L1 and

C2. The LM2575 has an ON/OFF pin that is switched on by pulling it to ground. T1, R2, and R1 (pull-up resistor) pull the ON/OFF pin to ground when power signal from PC or +5 V is received. D2, a red LED with current resistor R3, serves to indicate 'good' power condition or stable 5V. C3 is a high-frequency decoupling capacitor. The author managed to cut a USB cable in half without actually cutting data wires. It is advisable to look at the USB cable pin assignment for safety.

Myo Min   (040453-1)

177
Resistor-Programmable Temperature Switch

The switching threshold of the MAX6509 temperature switch from Maxim (www.maxim-ic.com) can be programmed over a range of –40 °C to +125 °C using an external resistor. The IC needs only two external components (see Figure 1). A hysteresis value of 2°C (typical) or 10 °C (typical) can be selected by connecting the HYST pin to ground or Vcc, respectively. The standard version of the IC is the MAX6509C, which pulls its open-drain output to ground when the temperature is below the threshold value set using the resistor. As shown in Figure 2, this ver-

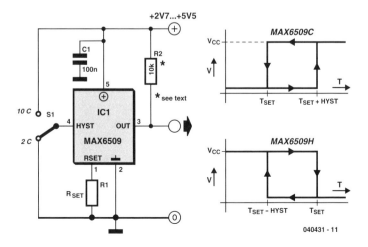

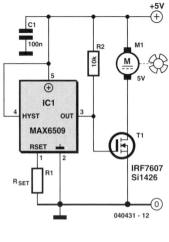

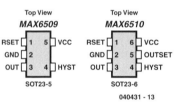

040431 - 13

T in°C	R_{SET}	T in°C	R_{SET}
−40	150 k	35	62 k
−30	136 k	40	57.6 k
−20	122 k	45	53.3 k
−10	108 k	50	49.2 k
−5	102 k	60	41.3 k
0	96.5 k	70	33.9 k
5	92 k	80	27 k
10	86.2 k	90	20.4 k
15	81 k	100	14.1 k
20	76 k	110	8.2 k
25	71 k	120	2.6 k
30	66.5 k	125	0

sion of the IC can be used to control a fan via an external MOSFET. The MAX6509H version has an inverted output, which means it is switched to ground when the temperature is above the threshold. A possible application for this version is switching on a heater in an oven when the temperature drops below the set point. The IC is housed in a SOT232 SMD package (Figure 3). The MAX6509 operates over a supply voltage range of +2.7–5.5 V with a supply current of only approximately 40 µA. The resulting self-heating is thus small enough to avoid corrupted temperature measurements (as long as the out-

put transistor is not required to switch high currents to ground). The table lists suitable values of resistor RSET for various threshold temperatures. The companion MAX6510 has an output stage that can be configured via a separate pin. The options are active high, active low, and open drain with an internal pull-up resistor. Like the MAX6510, MAX6510 is available in a -C version (open drain when the threshold temperature is exceeded) and an -H version (output pulled to ground when the threshold temperature is exceeded).

Gregor Kleine (040431-1)

178
MSP430 Programmer

For many applications, programming a microcontroller after it has been soldered

to the circuit board in the target application is more convenient than using a separate

programmer. With the Texas Instruments MSP430F11x1, this can be done quite easily using the JTAG pins. The Flash Emulation Kit makes it very easy to develop programs for the MSP430, debug the programs and program them into the microcontroller.

However, prototype testing usually reveals a need for minor improvements to the software. The MSP430 has a JTAG port that can be enabled by applying a High level to the TEST pin. The registers, RAM and Flash memory can be read and written via this interface. Naturally, this feature can also be used in the target application. However, it's important to bear in mind that the associated pins have dual functions. For in-circuit programming, you will need a 20-way SOJ test clip (available from 3M, for example) that can grip the pins of the SO IC package in the soldered-in state. A total of eight pins must be connected to the Flash Emulation Kit to allow the microcontroller to be programmed. It's important to ensure that a High level is applied to the /RST pin for the duration of the programming process, and a supplementary 30 kΩ resistor must be connected to the TEST pin to ensure a well-defined Low level.

Dirk Gehrke   (040458-1)

References and software

[1] IAR Embedded Workbench Kickstart Version 3 Rev. D Document ID: slac050d.zip
[2] MSP430F11X(1) Flash Emulation Tool (US $49)
[3] MSP-FET430 Flash Emulation Tool [http://focus.ti.com/lit/ug/slau138a/ slau138a.pdf]
[4] http://www.msp430.com

179
Code Lock with One Button

The unusual feature of this code lock is that it can be operated with just one push-button. In situations where a tamper-proof solution is required this circuit has a great financial advantage: only one robust pushbutton needs to be bought. The disadvantage of this solution is that it takes a little longer to enter the code. The operation of the code lock is as follows. After pressing the button, the PIC16F84

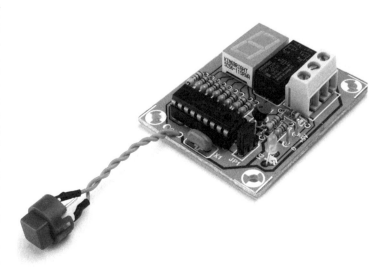

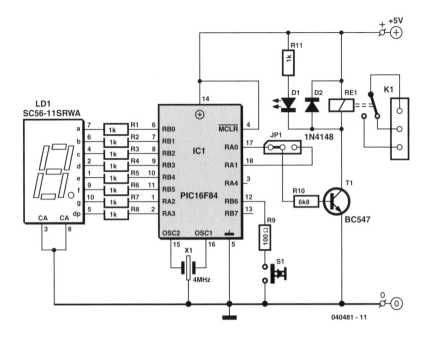

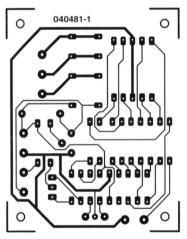

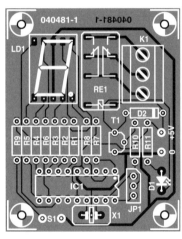

starts to count at a rate of one hertz. The numbers are visible on the LED display. The button is released once the correct number is displayed. This operation is repeated for the other digits in the code. The time between releasing the button and pressing it again for the next digit may not be more than 15 seconds. After the last digit, the letter 'E' (Enter) needs to be entered. If the entered code is incorrect, the display will show an 'F' (Fault) for 15 seconds. If the code has been entered incor-

rectly three times in a row, the lock will block further input for 1 minute. During this time the display shows a flashing 'F'. To alter the code, it is necessary to hold the button until the letter 'C' appears (Change code). The display subsequently shows an 'o' (Old code). Now enter the (valid) old code, followed by an 'E'. The PIC now asks for the new code by displaying an 'n' (New code). Now enter the new code, also followed by 'E'. The display now shows 'c' (Confirm). Repeat entry of the new

Components list

Resistors:
R1-R8, R11 = 1kΩ
R9 = 100Ω
R10 = 6kΩ8

Capacitors:
C1 = 100nF

Semiconductors:
D1 = LED, green, low-current
D2 = 1N4148
IC1 = PIC16F84 (programmed, order code 040481-41 *)
LD1 = 7-segment display, red, common-anode (e.g., Kingbright Sc56-11SRWA)
T1 = BC547B

Miscellaneous:
JP1 = 3-way SIL pinheader with jumper
K1 = 3-way PCB terminal block, lead pitch 5mm
Re1 = relay, 5V coil, (e.g., Omron G6A-234P-ST-US-DC5)
S1 = switch, 1 make contact, tamper proof (see text)
X1 = 4MHz ceramic resonator
PCB, ref. 040481-1 from The PCBShop Disk, source and hex code, order 040481-11 *

* see Elektor SHOP pages or www.elektor-electronics.co.uk

code (and once again an 'E' as confirmation). The code is now changed. A maximum of 10 digits can be used for the code. JP1 is used to set whether the relay is energised or de-energised when the correct code has been entered. You can use a standard 5 V regulated mains adapter for the power supply.

Zorislav Miljak   (040481-1)

180
Valve Sound Converter

'Valve sound' is not just an anachronism: there are those who remain ardent lovers of the quality of sound produced by a valve amplifier. However, not everyone is inclined to splash out on an expensive valve output stage or complete amplifier with a comparatively low power output. Also, for all their aesthetic qualities, modern valve amplifiers burn up (in the full sense of the word!) quite a few watts even at normal listening volume, and so are not exactly environmentally harmless. This valve sound converter offers a cunning way out of this dilemma. It is a low-cost unit that can be easily slipped into the audio chain at a suitable point and it only consumes a modest amount of energy.

A valve sound converter can be constructed using a common-or-garden small-signal amplifier using a readily available triode. Compared to using a pentode, this simplifies the circuit and, thanks to its less linear characteristic, offers even more valve sound. For stereo use a double triode is ideal. Because only a low gain is required, a type ECC82 (12AU7) is a better choice than alternatives such as the ECC81 (12AT7) or ECC83 (12AX7). This also makes things easier for homebrewers only used to working with semiconductors, since we can avoid any difficulties with high voltages, obscure transformers and the like: the amplifier stage uses an anode voltage of only 60 V, which is generated using a small 24 V transformer and a voltage doubler (D3, D4, C4 and C5). Since the double triode only draws about 2 mA at this voltage, a 1 VA or 2 VA transformer

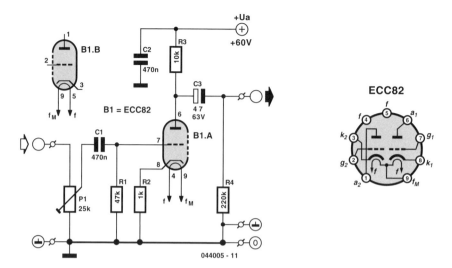

044005 - 11

will do the job. To avoid ripple on the power supply and hence the generation of hum in the converter, the anode voltage is regu-lated using Zener diodes D1 and D2, and T1. The same goes for the heater sup-ply: rather than using AC, here we use a DC supply, regulated by IC1. The 9 V transformer needs to be rated at at least 3 VA. As you will see, the actual amplifier circuit is shown only once. Components C1 to C3, R1 to R4, and P1 need to be du-plicated for the second channel. The inset valve symbol in the circuit diagram and the base pinout diagram show how the anode, cathode and grid of the other half of the double triode (V1.B) are connected. Con-struction should not present any great diffi-culties. Pay particular attention to screen-ing and cable routing, and to the placing of the transformers to minimise the hum in-duced by their magnetic fields. Adjust P1 to set the overall gain to 1 (0 dB). The out-put impedance of 47 kΩ is relatively high, but should be compatible with the inputs of most power amplifiers and preamplifiers. For a good valve sound, the operating point of the circuit should be set so that the audio output voltage is in the region of a few hun-dred millivolts up to around 1.5 V. If the valve sound converter is inserted between a preamplifier and the power amplifier, it should be before the volume control poten-tiometer as otherwise the sound will change significantly depending on the vol-ume. As an example, no modifications are needed to an existing power amplifier if the converter is inserted between the output of a CD player and the input to the amplifier.

Stefan Dellemann   (044005-1)

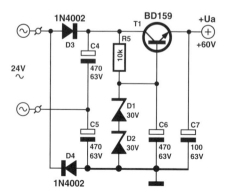

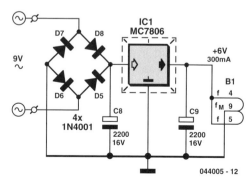

044005 - 12

181
Hard Drive Switch

Readers of Elektor Electronics who have both a PC and children face a particular problem. Since the young tend not to be too circumspect in their surfing habits, parents' files are in permanent danger of being infected with viruses or deleted. There is also the risk of children or their friends gaining unauthorised access to files not intended for their eyes. Perhaps a separate PC for the children would take up too much space or is ruled out for pedagogical reasons; in that case, the solution is to install two separate hard drives in the PC, one for the children and one for the adults. Ideally the two hard drives will each carry their own operating system and have their own software installed. As long as things are arranged so that the children can only boot their own drive, the parents' data will remain secure. All that is required, besides a second drive, is a specially-designed hard drive switch. This can be achieved, as has

already been described in Elektor Electronics, by switching the two drives between master and slave modes using the IDE cable, and only activating the master drive in the BIOS. However, the IT skills of children should not be underestimated: the BIOS is easily changed back. The solution described here is a bit more secure. Both drives are bootable and configured as masters. One is connected to IDE bus 1, and the other to IDE bus 2. The power supply voltages (+12 V and +5 V) are, however, only applied to one drive at a time. In principle, a simple double-pole change-over switch would do the job, but that has the disadvantage that it is possible to forget to reset the switch to child mode after use, especially if the switch is hidden. A better solution is to have to press a (hidden) button during boot to put the machine into parent mode. We will now see how this is done. If the button is not pressed when the

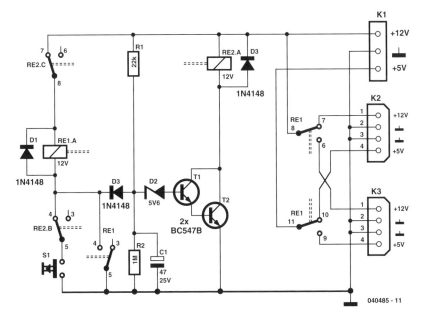

040485 - 11

PC is switched on, then, after a short delay of about 0.7 s (determined by R1, C1, D2 and the base-emitter junction threshold voltage of the Darlington pair formed by T1 and T2) RE2 pulls in. RE1 remains unenergised and hence the children's drive connected to K2 is active. Subsequently pressing S1 has no effect since RE1 has been isolated by the contacts of RE2. If the secret button is pressed, either briefly or continuously, during the 0.7 s sensitive period after the computer is switched on, RE1 pulls in immediately and holds itself in this state. D3 now prevents RE2 being subsequently activated. Since the contacts of RE1 have changed over, the PC now boots

from the parents' drive. It is impossible to forget to return the computer to child mode, since the computer will always start up in this mode if the secret button is not pressed. A 12 V miniature relay with contacts rated for 100 mA is suitable for RE2. The contacts of RE1 should be rated for the currents taken by a typical hard disk drive (say 2 A to 3 A). A key switch can be used instead of a secret button as a last resort against resourceful children, since the circuit will continue to operate correctly if S1 is left in parent mode permanently while the computer is on.

Dieter Brunow   (040485-1)

182
On/Off Button

It features at least once in every Small Circuits collection: the 555 timer. In this simple circuit we give the chip a little more attention than usual (refer to 'The Eternal 555' in the July/August 2004 issue). It is astonishing what can be built with a 555. Here at *Elektor Electronics* we are always infatuated with simple circuits using this IC, such as the one shown here. The 555 is used here so that a single pushbutton can operate a relay. If you press the button once, the relay is energised. When you press it again the relay turns off. In addition, it is possible to define the initial state of the relay when the power supply is switched on.

The design is, as previously mentioned, very simple. Using R1 and R2, the threshold and trigger inputs are held at half the power supply voltage.

When the voltage at the threshold pin becomes greater that 2/3 of the power supply voltage, the output will go low. The output goes high when the voltage at the trigger input is less than 1/3 of the power supply voltage. Because C2, via R3, will eventu-

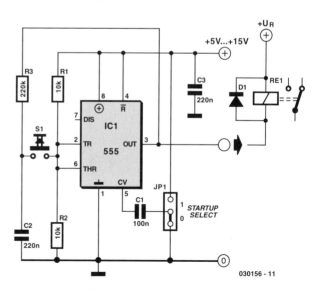

030156 - 11

ally have the same level as the output, the output will toggle whenever the pushbutton is pressed. If, for example, the output is low, the level of the trigger input will also become low and the output will go high! C1 defines the initial state of the re-

lay when the power is applied. If the free end of C1 is connected to Vcc, then the output is high after power up; the output is low when C1 is connected to ground.

Ger Langezaal   (030156-1)

183
Overcurrent Cutout Switches

Overcurrent sensors using external low-resistance sense resistors are fairly common. However, the members of a new family of ICs from Maxim (www.maxim-ic.com) feature an internal sensor resistor and a switch for disconnecting the load if the current limit is exceeded. The members of this IC family are listed in the table. There are two sorts of overcurrent switches in the family. The types listed in the 'Latching' column store any occurrence of an overcurrent condition and indicate it at the FLAG output until they are switched off and then on again by a pulse on the ON input. The types in the 'Auto-Retry' column automatically attempt to reconnect the load after a delay time. When the delay time expires, they check whether an overcurrent recurs, and if necessary they immediately switch off again.

The auto-retry types do not have a /FLAG output. They switch on for approximately 40 ms every 300 ms (typical) to measure the current. During this 40 ms 'blanking time', the IC checks whether the current is less than the selected limit level. The latching types have the same time delay before the switch opens and the FLAG output is asserted. The FLAG output can act as signal for a microcontroller or simply drive an LED. In the latter case, the input voltage must be greater than

the forward voltage of the LED. R1 must be dimensioned for the desired current through the LED. Capacitors C1 and C2 provide decoupling and prevent false triggering of the IC by spurious voltage spikes. The MAX47xx family of ICs operates over a supply voltage range of +2.3–5.5 V. The ICs have undervoltage lockout (UVLO) and reliably switch off when the current exceeds the type-specific limit, even if the current flows in the reverse direction (from the load to the input).

The table indicates the possible range of the overcurrent threshold for each type. For instance, a given MAX2791 might switch off at a current as low as 250 mA. However, other examples of the same type will not switch off until the current reaches 350 mA. The same threshold values apply to reverse currents. An overtemperature cutout circuit protects the IC against ther-

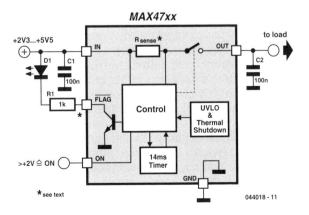

044018 - 11

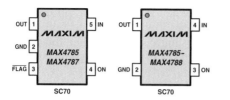

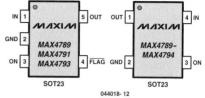

044018- 12

Latching	Auto-Retry	min. Limit	max. Limit	Package
MAX 4785	MAX 4786	50 mA	120 mA	SC70
MAX 4787	MAX 4788	100 mA	240 mA	SC70
MAX 4789	MAX 4790	200 mA	300 mA	SC70
MAX 4791	MAX 4792	250 mA	375 mA	SOT23/143
MAX 4793	MAX 4794	300 mA	450 mA	SOT23/143

mal destruction. The latching types come in a 5-pin SMD package, while the auto-retry types without a /FLAG output manage with only four pins. The 50 mA and 100 mA versions fit into the tiny SC70 package. The types for higher current levels require an SOT23 or SOT143 package. There are also other Maxim ICs with similar functions, such as the MAX4795–MAX4798 series with typical cutoff thresholds of 450 mA and 500 mA. Finally, there are the MAX4772 and MAX4773, which have a programmable threshold that can be set to 200 mA or 500 mA using a Select input. However, the IC types mentioned in this paragraph require a different circuit arrangement than what is shown here.

Gregor Kleine

(044018-1)

184
Negative Auxiliary Voltage

Some circuits need a negative supply voltage that only has to supply a small current. Providing a separate transformer winding for this (possibly even with a rectifier and filter capacitor) would be a rather extravagant solution. It can also be done using a few gates and several passive components.

The combination of gate IC1a and the other three gates (wired in parallel) forms a square-wave generator. D1 and D2 convert the ac voltage into a dc voltage. As a CMOS IC is used here, the load on the neg-

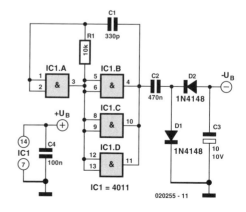

020255 - 11

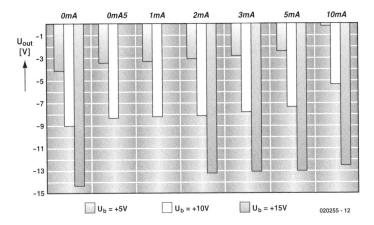

Ludwig Libertin

(020255-1)

ative output is limited to a few milliampères, depending on the positive supply voltage (see chart), despite the fact that three gates are connected in parallel. However, as the figure shows, the negative voltage has almost the same magnitude as the positive input voltage, but with the opposite sign. If a clock signal in the range of 10–50 kHz is available, it can be connected to the input of IC1a, and R1 and C1 can then be omitted.

185
Converter IC with Integrated Schottky Diode

Conventional step-up switching regulator ICs need at least one external Schottky diode and have the disadvantage that there is no effective output short-circuit current limiting. This means that very large currents can flow via coil L and Schottky diode D. Such currents can overload upstream components or destroy circuit board tracks. This situation is now remedied by the new LT3464 step-up switching regulator from Linear Technology (www.linear.com) in an 8-pin SOT23 package. Not only does it have an integrated Schottky diode, it also has an internal switching transistor that isolates the output from the input voltage in the shutdown mode. The switching transistor also has short-circuit current limiting that becomes effective at around 25 mA. The IC operates with input voltages between +2.3 V and +10 V and can supply an output voltage as high as +34 V. The amount of output current that can be drawn increases as the input voltage increases. For example, the maximum output current is 15 mA with an input voltage of +9 V and an output voltage of +20 V. A quite common application is generating +12 V from a +5 V source, for which the maximum output current is 20 mA. The output voltage is regulated via the Feedback pin (pin 2), with the voltage being determined by resistors R1 and R2 according to the formulas

$$V_{out} = 1.25 \cdot \left(1 + \frac{R2}{R1} \right)$$

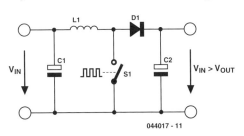

044017 - 11

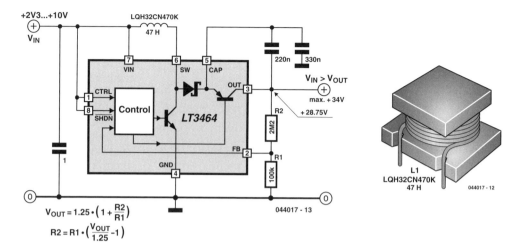

$$V_{OUT} = 1.25 \cdot \left(1 + \frac{R2}{R1}\right)$$

$$R2 = R1 \cdot \left(\frac{V_{OUT}}{1.25} - 1\right)$$

044017 - 13

$$R_2 = R_1 \cdot \left(\frac{V_{out}}{1.25} - 1\right)$$

Voltage divider R1/R2 can also be connected to the CAP pin (pin 5) ahead of the switching transistor. This avoids having an open-loop condition when the out-put is switched off, but it reduces the accuracy of the output voltage setting. The circuit shown here generates a voltage spike when the IC is switched back on, since the feedback loop drives up the voltage on the CAP

pin when the loop is open. A Murata type LQH32CN470K coil (47 µH) is used here as a storage choke due to its very compact construction. Other types of storage chokes in the range of 10–100 µH can also be used. The input capacitor, the capacitor connected to the CAP pin (pin 5), and the capacitor connected to the OUT pin (pin 3) are multilayer ceramic types (X5R and X7R).

Gregor Kleine   (044017-1)

186
Passive 9th-order Elliptical Filter

Steep filters can be realised in many different ways, for example by connecting active 2nd to 5th-order sections in series and calculating the component values for the higher order. They can also be made passive, but in practice this has a few difficulties associated with it. You cannot avoid the need for inductors with values that deviate from the standard se-

ries. You will have to wind them yourself on a specially selected core. The filter pre-

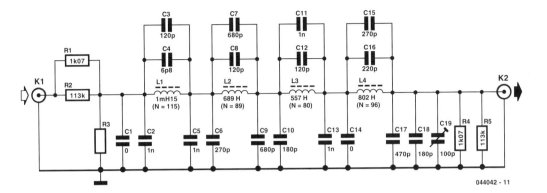

044042 - 11

sented here was originally designed to enable measurements to be made on the Class-T Amplifier (yes indeed, the one in Elektor Electronics, June 2004). When designing and testing audio equipment, we use a System Two Cascade Plus Analyzer made by Audio Precision. The accuracy of the measurements with this instrument is reduced if frequency components above 200 kHz are present at significant levels. This is the case with our amplifier, particularly at low signal levels. We immediately went for large artillery, namely a 9th-order elliptical filter. During the design of the filter we made use of normalised tables. In the end it became a filter with identical termination impedances, which unfortunately means an attenuation of two times within the pass band. When converting to realistic values we selected pure E12-series values for C1(+C2). All capacitors are arranged as two in parallel in order to closely approximate the calculated value. This applies to the resistors as well. With the inductors there is no way to avoid 'funny' values and series or parallel connections don't make much sense because to achieve a certain

quality, standard coils are not appropriate. So we had to think of a solution ourselves. The input and output impedance are theoretically 1.060 kΩ and are approximated quite well with components in parallel (1.05996 kΩ). By making use of a voltage divider it becomes possible for R3 to handle a higher voltage (otherwise note the

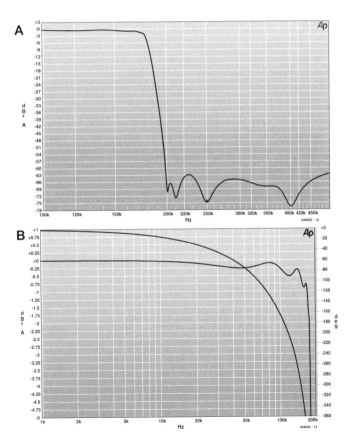

dissipation of R1!). Any voltage divider needs to have an output impedance of 1.06 kΩ (R1//R2//R3). In the last section, the parasitic capacitance of the connecting cable and input impedance of the analyser has been taken into account. Trimmer C19 can be used to compensate the attached capacitance and R5 can be omitted if the input impedance is about 100 kΩ. A deviation of about 50 pF makes little difference to the amplitude characteristic in the pass-band. The advantage of an elliptical filter compared to, for example a Chebyshev filter, is to trade off a limited attenuation in the stop-band to a much steeper transition from pass- to stop-band. It suffices to mention that the curve from 180 kHz to 200 kHz falls by more than 60 dB, quite steep and certainly not bad for a passive filter! In practice the attenuation in the stop-band at –63 dB was a little lower than the theoretical value of 60.2 dB, which was the design value. Frequency characteristic A shows mainly the stop-band and the

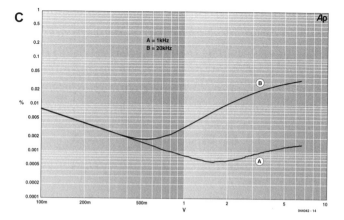

characteristic behaviour of an elliptical filter can be clearly seen. Frequency characteristic B shows an enlarged version of the ripple in the pass-band, which also shows the phase behaviour of the filter (scale on the right). At 20 kHz the attenuation is only 0.1 dB and the phase shift only –30°. The first dip of only –0.263 dB occurs at about 46 kHz and the attenuation at 100 kHz is only 0.276 dB. Above that, the non-ideal behaviour of the components becomes noticeable and the curve starts to drop a little too soon, but the characteristic elliptical behaviour is still clearly visible at

COMPONENTS LIST

Resistors:
R1,R4 = 1kΩ07
R2,R5 = 113kΩ
R3 = not fitted *

Capacitors:
C1,C14 = not fitted *
C2,C5,C11,C13 = 1nF 500V 1% silvered mica (Farnell 868-012)
C3,C8,C12 = 120pF 500V 1% silvered mica (Farnell 867-901)
C4 = 6pF8 500V 1% silvered mica (Farnell 867-779)
C6,C15 = 270pF 500V 1% silvered mica (Farnell 867-949)
C7,C9 = 680pF 500V 1% silvered mica (Farnell 867-998)
C10,C18 = 180pF 500V 1% silvered mica (Farnell 867-925)
C16 = 220pF 500V 1% silvered mica (Farnell 867-937)

C17 = 470pF 500V 1% silvered mica (Farnell 867-974)
C19 = 100pF trimmer

Inductors:
L1 = 1mH15, 115 turns of 0.5mm dia. ECW on core TN23/14/7-4C65 from BCcomponents (Farnell # 180-009)
L2 = 689μH, 89 turns of 0.5mm dia. ECW on core TN23/14/7-4C65 from BCcomponents (Farnell # 180-009)
L3 = 557μH, 80 turns of 0.5mm dia. ECW on core TN23/14/7-4C65 from BCcomponents (Farnell 1# 80-009)
L4 = 802μH, 96 turns of 0.5mm dia. ECW on core TN23/14/7-4C65 from BCcomponents (Farnell # 180-009)

Miscellaneous:
K1,K2 = cinch socket, PCB mount, e.g., T709G (Monacor/Monarch)
PCB, available from The PCBShop

* see text

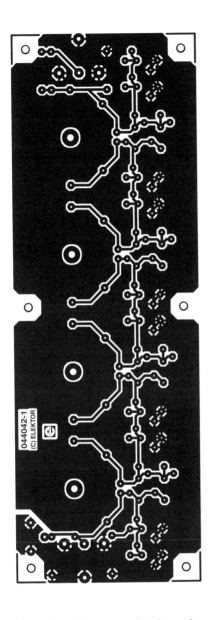

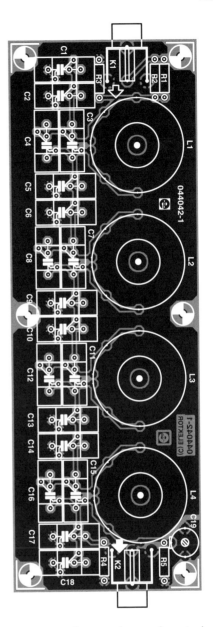

180 kHz. The filter proved to be quite useful in filtering the PWM signal and analyse the LF- amplitude. The only disadvantage is the increasing distortion at 20 kHz (from 0.5 V input signal) so that good THD+N measurements can only be done at 1 kHz. This can be seen clearly in Graph C. With 1 W into 8 Ω (2.828 V) the distortion at 1 kHz is less than 0.001%, but at 20 kHz the distortion is 20 times larger. In this

measurement the maximum input signal was 13.33 V (maximum from the analyser). For those who love to experiment and wind inductors, we have also designed a PCB. A low permeability core material (TN23/14/7-4C65) was selected for the inductors, so that saturation and material properties are less of a problem. Unfortunately this results in a higher number of turns, but also means that the inductor

Theoretical component values

R1//R2 = 1.060 kΩ
R4//R5 = 1.060 kΩ
C1+C2 = 1.000 nF
C3+C4 = 128.0 pF
C5+C6 = 1.277 nF
C7+C8 = 809.0 nF
C9+C10 = 860.4 nF
C11+C12 = 1.125 nF
C13+C14 = 996.8 pF
C15+C16 = 492.7 nF
C17+C18+C19 = 742.4 pF
L1 = 1.148 mH
L2 = 693.3 µH
L3 = 556.4 µH   L4 = 809.6 µH

value can be made more accurate. A larger core may have resulted in a lower distortion, but it would have been harder to obtain an accurate value. Toroids were selected to minimise mutual coupling – that this was successful is shown in Graph A. It

is easiest when winding the cores to calculate the amount of wire required beforehand and then add 10 or more centimetres. You have to wind tightly and put the turns close together to prevent the second layer dropping in between the first layer. This applies to the inside of the ring core. When using 0.5 mm enamelled wire the second layer turns easily fit between the first layer turns. The PCB has been designed such that connections can be made in several places (3 inside a quarter circle). The capacitors are 1% silvered mica types with an operating voltage of 500 V. That way even extreme voltage peaks will not cause any harm. There is also room for 1% tolerance 'Styroflex' (polystyrene) capacitors from Siemens (that are not made any more), which we have used in the past. Other manufacturers also use this shape.

Ton Giesberts   (044042-1)

187
Three-component Li-ion Charger

The LTC4054 from Linear Technology (www.linear.com) is a very simple device for charging 4.2 V Li-ion batteries. This SMD IC comes in a five-pin SOT-23 package and needs just two external components (the LED is not absolutely essen-

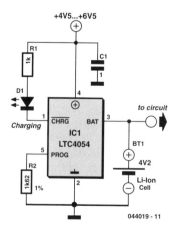

044019 - 11

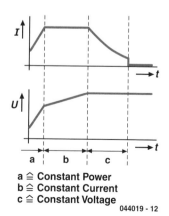

a ≅ **Constant Power**
b ≅ **Constant Current**
c ≅ **Constant Voltage**

044019 - 12

tial): a decoupling capacitor of at least 1 µF and a resistor connected to pin 5 (PROG) to set the charging current. The value of 1.62 kΩ shown here gives a charging current I_{CELL} of 600 mA when the device is in constant current mode. The formula

$$I_{cell} = \frac{V_{prog}}{R_{prog}} \cdot 1000$$

where $V_{PROG} = 1$ V, gives the charging current in terms of R_{PROG}.

The device works from a supply voltage of between 4.25 V and 6.5 V and is therefore suitable for connection to a USB port on a computer. To avoid risk of damage to the cells, the charging process is divided into three stages. In the first stage, which is brief, a constant power is delivered to the cell. In the next stage a constant current is delivered, and the cell voltage rises linearly. Finally the devices switches to a con-

stant voltage mode, and the current drops off sharply. The LTC4054 goes into a high-impedance state when the input voltage falls below a set value to ensure that the battery does not get discharged. The CHARGE pin (pin 1) indicates the charging state. It is an open drain output which is pulled down to ground via a low impedance during charging, allowing an LED to be connected. The pin sinks approximately 20 µA to ground when the Li-ion cell voltage is between 2.9 V and 4.05 V: this is the standby state. If the cell voltage falls below 2.9 V, the LTC4054 begins charging again. CHARGE goes into a high-impedance state if the input voltage is not at least 100 mV higher than the cell voltage. The under-voltage lockout circuit is then activated, with less than 2 µA being drawn from the cell.

Gregor Kleine   (044019-1)

188
LED Flasher for 230 V

The small circuit shown here could act as a power indicator for the 230 V mains supply and in terms of efficiency is the equal of classical neon bulbs. Note first that the LED in this circuit flashes rather than lighting continuously, and is therefore also suitable for applications where a flashing light is wanted for decorative purposes or as a gimmick. Diode D1 rectifies the input voltage, and C1 is charged by the rectified mains voltage via R1. When, after a number of half-cycles of the mains, the voltage on C1 exceeds the breakover voltage of the diac

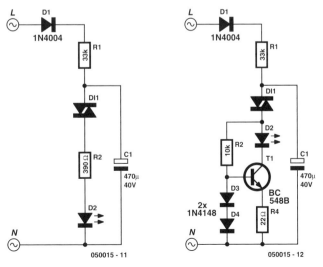

DI1, the diac conducts and C1 discharges via R2 and light-emitting diode D3. This discharge results in a brief flash of light. A 470 µF/40 V capacitor is suitable for C1. For the diode a 1N4004 can be used, and R1 should have a value of 33 kΩ, be rated at 0.6 W and be suitable for use at 350 V. As an alternative, the value of 33 kΩ can be made up from two (or more) resistors wired in series: for example, 15 kΩ + 18 kΩ or 2 × 10 kΩ and 1 × 12 kΩ. R2 should be 390 Ω. firing voltage of the diac should be 30 V. Using these values the LED flashes for 0.3 s every second.

Matthias Haselberger   (050015-1)

Caution: high voltage

This circuit is connected directly to the mains, and it is therefore dangerous to touch any part of it. This goes for the LED itself as well. It is absolutely essential to build the circuit in an insulating touchproof plastic enclosure; see also the safety advice pages.

189
GigaBit Crossover Cable

We previously described a crossover cable in last year's Summer Circuits issue ('Home Network for ADSL', p. 142). However, that cable cannot be used for GigaBit links, since all eight conductors are important in that case. A GigaBit link uses four signal pairs. If the cable is longer than 8 metres, it is also necessary for each paired set of conductors to be twisted together. Otherwise crosstalk will occur, which will cause data communication errors. A crossover cable that is suitable for 1000 Mbit networks must have all of the conductors connected differently at one

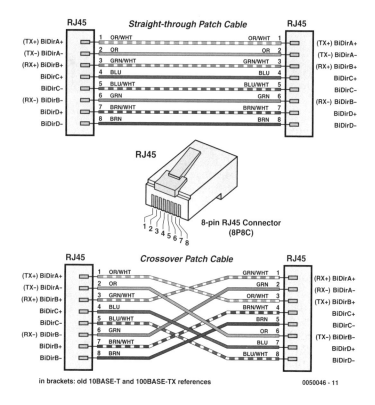

in brackets: old 10BASE-T and 100BASE-TX references   0050046 - 11

end. TX is thus connected to RX, and the two other pairs are also swapped. The connection scheme is shown clearly in the drawing.

Naturally, such a cable can also be used in a 10 Mbit or 100 Mbit network, but it is not suitable for use in combination with lines for an analogue telephone, such as was suggested in the article in the 2004 Summer Circuits issue.

Henri Derksen   (050046-1)

190
On-demand WC Fan

In most WCs with an extractor the fan is connected to the lighting circuit and is switched on and off either in sympathy with the light or with a short delay. Since toilets are sometimes used for washing the hands or just for a quick look in the mirror, it is not always necessary to change the air in the smallest room in the house. The following circuit automatically determines whether there really is any need to run the fan and reacts appropriately. No odour sensor is needed: we just employ a small contact that detects when and for how long the toilet seat lid is lifted. If the seat lid is left up for at least some presettable minimum time t1, the fan is set running for another presettable time t2. In the example shown

the contact is made using a small magnet on the lid and a reed switch mounted on the cistern. The rest is straightforward: IC2, the familiar 555, forms a timer whose period can be adjusted up to approximately 10 to 12 minutes using P2. This determines the fan running time. There are three CMOS NAND gates (type 4093) between the reed switch and the timer input which generate the required trigger signal. When the lid is in the 'up' position the reed switch is closed. Capacitor C1 charges through P1 until it reaches the point where the output of IC1a switches from logic 1 to logic 0. The output of IC1b then goes to logic 1. The edge of the 0-1 transition, passed through the RC network formed by

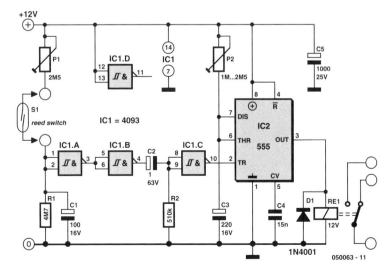

C2 and R2, results in the output of IC1c going to logic 0 for a second. This is taken to the trigger input on pin 2 of timer IC2, which in turn switches on the relay which causes the fan to run for the period of time determined by P2. The circuit is powered from a small transformer with a secondary winding delivering between approximately 8 V and 10 V. Do not forget to include a suitable fuse on the primary side. The circuit around IC1b and IC1c ensures that the

fan does not run continuously if the toilet seat lid is left up for an extended period. The time constant of P1 and C1 is set so that the fan does not run as a result of lavatorial transactions of a more minor nature, where the lid is opened and then closed shortly afterwards, before C1 has a chance to charge sufficiently to trigger the circuit.

Ton Giesberts   (050063-1)

191
Micropower Voltage Regulator

This circuit was developed to power an AVR microcontroller from a 12 V lead-acid battery. The regulator itself draws only 14 µA. Of course, there are dedicated ICs, for example from Linear Technology or Maxim, which can be used, but these can be very hard to get hold of and are frequently only available in SMD packages these days. These difficulties are simply and quickly avoided using this discrete circuit.

The series regulator component is the widely-available type BS170 FET. When power is applied it is driven on via R1. When the output voltage reaches 5.1 V, T2 starts to conduct and limits any further rise in the output voltage by pulling down the voltage on the gate of T1. The output voltage can be calculated as follows:

$$U_{out} = \left(U_{LED} + U_{BE}\right) \cdot \frac{R4 + R2}{R4}$$

where we can set U_{LED} at 1.6 V and U_{BE} at 0.5 V. The temperature coefficients of U_{LED} and U_{BE} can also be incorporated into the formula. The circuit is so simple that of course someone has thought of it before. The author's efforts have turned up an example in a collection of reference circuits

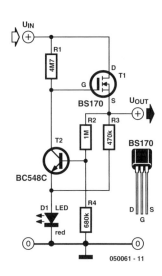

dating from 1967: the example is very similar to this circuit, although it used germanium transistors and of course there was no FET. The voltage reference was a Zener diode, and the circuit was designed for currents of up to 10 A.

Perhaps Elektor Electronics readers will be able to find even earlier examples of two-transistor regulators using this principle?

Reinhold Oesterhaus   (050061-1)

192
PIC PWM Controller

An efficient and economical method to control the power into a load (for example the speed of a motor or the temperature of a heating element) is to use PWM (Pulse Width Modulation). But things are a little bit more involved of we want an accurate adjustment from 0 to 100 % and an indication of the power.

A little 8-pin microcontroller can do these tasks easily: generating a PWM signal and indicating the power via a 4017 (see schematic). This configuration of the microcontroller does not require an external reset circuit (so that there is a spare nal reset circuit (so that there is a spare

pin), because we use the automatic internal one. A quartz crystal is not required either, since we're using the integrated 4 MHz oscillator, which despite being based on an RC network, is accurate to 1 % thanks to the calibration carried out at the factory. Once the microcontroller has been provided with a suitable program, it can carry out the instructions from the user by means of two pushbuttons.

A PWM signal with a frequency of 100 Hz is generated, while the power is indicated with a classical 4017. The circuit can be kept very simple because the

1

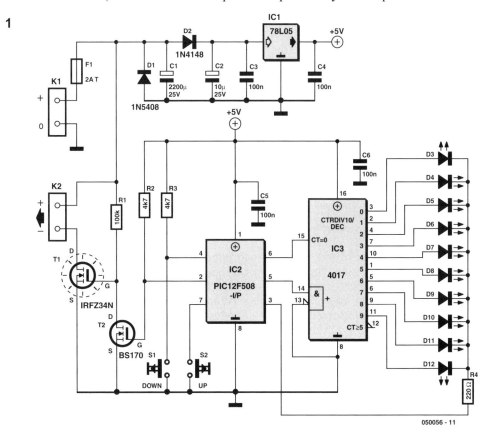

050056 - 11

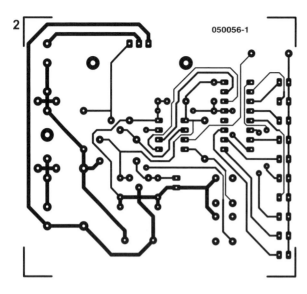

050056-1

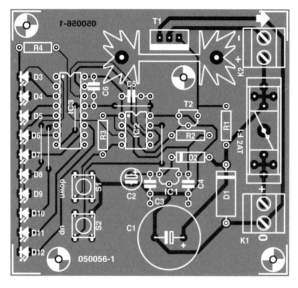

with the 4017 and 10 LEDs. When first powered up, a back-and-forth running light indicates that the circuit is powered but the output is not active. As soon as the load is switched on with pushbutton S1, the first LED (10%) will light up. The correct operation of the circuit is indicated in an eye-catching manner by flashing the LED that indicates the power, using pin 3 of the controller. The program is of a very simple design and the source code together with the hex file are available from the Elektor Electronics website or on a floppy disk (order code 050056-11). You are free to add improvements, because there is plenty of space left in the program memory of the controller. To compile the code (written in C) you can use the evaluation version of the CC5X-compiler (limited to 1024 words, which in our case is more than enough). This is available from the website www.bknd.com/cc5x (choose the Free Edition). Another handy piece of freeware is the ConText editor. It can be found at www.context.cx. If you would like to experiment with the circuit, it is recommended to use the reprogrammable 12F629. In this case R3 is required. It is not necessary when a 12x508 is used. When programming, don't forget to check that all the fuses are configured correctly:

- Oscillator: Internal_RC;
- WatchDog_Timer: ON;
- Master_Clear_Enable: Internal;
- Code_Protect: OFF.

This is particularly important when using the 12C508, because this is an OTP-type (can be programmed only once). Figure 2 shows the printed circuit board, which, because of the 4 wire links has remained sin-

microcontroller carries out all the complicated tasks. A 78L05 provides the power supply, a 4017 for the indicator and a pair of MOSFETs for the power stage. Connector K1 receives the input power supply voltage (9 to 12 V at 1 to 5 A). The load to be controlled is connected to K2. With 2 pushbuttons we can control the power in steps of 10 %. To stop immediately, both need to be pressed at the same time. That concludes the user manual! A special feature of the circuit is the power indication

COMPONENTS LIST

Resistors:
R1 = 100kΩ
R2,R3 = 4kΩ7
R4 = 220Ω

Capacitors:
C1, = 2200µF 25V
C2 = 10µF 25V
C3-C6 = 100nF

Inductors:
D1 = 1N5408
D2 = 1N4148
D3-D7 = LED, 3mm, green
D8-D12 = LED, 3mm, red

T1 = IRFZ34N
T2 = BS170
IC1 = 78L05
IC2 = PIC 12C508-1/P
IC3 = CD 4017

Miscellaneous:
K1,K2 = 2-way PCB terminal block, lead pitch 5mm
S1,S2 = pushbutton, 1 make contact, DTS6
F1 = 2 AT (time lag) fuse with PCB mount holder
Heatsink type SK104 (Fisher)
PCB, ref. 050056-1 from The PCBShop Disk, PIC source and hex code: order code 050056-11 or Free Download

gle-sided and provides enough space for all the parts. T1 only requires a heatsink of you intend to regulate currents greater than 2 A over extended periods of time. The choice of fuse depends on the current requirements of the load that is connected. T1 and the PCB traces can easily handle 5 A. As soon as the power supply is applied to the circuit, fitted with a programmed PIC, the LEDs should light up as a running light. Every button press of S2 increases the on/off ratio with 10 % to a maximum of 100 %. Thanks to the accurate oscillator in the 12C506, each step of 10 % corresponds to 1 ms and the entire cycle is repeated at a rate of 10 ms, which corresponds to a frequency of 100 Hz. Ideal for small motors. A final note is that the program uses the watchdog functionality of the 12C508. This generates an automatic reset within 20 ms if the program crashes for some reason (for example large voltage surges). So there is nothing left to be desired regarding the reliability of the circuit…

Jean-Marc Bühler   (050056-1)

193
Simple Stepper Motor Driver

There is hardly any field in the world of electromechanics that has not found an application for the stepper motor. They are used extensively in the world of model making and as actuators in remote control equipment. In industry, picture scanners and printers are probably the most obvious devices that simply would not function

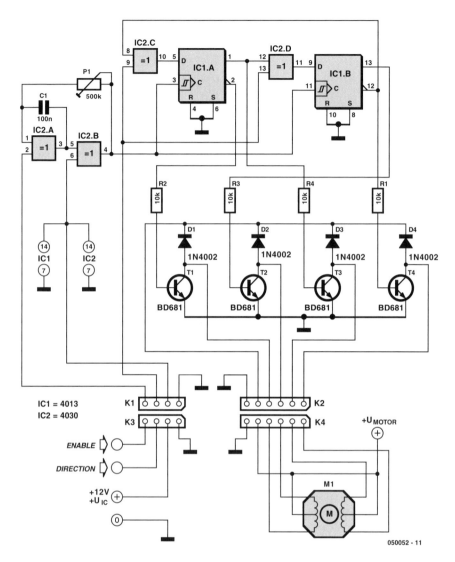

050052 - 11

without them, so no excuse is needed to include this very simple 4-phase stepper motor driver design in this collection of circuits. The circuit clock generator is formed from two exclusive OR (XOR) gates IC2A and IC2B together with C1 and P1. A logic '0' on the ENABLE input enables the clock generator and its output frequency is defined by 1.4 P1C1.

The output signal from the clock generator is connected to the clock inputs of the two D-type flip-flops IC1.A and IC1.B. These two flip flops are connected together in a ring to form a 2-bit shift register so that the

Q output of IC1.B is fed back to the D input of IC1.A and the Q output of IC1.A is fed into the D input of IC1.B. This configuration supplies the 4 phase impulses necessary to provide motor rotation. When the DIRECTION input is changed to logic zero IC2C and IC2D operate as non-inverting gates, reversing the phase sequence of the output signals and making the motor spin in the opposite direction. The actual rotation direction will depend on the sense of the motor windings. Swapping the outer two coil connections on one of the windings will reverse the direction if this is nec-

essary. With the components specified the circuit oscillates at a frequency of 10 Hz. The clock frequency can be adjusted between 0.2 and 100 Hz by substituting different values for P1 and C1. It is important to ensure that power drawn by the stepper motor is within the power handling capability of the driver transistors T1 to T4. Diodes D1 to D4 are necessary to conduct away the back-EMF produced each time a drive impulse to each of the motor coils is switched off.

Paul Goossens   (050052-1)

194
Extended Timer Range for the 555

Anyone who has designed circuits using the 555 timer chip will, at some time have wished that it could be programmed for longer timing periods. Timing periods greater than a few minutes are difficult to achieve because component leakage currents in large timing capacitors become significant. There is however no reason to opt for a purely digital solution just yet. The circuit shown here uses a 555 timer in the design but nevertheless achieves a timing interval of up to an hour! The trick here is to feed the timing capacitor not with a constant voltage but with a pulsed dc volt-

age. The pulses are derived from the unsmoothed low voltage output of the power supply bridge rectifier. The power supply output is not referenced to earth potential and the pulsing full wave rectified signal is fed to the base of T1 via resistor R1. A 100-Hz square wave signal is produced on the collector of T1 as the transistor switches. The positive half of this waveform charges up the timing capacitor C1 via D2 and P1. Diode D2 prevents the charge on C1 from discharging through T1 when the square wave signal goes low. Pushbutton S1 is used to start the timing

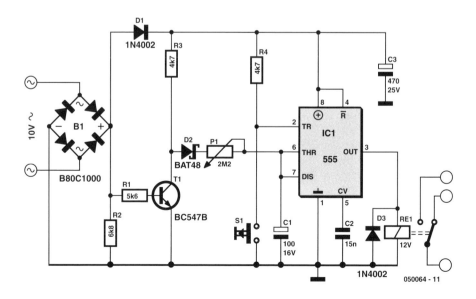

050064 - 11

period. This method of charging uses relatively low component values for P1 (2.2 MΩ) and C1 (100 to 200 µF) but achieves timing periods of up to an hour which is much longer than a standard 555 circuit configuration.

Ton Giesberts   (050064-1)

195
Transistor Dip Meter

The dip meter consists of a tunable RF oscillator whose resonant circuit is held in the vicinity of a resonant circuit to be checked. If the frequencies of the two circuits match, the circuit being measured draws energy from the oscillator circuit. This can be measured. This type of meter is also called a 'grid dip meter', since it was originally built using a valve. The amplitude of the voltage on the tuned circuit could be measured from the grid leakage current. Such meters typically have a set of interchangeable coils and several frequency scales. A meter that can manage with a low voltage of only 1.5 V can be built using a circuit with two transistors. In addition, a coil tap is not required in this design. That makes it easy to connect many different coils to cover a large number of frequency bands.

If a sufficiently sensitive moving-coil meter is not available, an acoustic signal can be used instead of a pointer display. This involves a sound generator whose frequency increases when its input voltage rises. A resonance dip is then indicated by a falling tone. This acoustic indicator draws less current from the measurement rectifier than a moving-coil instrument. That allows the amplitude of the oscillation to be decreased slightly by reducing the emitter current. This increases the sensitivity, so the dip meter can measure resonant circuits at greater distances.

Burkhard Kainka   (050097-1)

1

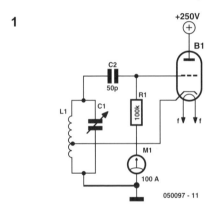

050097 - 11

2

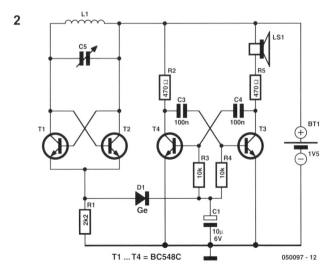

T1 ... T4 = BC548C   050097 - 12

196
Fridge Thermostat

What to do when the thermostat in your fridge doesn't work any more? Get it repaired at (too) much expense or just buy a new one? It is relatively simple to make an electronic variation of a thermostat yourself, while saving a considerable amount of money at the same time. However, be careful when working with mains voltages. This voltage remains invisible and can sometimes be fatal! This design allows for five temperatures to be selected with a rotary switch. By selecting suitable values

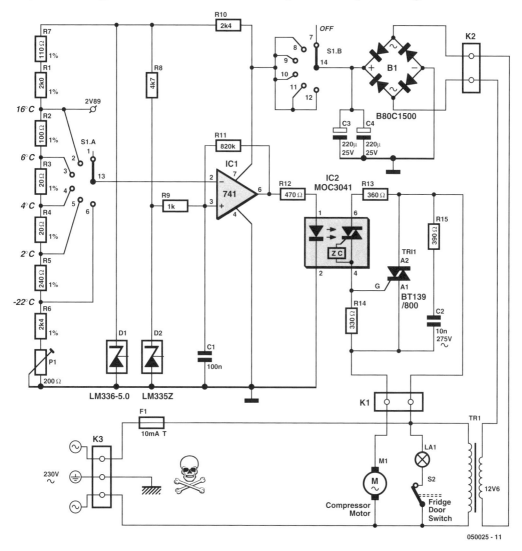

050025 - 11

for the resistors (R1 to R7), the temperatures at the various switch positions can be defined at construction time. With the resistance values shown here, the temperature can be adjusted to 16, 6, 4, 2 and –22 °C. 16° C is an ideal temperature for the storage of wine, while 6, 4, and 2 degrees are interesting for beer connoisseurs and the minus 22°degrees position transforms the fridge into a large freezer. Note for wine connoisseurs: to prevent mould on the labels, it is necessary to place a moisture absorber or bag of silica gel in the fridge.

The circuit is built around an old workhorse among opamps, the 741. D1 provides a stable reference voltage of 5 V across the entire resistor divider. P1 allows adjustment of the voltage at the node of R1 and R2. To use the above-mentioned temperatures as setpoints this voltage needs to be adjusted to 2.89 V. D2 is a precision temperature sensor, which can be used from –40 to +100 °C. The voltage across this diode varies by 10 mV per Kelvin. In this way D2 keeps an eye on the temperature in the fridge. The reference voltage derived from the voltage divider (selected with S1) is compared by IC1 with the voltage across the temperature sensor. Based on this, the 741 switches, via the zero voltage crossing driver (IC2), a triac that provides voltage to the compressor motor. The zero voltage crossing IC switches only at the zero crossings of the mains voltage, so that interference from the compressor motor is avoided when turning on. The power supply for the circuit is provided by a simple bridge rectifier and filtered with two electrolytic capacitors of 220 µF each. The design can also be used for countless other uses. You can, for example, make a thermostat for heating by swapping the inputs of the opamp. Keep in mind the safety requirements when building and mounting the circuit.

Tony Beekman   (050025-1)

197
Mobile Phone Operated Code Lock

Suitcase numberwheels and doorside keypads have evolved from well-known code locks, to hot topics now mainly due to Dan Brown's bestseller novel The daVinci Code. The main ideas behind the lock described here are minimum obtrusiveness and minimum user interface. A typical code lock is operated with a four-digit secret code and the lock can be opened by presenting this code. The lock described here has no buttons or keypad at all, a small hole or other hiding place for the microphone capsule is enough.

Nowadays practically everyone has a keypad in the pocket – it's on your mobile phone! The lock listens to mobile phone keytones (DTMF tones) and responds to the valid, preset four-digit code. No visible interface is needed as the microphone capsule can be located behind a small hole. Note that the mobile is used 'off-line', so no phone expenses are involved. Electret microphone M is connected via transistor amplifier stage T1 to the input pin (2) of DTMF-receiver/decoder IC7. The decoder's four-digit output word (on pins 11, 12, 13, 14) and 'valid digit present' flag (pin 15) are connected to two shift registers, IC1 and IC2. A rising edge on pin 15 of the '8870 chip triggers each shift register to read its input code and shift it by one increment. Shift register outputs are connected to BCD to Decimal-converters type 4028 (IC3-IC6). The register status is shown as a high signal level at certain pins of DIP switches S1-S4. Depending on the switch settings, one combination causes High levels at all AND gate inputs and the lock is opened for a moment. Due to the operation of the shift registers, the latest input digit appears on IC3's outputs. So if the desired code is, for example, 2748, S1 con-

tact 2, S2 contact 7, S3 contact 4 and S4 contact 8 are closed. When four digits in sequence match with the code set by the switches, relay RE1 is activated for a user-defined time. Upon a rising edge on input pin 8, monostable multivibrator (MMV) IC8 pulls pin 10 high for a moment, activating relay Re1 via transistor

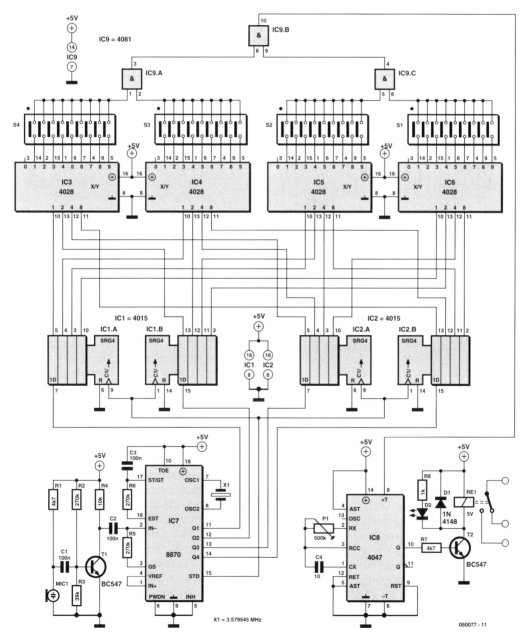

050077 - 11

T2. The duration of active time can be adjusted with the preset between pins 2 and 3. A green LED can be connected across the relay coil to indicate lock opening. The

minimum distance from the microphone is about 20 cm.

Heikki Kalliola   (050077-1)

198
EE-ternal Blinker

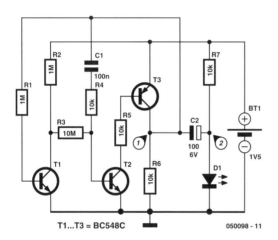

T1...T3 = BC548C   050098 - 11

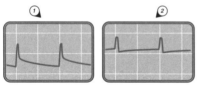

You occasionally see advertising signs in shops with a blinking LED that seems to blink forever while operating from a single battery cell. That's naturally an irresistible challenge for a true electronics hobbyist... And here's the circuit. It consists of an astable multivibrator with special properties. A 100 µF electrolytic capacitor is charged relatively slowly at a low current and then discharged via the LED with a short pulse. The circuit also provides the necessary voltage boosting, since 1.5 V is certainly too low for an LED. The two oscillograms demonstrate how the circuit works. The voltage on the collector of the

PNP transistor jumps to approximately 1.5 V after the electrolytic capacitor has been discharged to close to 0.3 V at this point via a 10 kΩ resistor. It is charged to approximately 1.2 V on the other side. The difference voltage across the electrolytic capacitor is thus 0.9 V when the blink pulse appears. This voltage adds to the battery voltage of 1.5 V to enable the amplitude of the pulse on the LED to be as high as 2.4 V. However, the voltage is actually limited to approximately 1.8 V by the LED, as shown by the second oscillogram. The voltage across the LED automatically matches the voltage of the LED that is used. It can theo-retically be as high as 3 V. The circuit has been optimised for low-power operation. That is why the actual flip-flop is built using an NPN transistor and a PNP transistor, which avoids wasting control current. The two transistors only conduct during the brief interval when the LED blinks. To ensure stable operating conditions and reliable oscillation, an additional stage with negative DC feedback is included. Here again, especially high resistance values are used to minimise current consumption. The current consumption can be estimated based on the charging current of the electrolytic capacitor. The average voltage across the two 10 kΩ charging resistors is 1 V in total. That means that the average charging current is 50 µA. Exactly the same amount of charge is also drawn from the battery during the LED pulse. The average current is thus around 100 µA. If we assume a battery capacity of 2500 mAh, the battery should last for around 25,000 hours. That is more than two years, which is nearly an eternity.

As the current decreases slightly as the batter voltage drops, causing the LED to blink less brightly, the actual useful life could be even longer. That makes it more than (almost) eternal.

Burkhard Kainka (050098-1)

199
Short-Wave
Superregenerative Receiver

Superregenerative receivers are characterised by their high sensitivity. The purpose of this experiment is to determine whether they are also suitable for short-wave radio. Superregenerative receivers are relatively easy to build. You start by building a RF oscillator for the desired frequency.

The only difference between a super-regenerative receiver and an oscillator is in the base circuit. Instead of using a voltage divider, here we use a single, relatively high-resistance base resistor (100 kΩ to 1MΩ). Superregenerative oscillation occurs when the amplitude of the oscillation is sufficient to cause a strong negative

charge to be applied repeatedly to the base. If the regeneration frequency is audible, adjust the values of the resistors and capacitors until it lies somewhere above 20 kHz. The optimum setting is when you hear a strong hissing sound. The subsequent audio amplifier should have a low upper cut-off frequency to strongly attenuate the regeneration signal at its output while allowing signals in the audio band to pass through. This experimental circuit uses two transistors. A Walkman headphone with two 32 Ω earphones forms a suitable output device.

The component values shown in the schematic diagram have proven to be suitable for the 10–20 MHz region. The coil consists of 27 turns wound on an AA battery serving as a winding form. The circuit produces a strong hissing sound, which diminishes when a station is received. The radio is so sensitive that it does not require any antenna to be connected. The tuned circuit by itself is enough to receive a large number of European stations. The circuit is usable with a supply voltage of 3 V or more, although the audio volume is greater at 9 V.

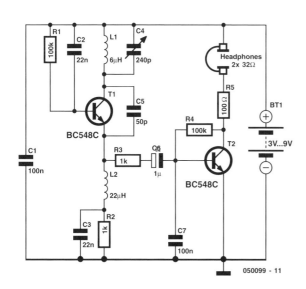

050099 - 11

One of the major advantages of a superregenerative receiver is that weak and strong stations generate the same audio level, with the only difference being in the signal to noise ratio. That makes a volume control entirely unnecessary. However, there is also a specific drawback in the short-wave bands: interference occurs fairly often if there is an adjacent station separated from the desired station by something close to the regeneration frequency. The sound quality is often worse than with a simple regenerative receiver. However, this is offset by the absence of the need for manual feedback adjustment, which can be difficult.

Burkhard Kainka   (050099-1)

200
9-in-1 Logic Glue / Level Translators

Two logic devices from revered Texas Instruments, the SN74AUP1T97 and '98 are logic glue components in tiny SOT23 cases. As shown in the diagrams, both the AUP1T97 and its inverting counterpart the AUP1T98 can be configured as voltage level translators with nine different logic functions marked by Schmitt trigger inputs. Apart from their logic function, these devices can also act as voltage level translators between low-voltage logic systems, as follows:

in	out	at V_{cc}
1.8 V	3.3 V	3.3 V
2.5 V	3.3 V	3.3 V
1.8 V	2.5 V	2.5 V
3.3 V	2.5 V	2.5 V

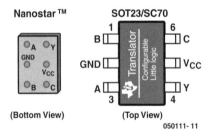

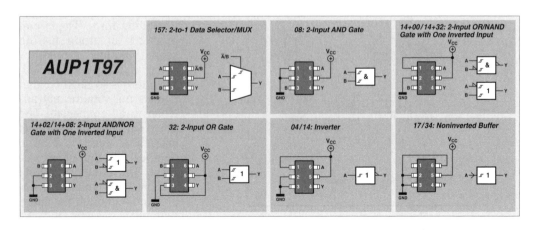

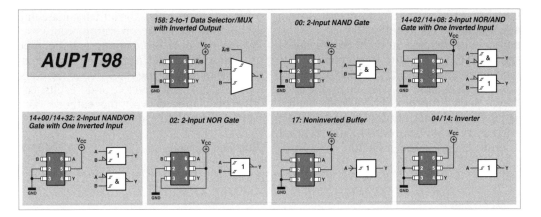

An example of a level translator could be one converting from 1.8 V LVCMOS to 3.3 V LVTTL or LVCMOS. input transitions and noisy signals. The devices, says TI, are tolerant to slow input transitions and noisy signals.

Dirk Gehrke

(050111-1)

201
DRM Direct Mixer Using an EF95/6AK5

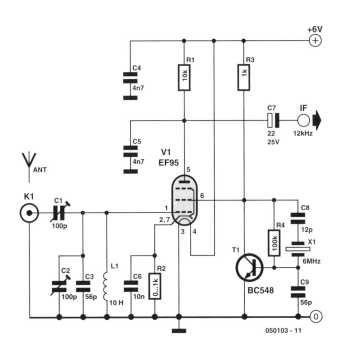

050103 - 11

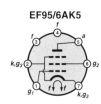

EF95/6AK5

This hybrid DRM receiver with a single valve and a single transistor features good large-signal stability. The EP95 (US equivalent: 6AK5) acts as a mixer, with the oscillator signal being injected via the screen grid. The crystal oscillator is built around a single transistor. The entire circuit operates from a 6 V supply. The receiver achieves a signal-to-noise ratio of up to 24 dB for DRM signals. That means the valve can hold its own against an NE612 IC mixer. The component values shown in the schematic have been selected for the RTL2 DRM channel at 5990 kHz. That allows an inexpensive 6 MHz crystal to be used. The input circuit is built using a fixed inductor. Two trim-

mer capacitors allow the antenna matching to be optimised. The operating point is set by the value of the cathode resistor. The grid bias and input impedance can be increased by increasing the value of the cathode resistor. However, good results can also be achieved with the cathode connected directly to ground.

Burkhard Kainka   (050103-1)

202
Digital VU Meter

The circuit presented here has the same functionality as the renowned LM3914, an LED driver that can control 10 LEDs. The circuit shown here can even drive 12 of them, with only 4 outputs. A necessary requirement is that low-current LEDs are used. The audio signal is amplified by (half of) an LM358 opamp. P1 allows adjust-

ment of the gain and therefore the sensitivity. One input of the opamp is connected to ground so that only the positive half of the input signal is amplified and is effectively rectified at the same time. The combination of R2/C4 then performs an averaging function after which the signal is converted to a digital value by the 8-bit ADC in the

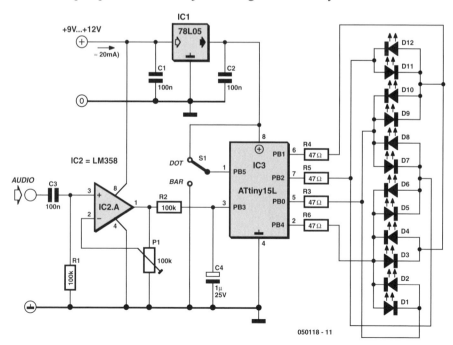

050118 - 11

Attiny15L microcontroller. Subsequently, the 12 LEDs are then controlled through a method called 'charlieplexing'. Switch S1 selects between dot and bar mode. The power supply is regulated with the usual 7805.

The maximum current consumption amounts to 20 mA at 13.8 V. Power supply voltages of up to 30 V are not a problem, since the LM358 can deal with a maximum of 33 V at least (a 7805 can typically deal with 35 V). The minimum voltage at which the circuit will operate is around 7 V. The software can be downloaded in the form of a hex file from our website at www.elektor-electronics.co.uk, The file number is 050118-1.zip. After the code has been programmed into the ATtiny, it is necessary the disable the external reset with the appropriate fuse, since the reset pin in this circuit is used as an output. Those among you who would like to know more on the subject of charlieplexing are referred to www.maxim-ic.com.

Robin van Arem   (050118-1)

The chip is also available ready-programmed as no. 050118-41.

203
Short-Wave Regenerative Receiver for AM and DRM

Is it possible to make a short-wave regenerative valve receiver so stable that it is even suitable for DRM? And is it possible to do this using a 6 V supply, so only a single voltage is necessary for the filament and anode supply? It looks like it might be possible with an EL95, which has good transconductance even at low anode voltages, although it is actually an output-stage pentode instead of an RF valve. Besides that, it draws only a modest 200 mA of heater current. Everything can be operated from a small battery, which eliminates any problems with 50 Hz hum. The stability depends entirely on the tuned circuit. Consequently, a robust coil with 20 turns of 1.5 mm wire was wound on a PVC pipe with a diameter of 18 mm. With short leads to the air-dielectric variable

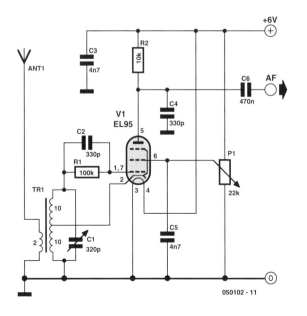

050102 - 11

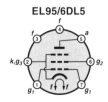

EL95/6DL5

capacitor, this yields an unloaded Q factor considerably larger than 300. The schematic shows a regenerative receiver with feedback via the cathode. The amount of feedback is adjusted using the screen grid voltage. The audio signal across the anode resistor is coupled out via a capacitor. No additional gain is necessary, since the voltage level is sufficient for a direct connection to the Line input of a PC sound card. A screened cable should be used for this connection.

A two-turn antenna coil is located at the bottom end of the tuned circuit. It provides very loose coupling to the antenna, which is important for good stability. Now it's time to see how it works in practice. Despite the open construction, the frequency

drift is less than 1 Hz per minute. That's the way it should be if you want to receive DRM. Quite strong feedback should be used, so the regenerative circuit acts like a direct mixer or a self-oscillating mixer stage. Every strong DRM signal could be seen using DREAM and tuned to 12 kHz. A total of six different DRM frequencies could be received in the 40 m and 41 m bands. If no good DRM stations are available, the receiver can also pick up AM transmitters. In this case, the amount of feedback should be reduced. The PC can be set aside for AM reception, since all you need is a direct connection to an active PC speaker.

Burkhard Kainka   (050102-1)

204
Modifying Stripboard for SO Packages

This technique for modifying standard predrilled prototyping stripboard ('perfboard' or veroboard') to accept SO IC packages is intended to be used for inex-pensive breadboard constructions. It provides an economical alternative to using special circuit boards that accept SO packages and increase the lead spacing

from 1.27 mm to the 2.54 mm grid spacing. The standard dimensions of SO and plastic DIP packages are shown for comparison in Figure 1 (with metric dimen-sions in parentheses).

The grid spacing of predrilled stripboard is 2.54 mm. Predrilled stripboard is available in various sizes with glass–epoxy or pa-

1

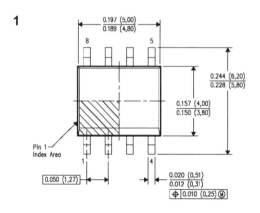

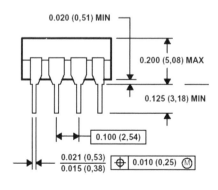

per–phenolic substrate material. It is commonly sold in Eurocard format with dimensions of 100 × 160 mm To allow an IC in a SO package to be soldered to the predrilled stripboard, a utility knife is used to separate a section of a strip down the middle through the holes, thus reducing the grid spacing from 2.54 mm to 1.27 mm. After being prepared in this manner, the board can accept SO packages. Figure 2 shows a diagram of simple example application using a TL5001 in a SO package. Here the green points and lines

2

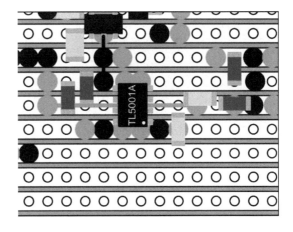

represent cuts that must be made using a strip cutter and/or a utility knife.

Dirk Gehrke   (050119-1)

205
Medium-Wave Modulator

If you insist on using a valve radio and listening to medium-wave stations, you have a problem: the existing broadcasters have only a limited number of records. Here there's only one remedy, which is to build your own medium-wave transmitter. After that, you can play your own CDs via the radio. The transmitter frequency is stabilised using a 976 kHz ceramic resonator taken from a TV remote control unit. Fine tuning is provided by the trimmer capacitor. If there's another station in the background, which will probably be weak, you can tune it to a heterodyne null, such as 981 kHz. As an operator of a medium-wave transmitter,

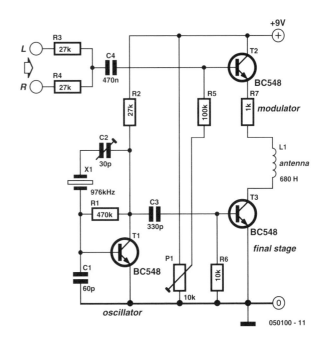

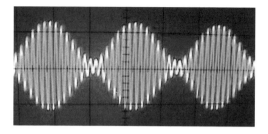

that's your obligation with respect to the frequency allocations. And that's despite the fact that the range of the transmitter is quite modest. The small ferrite coil in the transmitter couples directly into the ferrite rod antenna in the radio. The modulator is designed as an emitter follower that modulates the supply voltage of the output amplifier. As the medium-wave band is still mono, the two input channels are merged. The potentiometer can be adjusted to obtain the least distortion and the best sound. The RF amplifier stage has intentionally been kept modest to prevent any undesired radiation. The quality of the output signal can also be checked using an oscilloscope. Clean amplitude modulation should be clearly visible. The medium-wave modulator can simply be placed on top of the radio. A signal from a CD player or other source can be fed in via a cable. Now you have a new, strong station on the radio in the medium-wave band, which is distinguished by good sound quality and the fact that it always plays what you want to hear.

Burkhard Kainka (050100-1)

206
Simple Short-Circuit Detection

This circuit is suitable in every situation where over-current protection is required. Here we give an example from the model train world. Every seasoned model train enthusiast knows that there is nothing worse than having to find the cause of a short-circuit. On a small model railway with one locomotive it is obviously fairly easy, but on large layouts all locomotives stand still when there is a short and then you have to check each one in turn to find the culprit. If the track is divided into sections then we can use this super simple circuit to make our lives a lot easier. A multifuse is inserted into one of the supply lines for each of the sections. (A multifuse is also called a multiswitch, polyfuse or polyswitch, depending on the manufacturer). This is a type of fuse that cools down and conducts normally again once the short has been removed. The advantage is that only the section with the short becomes isolated. All the other locomotives in the other sections continue to move. The stationary locomotive is in principle the

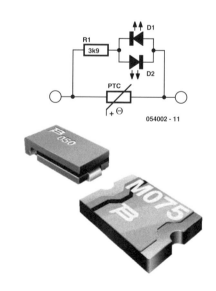

054002 - 11

Model	V_max (V)	I_hold	I_trip	Initial resistance (Ω)		Max. time to trip	
				Min.	Max.	I (A)	t (s)
MF-R005	60	0.05	0.10	7.3	11.1	0.5	5.0
MF-R010	60	0.10	0.20	2.50	4.50	0.5	4.0
MF-R017	60	0.17	0.34	2.00	3.20	0.85	3.0
MF-R020	60	0.20	0.40	1.50	2.84	1.0	2.2
MF-R025	60	0.25	0.50	1.00	1.95	1.25	2.5
MF-R030	60	0.30	0.60	0.76	1.36	1.5	3.0
MF-R040	60	0.40	0.80	0.52	0.86	2.0	3.8
MF-R050	60	0.50	1.00	0.41	0.77	2.5	4.0
MF-R065	60	0.65	1.30	0.27	0.48	3.25	5.3
MF-R075	60	0.75	1.50	0.18	0.40	3.75	6.3
MF-R090	60	0.90	1.80	0.14	0.31	4.5	7.2
MF-R090-0-9	30	0.90	1.80	0.07	0.12	4.5	5.9
MF-R110	30	1.10	2.20	0.10	0.18	5.5	6.6
MF-R135	30	1.35	2.70	0.065	0.115	6.75	7.3
MF-R160	30	1.60	3.20	0.055	0.105	8.0	8.0
MF-R185	30	1.85	3.70	0.040	0.07	9.25	8.7
MF-R250	30	2.50	5.00	0.025	0.048	12.5	10.3

culprit, but it's quite likely that several locomotives aren't moving since not all of them would be travelling in the first place. For this reason we connect an LED indicator across each multifuse, making it clear which section caused the problem. You can choose any colour LED, but we recommend that you use low-current types that emit a lot of light at only a few mA. The value of the current limiting resistor may be changed to give an acceptable LED brightness. As long as the current is small, the resistance of the multifuse is also low and there will barely be a voltage drop. At high currents the resistance increases, which causes a voltage drop across the multifuse that is large enough to light up the LED. As we don't know the direction of the current flow (the train could be moving either forwards or backwards and digital controls use an alternating current) we connected two LEDs in parallel with opposite polarities. Multifuses are available for many different trip currents. Choose a value that is slightly higher than the maximum current consumption of a locomotive in a section. The table below shows the characteristics of several types from the MF-R series made by Bourns. (Raychem is another well-known manufacturer of polyswitches.) Ihold is the current at which the multifuse still conducts normally, Itrip is the short-circuit current.

Karel Walraven (054002-1)

207
DRM Double Superhet Receiver

Using an EF95/6AK5

This receiver arose from the desire to demonstrate that valves are fully capable of holding their own against modern semi-conductor devices. Valves often have better large-signal characteristics and less noise. The decisive difference is that the circuit must be designed with higher input and output impedances. This circuit is built using four EF95 (US equivalent: 6AK5) valves, since this type of valve is small and has proved to operate well with low anode voltages. All four heaters are connected in series and operated from a 24 V supply. That makes it attractive to use the same

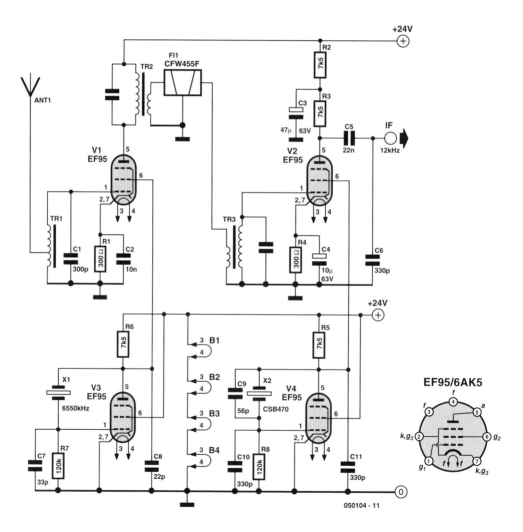

050104 - 11

supply for the anode voltage. The achievable gain is fully adequate. The receiver is designed for the RTL DRM channel at 6095 kHz. It consists of two mixer stages with two crystal oscillators. A steep-edged ceramic filter (type CFW455F) with a bandwidth of 12 kHz provides good IF selectivity. Thanks to the high-impedance design of the circuit, the valves achieve a good overall gain level. The receiver performance is comparable to that of the Elektor Electronics DRM receiver design published in the March 2004 issue, and it can even surpass the performance of the latter circuit with a short antenna, since the tuned input circuit allows better matching to be obtained.

The key difficulty is obtaining a suitable crystal with a frequency of 6550 kHz. Old-style FT243 crystals with exactly this frequency are available from American army radio units. However, it takes a bit of luck to obtain a crystal with exactly the right frequency. It's also possible to use a standard crystal with a frequency of 6553.6 kHz, which yields an IF that is 3.6 kHz too high. However, that shouldn't be a problem if a relatively wide-bandwidth ceramic filter is used. One possibil-ity is the CFW455C, which has a bandwidth of 25 kHz. If the frequency of the second crystal oscillator remains unchanged, the DRM baseband signal will appear at around 9 kHz, approximately 3 kHz below the nominal value. Nevertheless, the signal can easily be decoded by the DREAM software, since it does not depend on the signal being at 12 kHz. Another possibility would be to use the programmable crystal oscillator design published in the March 2005 issue of Elektor Electronics.

Burkhard Kainka   (050104-1)

208
Virtual Prototyping Board

It is often very useful to test a small circuit before designing the printed circuit board. Often a small piece of prototyping board, also known as breadboard, perfboard or veroboard, is used for this. An alternative approach is to use simulation software. This is usually faster and more convenient. If the circuit doesn't work as expected, then such software makes it very easy to try something different. Most of the software available for this purpose is pretty ex-

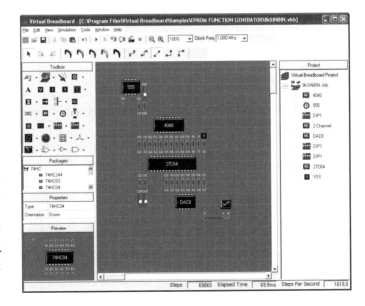

pensive. However, there are fortunately also a number of freeware programs available. On of those is 'Virtual Breadboard'. This program can be downloaded from www.muvium.com. This software allows for the simulation of digital circuits. A library with a number of standard parts is also provided. A small disadvantage is that the ICs are shown as they appear in reality, instead of a schematic block that indicates which functions each IC has. What is very practical however is that the simulator can also simulate PIC-microcontrollers and the

BASIC-stamp. This allows not only the hardware to be tested, but the software as well. Reprogramming is done in no time. You only need to load the new HEX file and can test immediately. This program is not as extensive as the expensive, professional software packages, of course, but if you would like to test some small circuits (possibly including a PIC microcontroller or BASIC Stamp) then it is certainly worth it to check this program out!

Paul Goossens   (054003-1)

209
Short-Wave Converter

This short-wave converter, which doesn't have a single coil requiring alignment, is intended to enable simple medium-wave receivers to be used to listen to short-wave signals. The converter transforms the 49 m short-wave band to the medium-wave frequency of 1.6 MHz. At the upper end of the medium-wave band, select an unoccupied frequency that you want to use for listening to the converted short-wave signals. Good reception performance can be obtained using a wire antenna with a length of one to two metres. The converter contains a free-running oscillator with a frequency of around 4.4 MHz, which is tuned using two LEDs (which act as variable-capacitance diodes!) and a normal potentiometer. The frequency range is set by adjusting the emitter current using a 1 kΩ trimpot. The oscillator frequency depends strongly on the operating point. This is

due to the combination of using an audio transistor and the extremely low supply voltage. Under these conditions, the transistor capacitances are relatively large and strongly dependent on the operating point. The second transistor forms the mixer stage. If you calculate the resonant frequencies of the tuned circuits, you will ob-

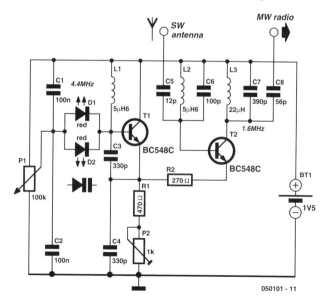

050101 - 11

tain 6.7 MHz for the antenna circuit and 1.7 MHz for the output circuit. Additional transistor capacitances and the effects of the coupling capacitors shift each of the resonant frequencies down-ward. The tuned circuits are relatively heavily damped to obtain bandwidths that are large enough to allow the circuit to be used with-out any specific alignment. The results are good despite the low collector–emitter voltage of around only 0.6 V, due to the fact that only a modest amount of mixer gain is necessary. The entire circuit also draws less than 1 mA.

Burkhard Kainka (050101-1)

210
Toslink Repeater/Splitter

The purpose of this circuit is not only to extend an optical digital audio connection (Toshiba Toslink) but also to function as a splitter, so that two devices can be connected to one Toslink output. The circuit consists of the standard application for the Toslink receiver module TORX173 (IC1) and the Toslink transmitter module TOTX173 (IC2 and IC3), that we have used on previous occasions. Inverters from a 74HCU04 are connected between the input and output and act as pulse restorers. To keep the quality as high as possible, each of the transmitter modules is driven

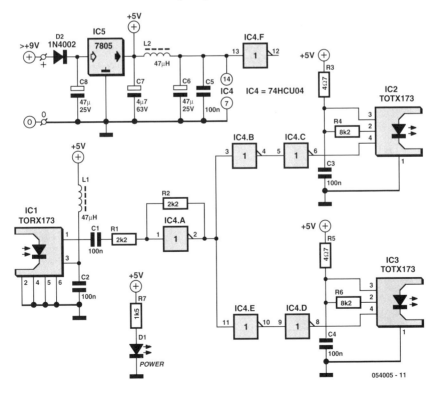

054005 - 11

COMPONENTS LIST

Resistors:
R1,R2 = 2kΩ2
R3,R5 = 4Ω7
R4,R6 = 8kΩ2
R7 = 1kΩ5

Capacitors:
C1-C5 = 100nF
C6,C8 = 47µF 25V radial
C7 = 4µF7 63V radial

Inductors:
L1,L2 = 47µH

Semiconductors:
D1 = low-current LED, red
D2 = 1N4002
IC1 = TORX173
IC2,IC3 = TOTX173
IC4 = 74HCU04
IC5 = 7805
PCB, ref. 054005-1 from The PCBShop

individually. The first inverter is AC-coupled and ensures that the received S/PDIF signal is centred on the threshold of the inverter. The other inverters (except one) amplify the signal to the maximum output voltage before presenting it to the transmitter modules. For the power supply a standard 7805 ply voltage of IC4 some more. This does was selected with D2 as reverse polarity L1 for IC1 and R3 and R5 for IC2 and protection. L2 decouples the power sup-IC3 respectively. R4 and R6 are required for the internal adjustment of the transmitter modules. The required power supply voltage amounts to a minimum of 9 V and can be provided by a mains

adapter. A panel mount socket can be fitted in the enclosure (that you can choose yourself) to suit the plug on the mains adapter. A printed circuit board has been designed for the circuit that contains all the components. Two pins are provided to connect the power supply. LED D1 indicates that the power supply is present. The current consumption without optical signal is

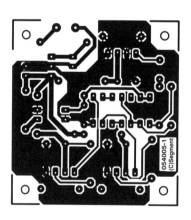

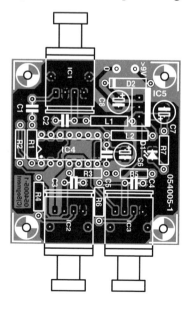

103 mA. When the cables are connected and with a frequency of 48 kHz, the current is 70 mA.

This is because the transmitter modules will now turn the internal LEDs on and off (when there are no cables connected, the LEDs are on continuously). At 96 kHz the current is about 3 or 4 mA higher. An FFT (fast fourier transform) analysis shows that the noise floor with a 16-bit PCM signal is about 40 to 45 dB higher compared to a 24-bit PCM signal.

Ton Giesberts   (054005-1)

211
Servo Points Actuator

Servos for model making can also be used for completely different purposes besides model airplanes, boats and cars. As an example, here a servo is used to operate a set of points for a model railway system. The points are directly actuated by a length of steel wire connected to the servo arm, and they move quietly and smoothly, so the illusion is no longer spoiled by that irritating click-clack sound. The voltage on the point tongues and nose is also switched over by the servo. As an example for DIY construction, we have designed two small PCBs that can be attached to a servo with a bit of glue. The larger board should be glued to the case of the servo, while the smaller (round) board should be glued to the arm. Several contact strips made from phosphor bronze must be soldered to the larger board. Phosphor bronze is a copper alloy with good spring characteristics. Almost all relay contacts arms are made from this material. It can be obtained from Conrad,

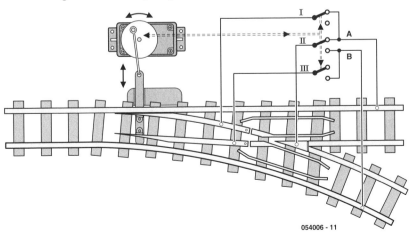

054006 - 11

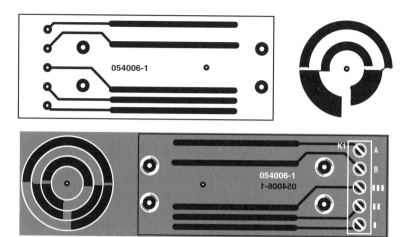

among other sources, in the form of small sheets that can be cut or sawn into strips. Any desired switching scheme can be created by devising a suitable pattern of copper tracks. The actuator works as follows: in the straight-through state, one tongue and the nose are connected to rail B. When the points are moving, both tongues and

the nose are floating, since there is a gap between the segments with no copper. Just before the points reach the turn-out state, the other tongue and the nose pick up the voltage from rail A. The PCB shown here is available through The PCBShop.

Karel Walraven

(054006-1)

212
Comparing Signed Integers

Every once in a while it is necessary to compare two signed integers with each other. Unfortunately, some programming languages do not support signed integers. This problem presented itself with a design in Verilog. This language has a direct method of comparing two unsigned integers. With comparing we mean determining whether integer A is more than or less then integer B, or equal. After some thought for an efficient solution we found the following: By inverting the MSB (Most Significant Bit) of both signed integers, both can be compared as unsigned integers with the correct result. "How can this be?", you will ask. The solution is simple. The difference

between an unsigned inte-ger and a signed integer is that the MSB of an unsigned integer has a value of 2n, while that same MSB of a signed integer has the value –2n. With positive numbers nothing special happens, that means, the value is the same whether they are treated as signed or unsigned. With a negative number (where the MSB=1 and is therefore significant) the value increases by 2*2n (instead of –2n–1 the weight of the MSB becomes 2n–1). By inverting the MSB, 2n–1 is added to both negative and positive numbers. A necessary condition is that the MSB of a signed value is equal to '1' (thus indicating a negative value) and zero for an unsigned value.

Decimal value of bits within an 8-bit integer								
7	6	5	4	3	2	1	0	
−128	64	32	16	8	4	2	1	signed
128	64	32	16	8	4	2	1	unsigned
Example:	10011100 = signed −100 inverting MSB: 00011100 = unsigned 28 00001111 = signed 15 10001111 = signed 143							

In this way the relative difference between the two numbers remains exactly the same. In the example you can see clearly that after the operation the value of each has been increased by exactly 128, provided they are both considered as unsigned integers.

This is independent on whether the original integer was positive or negative. Now both numbers can be compared as unsigned integers with (of course) the correct result!

Paul Goossens

(054004-1)

213
Converting a DCM Motor

We recently bought a train set made by a renowned company and just couldn't resist looking inside the locomotive. Although it did have an electronic decoder, the DCM motor was already available 35 (!) years ago. It is most likely that this motor is used due to financial constraints, because Märklin (as you probably guessed) also has a modern 5-pole motor as part of its range. Incidentally, they have recently introduced a brushless model.

The DCM motor used in our locomotive is still an old-fashioned 3-pole series motor with an electromagnet to provide motive power. The new 5-pole motor has a permanent magnet. We therefore wondered if we couldn't improve the driving characteristics if we powered the field winding separately, using a bridge rectifier and a 27 Ω current limiting resistor. This would effectively create a permanent magnet. The result was that the driving characteristics improved at lower speeds, but the initial acceleration remained the same. But a constant 0.5 A flows through the winding, which seems wasteful of the (limited) track power.

A small circuit can reduce this current to less than half, making this technique more acceptable. The field winding has to be disconnected from the rest (3 wires). A freewheeling diode (D1, Schottky) is then connected across the whole winding. The centre tap of the winding is no longer used. When FET T1 turns on, the current through the winding increases from zero until it reaches about 0.5 A. At this current the voltage drop across R4-R7 becomes greater than the reference voltage across D2 and the opamp will turn off the FET. The current through the winding continues flowing via D1, gradually reducing in strength. When the current has fallen about 10% (due to hysteresis caused by R3), IC1

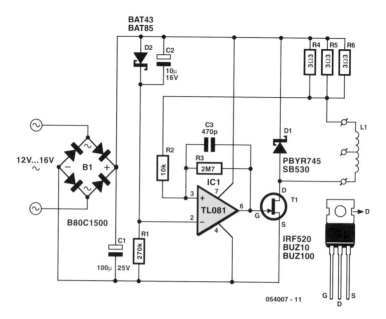

054007 - 11

will turn on T1 again. The current will increase again to 0.5 A and the FET is turned off again. This goes on continuously. The current through the field winding is fairly constant, creating a good imitation of a permanent magnet. The nice thing about this circuit is that the total current consumption is only about 0.2 A, whereas the current flow through the winding is a continuous 0.5 A. We made this modification because we wanted to convert the locomotive for use with a DCC decoder. A new controller is needed in any case, because the polarity on the rotor winding has to be reversed to change its direction of rotation. In the original motor this was done by using the other half of the winding. There is also a good non-electrical alternative: put a permanent magnet in the motor. But we didn't have a suitable magnet, whereas all electronic parts could be picked straight from the spares box.

Karel Walraven   (054007-1)

214
Compact 200 W Output Stage

There is no doubt that this small power amp packs a punch. It is capable of delivering a healthy 200 W into 4 Ω. Into 8 Ω it can still output 125 W (see Figure 2). These large power outputs are made possible through the use of Darlington transistors made by Sanken, the SAP16N and its opposite number, the SAP16P (in our prototype we used their predecessors, the SAP15N and P, because the SAP16 versions were not available at that time). These power transistors have an emitter resistor built in, as well as a diode for temperature compensation. Because of this, the whole emitter follower stage has just two components (and a preset for setting the quiescent current, shown in the circuit in Figure 1). One small disadvantage is that the transistors have to operate at a relatively low quiescent current, according to the datasheet. This causes an increase in distortion and a reduction in bandwidth. The current through the diodes has to be set to 2.5 mA, when the quiescent current will be 40 mA. This has the advantage that the driver transistors (T9, T10) do not need heat sinks, which helps to keep the circuit small. The amplifier is of a standard design and doesn't require much explanation. The input is formed by two differential amplifiers (T1, T2), which are each followed by a

1

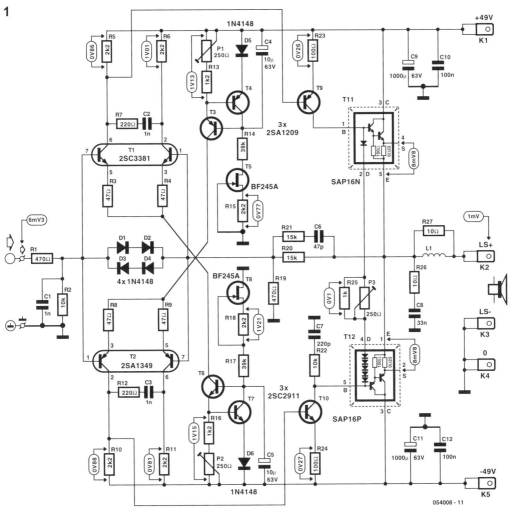

054008 - 11

Specifications

Input sensitivity		1Veff
Input impedance		10 kΩ
Sine-wave power	8Ω	125 W, THD+N = 1 %
	4Ω	200 W, THD+N = 1 %
Bandwidth		135 kHz (1 W/8Ω)
Slew rate		20 V/µs
Signal/noise ratio		101 dB (1 W/8 Ω, 22 Hz to 22 kHz)
		104 dBA
THD+noise		0.014 % at 1 kHz (60 W/8Ω)
Damping factor		>700 (1 kHz)
		>400 (20 kHz)

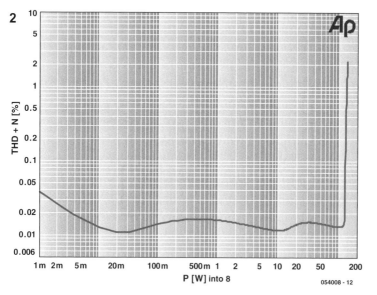

2

054008 - 12

buffer transistor (T9, T10). T9 and T10 together make a push-pull stage that drives the output transistors. For T1 and T2 we've used special complementary dual transistors made by Toshiba. These, along with the driver transistors, have been used previously in the High-End Power Amp in the March 2005 issue. The driver transistors are a complementary pair made by Sanyo, which have been designed specifically for these applications. Compensation in the amplifier is provided by R7/C2, R12/C3, R21/C6, R22/C7 and R26/C8. The dual transistors are protected by D1 to D4. The output inductor consists of 8 turns of 1.5 mm diameter enamelled copper wire (ECW). Since the current through the diodes is just 2.5 mA, the operating point of

T9 and T10 has to be set precisely. This operating point is determined purely by the operating point of the differential input amplifiers. Since the ambient temperature affects the operating point, any potential drift in the operating point of T9 and T10 is compensated for by the current sources of the differential amplifiers. The voltage drop across D5 (D6) and the base-emitter voltage of T4 (T7) determine the current through P1 (P2) and R13 (R16). T4 (T7) controls the voltage at the base of T3 (T6) and creates a constant current that is independent from the supply voltage.

Since the voltage across D5 (D6) and T4 (T7) depends on the temperature, the voltage at the base of T10 (T9) has been temperature-compensated as well as possible. T3 and T4 (T6 and T7) are fed by a simple constant current source built around JFET T5 (T8), which makes the differential amplifier around T2 (T1) even less dependent on the supply voltage. R14 (R17) restricts the maximum voltage across T5 (T8), which may not exceed 30 V. According to the datasheet the JFET current should be about 0.5 mA, but in practice a deviation of up to 50% is possible. The actual value is not critical, but the voltage across the JFET must always stay below the maximum value (also take into account any possible mains voltage variations). If you do need to reduce the voltage, that should be done by lowering the value of R15 (R18). This causes the voltage across R14 (R17) to increase and hence the

3

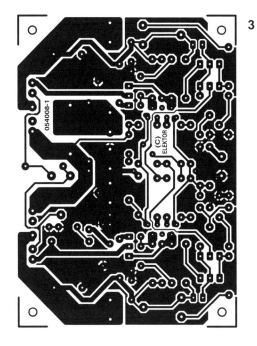

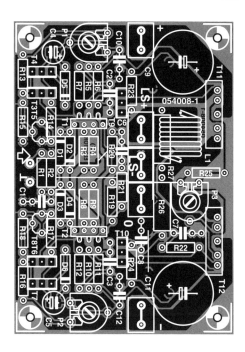

COMPONENTS LIST

Resistors:
R1, R19 = 470Ω
R2, R22 = 10kΩ
R3, R4, R8, R9 = 47Ω
R5, R6, R10, R11, R15, R18 = 2kΩ2
R7, R12 = 220Ω
R13, R16 = 1kΩ2
R14, R17 = 39kΩ
R20, R21 = 15kΩ
R23, R24 = 100Ω
R25 = 1kΩ
R26, R27 = 10Ω 1W
P1,P2,P3 = 250Ω preset

Capacitors:
C1,C2,C3 = 1nF
C4, C5 = 10µF 63V radial
C6 = 47pF
C7 = 220pF
C8 = 33nF
C9, C11 = 1000µF 63V radial
C10, C12 = 100nF

Inductors:
L1 = 8 turns 1.5mm dia. ECW, inside diameter 10mm.

Semiconductors:
D1-D6 = 1N4148
T1 = 2SC3381 (Toshiba) (Huijzer; Segor
Conrad Electronics)
T2 = 2SA1349 (Toshiba) (Huijzer; Segor
Electronics)
T3,T4,T9 = 2SA1209 (Sanyo) (Farnell #
410-3841)
T5, T8 = BF245A
T6, T7, T10 = 2SC2911 (Sanyo) (Farnell #
410-3853)
T11 = SAP16N (Sanken) or SAP15N (Farnell
# 410-3749)
T12 = SAP16P (Sanken) or SAP15P (Farnell
# 410-3750)

Miscellaneous:
K1-K5 = 2-way spade terminal, PCB mount,
vertical
Heatsink <0.5 K/W
Mica washers for T11 en T12, e.g., Electron-
ics # 189049.
4 wire links on PCB
PCB, ref. 054008-1 from The PCBShop

voltage across the JFET will drop. P1 and P2 are required to compensate for various tolerances. With the input open circuit you should set the output to zero, while keeping the current through T9 and T10 as close as possible to 2.5 mA. This can be measured across R23 and R24. It is not a problem if the current is a few tenths of a mA more than this. The quiescent current is set by P3. In the reference design a 200 Ω preset is used. We have put this together using a (standard) preset of 250 Ω in parallel with a 1kΩ resistor. An incidental advantage of this parallel resistor is that it limits any pos-sible current spikes when the wiper of the potentiometer makes a bad contact during the adjustment of the quiescent current. This amplifier provides a good opportunity to experiment with the Sanken tran-sistors. If you want to use the output stage in a complete power amplifier (refer to the print layout in Figure 3), you will need to add an input decoupling capacitor, a power-on delay with a relay for the loudspeaker and a beefy power supply. The input decoupling capacitor is certainly a necessity, since the offset is determined by the various tolerances and differences between the complementary transistors. In our prototype the input offset was 6.3 mV for a 0 V output voltage. This is amplified by a factor of 33, which would result in an output offset of over 200 mV if the input was shorted by, for example, a volume control. Elsewhere in this issue is a design for a small board, which contains an input decoupling capacitor (MKT or MKP) and a relay with a power-on delay.

Ton Giesberts   (054008-1)

215
Low-drop Regulator with Indicator

Even today much logic is still powered from 5 volts and it then seems obvious to power the circuit using a standard regulator from a rectangular 9 V battery. A disadvantage of this approach is that the capacity of a 9 V battery is rather low and the price is rather high. Even the NiMH revolution, which has resulted in considerably higher capacities of (pen-light) batteries, seems to have escaped the 9 V battery generation. It would be cheaper if 5 volts could be derived from 6 volts, for example. That would be 4 'normal' cells or 5 NiMH-cells. Also the 'old fashioned' sealed lead-acid battery would be appropriate, or two lithium cells. Using an LP2951, such a power supply is easily realised. The LP2951 is an evergreen from National Semiconductor, which you will have encountered in numerous Elektor Electronics designs already. This IC can deliver a maximum current of 100 mA at an input voltage of greater than 5.4 V. In addition to this particular version, there are also versions available for 3.3 and 3 V output, as well as an adjustable version. In this design we have added a battery indicator, which also protects the battery from too deep a discharge. As soon as the IC has a problem with too low an input voltage, the ERROR output will go low and the regulator is turned off via IC2d, until a manual restart is provided with the RESET-pushbutton. The battery voltage is divided with a few resistors and compared with the reference voltage (1.23 V) of the regulator IC. To adapt the indicator for different voltages you only need to change the 100 k resistor. The comparator is an LP339. This is an energy-friendly version of the LM339. The LP339 consumes only 60 µA and can sink 30 mA at its output. You can also use the

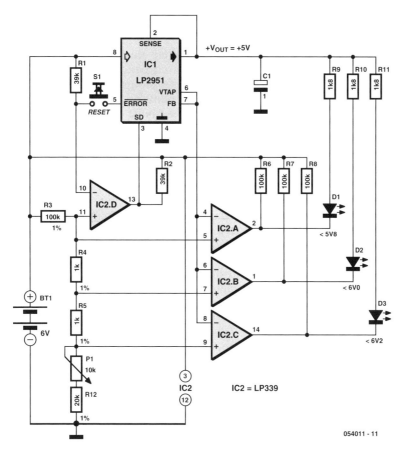

054011 - 11

LM339, if you happen to have one around, but the current consumption in that case is 14 times higher (which, for that matter, is still less than 1 mA). Finally, the LP2951 in the idle state, consumes about 100 μA and depending on the output current to be delivered, a little more.

Karel Walraven (054011-1)
(National Semiconductor application note)

216
Universal I/O for Power Amps

This two-part circuit was designed as an addition to the 'Compact 200 W Output Stage' in this issue. But it is also suitable for use with IC amplifiers that don't have their own power-on delay. The circuit consists of an input stage (nothing more than a resistor and capacitor) and a power-on delay with a relay for the amplifier output. A small PCB has been designed for the input signal. This board contains a phono socket, capacitor and a resistor that is connected directly across the input. This resistor

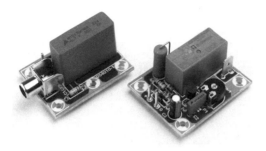

keeps the input side of the capacitor at ground level. It prevents any offset voltage at the amplifier input from appearing at the input of this circuit when there is no connection. This would otherwise cause a loud bang from the loudspeaker when a connection was made. We have taken account of the larger size of MKP and MKT capacitors. There are several mounting holes on the board to cater for the various sizes of capacitor. The maximum size is $18 \times 27.5 \times 31.5$ mm (WxHxL). To protect a loudspeaker from a DC off-set during the switch-on period a relay needs to be connected in series that turns on after a short delay. This circuit also makes use of the

falling supply voltage of the amplifier when it is switched off. This concept also makes this circuit suitable to protect against overloads. The circuit itself is relatively straightforward. MOSFET T1 turns on relay RE1 when the gate-source voltage rises above 2.5 V. The gate voltage is derived directly from the amplifier supply via potential divider R3/R4/P1. P1 is used to adjust the exact level at which the relay turns on. This is also used to compensate

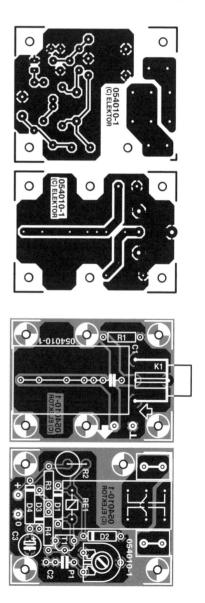

COMPONENTS LIST

Resistors:
R1 = 270kΩ
R2 = 1kΩ2
R3,R4 = 1MΩ
P1 = 250kΩ preset

Capacitors:
C1 = 4µF7 MKT or MKP (see text)
C2 = 47nF
C3 = 10µF 63V radial

Semiconductors:
D1,D3 = 1N4148
D2 = zener diode 24V 1.3W
D4 = 1N4002
T1 = BS170

Miscellaneous:
K1 = cinch socket, PCB mount, e.g.
Monacor / Monarch T-709G
K2, K3 = spade terminal, PCB mount, verti-
cal, 2 pins
RE1 = PCB relay 24 V/16 A (e.g., Omron
G2R-1-24, 1100 Ω, or Schrack RT314024,
1440 Ω)
PCB, ref. 054010-1 from The PCBShop

for the tolerance in the MOSFET threshold voltage. The relay operating voltage is limited to about 24 V by resistor R2. D2 restricts the voltage across T1 from rising above 24 V, stopping this voltage from rising too much when the relay is not actuated

It also keeps the voltage at 24 V when a relay with a higher coil resistance is used (see parts list). When a relay with a lower coil resistance is used, the value of R2 needs to be adjusted accordingly. To calculate the value for R2 you should take the minimum acceptable supply voltage, subtract 24 V and divide the result by the current through the relay. This value should then be rounded to a value in the E12 series. There is room on the PCB for a vertically mounted 5W resistor for R2, so it's even possible to use a 12 V relay (D2 should then be changed to a 12 V type as well). D1 protects T1 from induced voltages when the relay is turned off. The voltage at the gate only reaches its end value slowly due to the addition of D3 and C3. D3 prevents C3 from keeping the MOSFET conducting when the voltage drops. When the voltage has fallen below half the normal supply voltage, C3 starts to discharge via D4. In this way the switch-on delay is at a maximum even when the power is turned off and on repeatedly. It should be clear that this circuit can only be used with amplifiers that are stable until the threshold voltage is reached and hence do not create an offset at the output before the relay turns off.

Ton Giesberts   (054010-1)

217
Remote Control Extension using RF

Transmitter

We have, over the years, published numerous variations of remote control extenders in Elektor Electronics, but not yet one using RF. These days, transmit and receive modules that operate on the well-known 433 MHz licence-free frequency are reasonably cheap and freely available at elec-

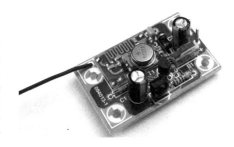

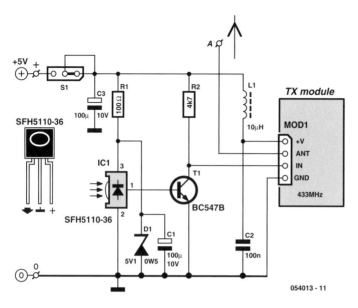

054013 - 11

system from Philips is being used. It is now a little dated but still has many applications, particularly for your own designs. The minimum pulse width of RC5 amounts to 0.89 ms. The maximum frequency is therefore 562 Hz $[1/(2 \cdot 0.89 \cdot 10^{-3})]$. This still passes reasonably well through the RF link. However, at the receiver some pulse stretching is required. The transmitter is simplicity itself. IC1 is an IR receiver for remote control systems. The output signal is active low. The pulses at the output, with RC5, have a minimum length of 0.89 ms. The transmitter is activated with an active high signal that is supplied by T1. When IC1 receives a pulse, T1 will leave conduction and the transmitter is turned on via R2. The little transmitter module has 4 connections: ground, power (3 to 12 V), data input and antenna output. The transmitter module is mounted on our PCB and connected with four pieces of wire. The overlay makes incorrect assembly just about impossible. You can use a piece of wire of about 15.5 cm long ($1/4 \lambda$) as the antenna. Current consumption in the idle state is about 4 mA and about 5.3 mA while transmitting. For the protection of IC1, a 5.1 V zener diode is connected in parallel with its power supply connections. An excessive power supply voltage is then turned into heat via decoupling resistor R1. That means that a power supply voltage of up to 10 V can be connected to the circuit without any problems (with a power rating of at least 0.25 W for R4). A disadvantage is that the circuit draws a little less than 50 mA additional current. You can increase the value of R1, of course, when the power supply voltage is always at the high end of the range. The transmitter module can deal with 3 to 12 V and that limits the maximum supply volt-

tronics stores. The circuit described here makes use of the transmitter/receiver combination available from Conrad Electronics, which stands out because of its low price. A disadvantage with this setup is the available bandwidth. At 2 kHz this is quite limited, but still sufficient for our purpose. We assume that the RC5 remote control

COMPONENTS LIST

Resistors:
R1 = 100Ω
R2 = 4kΩ7

Capacitors:
C1,C3 = 100µF 10V radial
C2 = 100nF ceramic

Inductors:
L1 = 10µH

Semiconductors:
D1 = zener diode 5V1 0.5W
T1 = BC547B
IC1 = IC1 = SFH5110-36

Miscellaneous:
S1 = miniature slide switch SX254 (Hartmann) (Conrad Electronics # 708062) or 3-way pinheader with jumper
MOD1 = 433.92 MHz AM transmitter module from Conrad Electronics set, # 130428
PCB, ref. 054013-1 from The PCBShop

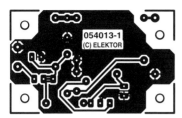

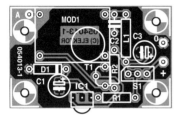

age to 12 V. According to the datasheet, the current consumption of IC1 is 5 mA. In our prototypes the current consumption was actually lower, less than 3 mA. R1 is then roughly $(U-5.1)/3.5 \cdot 10^{-3}$. However, just to be sure, measure the voltage across the

IR receiver and check that it is correct. You can use other devices for IC1, but keep in mind the current consumption and the pinout of the terminals. Another requirement is that the alternative IR-receiver has an active low output and an internal pull-up resistor. C1, L1 and C2 provide additional decoupling and C3 decouples the common power supply. In place of S1 you could just fit a wire link or a 3-way pin header with a jumper. The switch is really only useful when the circuit is powered from batteries. It is likely that the circuit will be placed in a fixed location and a regulated mains adapter as a power supply is more appropriate. On the PCB there is enough space so that IC1 can be fitted horizontally. The electrolytic capacitors can also be placed horizontally, so that the entire assembly can be quite thin, so that it can easily be mounted between, behind or in something.

Ton Giesberts (054013-1)

218
Extension for LiPo Charger

The 'Simple LiPo Charger' published in Elektor Electronics April 2005 is a small and handy circuit that allows you to quickly charge two or three LiPo cells. Especially in the model construction world are LiPo batteries used a lot these days, particularly model aeroplanes. It is usual to use a series connection of three cells with these models. Since working with these model aeroplanes usually happens in the field, it would be nice if the batteries could be charged from a car battery. We therefore designed a voltage converter for the

LiPo charger concerned, which makes it possible to charge three cells in series. The voltage per cell increases while charging to a value of about 4.2 V, which gives a total

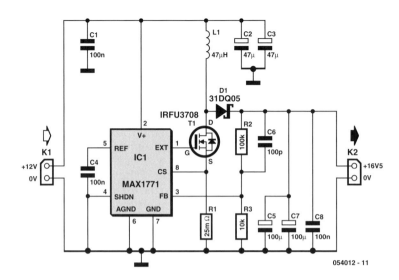

054012 - 11

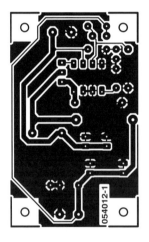

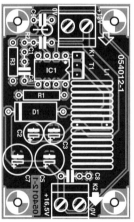

voltage of 12.6 V. The converter, therefore, raises the 12 V voltage from the car battery to 16.5 V, from which the LiPo charger can be powered. A step-up controller type MAX1771 in combination with an external FET carries out the voltage conversion. The IC operates at a moderately high switching frequency of up to 300 kHz, which means that quite a small coil can be

COMPONENTS LIST

Resistors:
R1 = 25mΩ (e.g., Digikey # 2ER025-ND)
R2 = 100kΩ
R3 = 10kΩ

Capacitors:
C1,C4,C8 = 100nF4
C2,C3 = 47μF 25V radial
C5,C7 = 100μF 25V radial
C6 = 100pF

Semiconductors:
D1 = 31DQ05 (e.g., Digikey # 31DQ05-ND)
IC1 = MAX1771-CPA (e.g., Digikey # MAX1771 EPA-ND)
T1 = IRFU3708 (e.g., Digikey # IRFU3708-ND)

Miscellaneous:
K1,K2 = 2-way PCB terminal block, lead pitch 5mm
L1 = 47μH high current suppressor coil, (e.g., Digikey # M9889-ND)
PCB, ref. 054012-1 from The PCBShop

used. Because the IC uses pulse frequency modulation (PFM) it combines the advantages of pulsewidth modulation (high efficiency at high load) with very low internal current consumption (110 μA). The IC is configured here in the so-called non-bootstrapped mode, which means that it is powered from the input voltage (12 V). The output voltage is adjusted with voltage divider R2/R3. This can be set to any required value, provided that the output voltage is greater than the input voltage. Finally, sense resistor R1 determines the maximum output current that the circuit can deliver. With the 25 mΩ value as indicated, this is 2.5 A.

(054012-1)

219
Improved DECT Battery Charger

After buying a number of DECT telephones, we noticed that these became quite warm while charging. That surprised us, because the manufacturer wrote in the manual that the batteries had to be charged for 14 hours. That would lead you to conclude that the batteries are charged at 1/10th of the nominal capacity. But we had the feeling that the batteries were getting rather warm for such a small charging current. That is why we quickly reached for a screwdriver and explored the innards of the charging station. The accompanying schematic reveals what is going on. The batteries are 'simply' charged from a 9 V mains adapter. In series with the output are a diode and a resistor. A quick calculation shows that a current of about 160 mA will flow and this was indeed the case when measured with a multimeter. That means that, for the AAA cells used here, rated at 650 mAh, the charging current isn't 1/10 C, but 1/4 C! This is rather high and certainly not good for the life expectancy of the batteries.

The remedy is simple: increase the value of the 25 Ω series resistor so that the charging current will be less. We chose a value of 68 Ω, resulting in a charging current of about 60 mA. You may ask what the purpose of the remainder of the circuit is. This is all required to turn on the LED, which indicates when charging is taking place.

During charging, there is a voltage drop of about 4 V across R1. T1 will then receive base current via R2 and the LED turns on. The resistor in series with the LED limits the LED current. The fact that the batteries are charged with an unregulated supply is not such a problem. The mains voltage will vary somewhat, but with a maintenance charge such

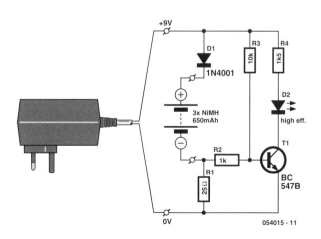

054015 - 11

as in this application it is not necessary to charge with an accurate constant current, provided that the current is not too high. For the curious, here is the calculation for the charging resistor: three cells are being charged.

The charging voltage of a NiMH cell is about 1.4 V. The voltage drop across D1 is about 0.7 V. That leaves a voltage across the resistor of:

$$9_{\text{mains adapter}} - 4.2_{\text{batteries}} - 0.7_{\text{diode}} = 4.1 \text{ V}$$

Therefore, there flows a current of $4.1/25 = 164$ mA.

Karel Walraven   (054015-1)

220
RF Remote ControlExtender

Receiver

This circuit works together with a transmitter described elsewhere in this issue to form a remote-control extender. Due to its limited bandwidth, the output of the RF receiver module does not exactly reproduce the original pulses from the IR receiver. That problem is remedied here by the circuit around T1. The modulator circuit for the data received from this stage is active low, so the primary function of T1 is to invert the received signal. However, the stage built around T1 also acts as a pulse stretcher in order to restore the original pulse length. Capacitor C2, which is connected in parallel with P1, charges quickly. The maximum current is limited by R3 to protect T1. The discharge time is determined by P1 and C2, and P1 can be adjusted to compensate for various deviations and tolerance errors. The timing should be matched to the RC5 code with P1 set to approximately its midrange position. In practice, the pulse length also proved to be slightly dependent on the signal strength. T1 provides the reset signal for a 74HC4060 IC (14-stage binary ripple counter with built-in oscillator). This counter restores the original modulation with a frequency of 36 kHz. Each received pulse starts the oscillator, and a frequency of exactly 36 kHz is present on pin 13 (division factor 29) if the crystal frequency is 18.432 MHz. Other oscillator frequencies can also be selected using J1, and various combinations are possible using other division factors (for 36 kHz, connect a jumper between pins 11 and 12 of J1). At frequencies below 12 MHz, C5 must be replaced by a resistor with a value of 1–2.2 kΩ (fitted vertically). The main reason for including J1 is to allow the remote con-

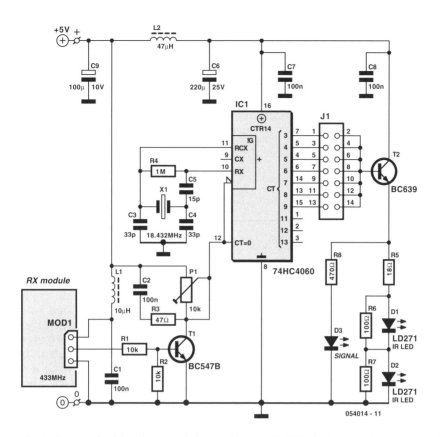

054014 - 11

trol extender to be used with other modulation frequencies. The counter output drives a transistor stage (T2) with two IR diodes (type LD271, but other types can also be used). The current through LEDs D1 and D2 is limited to approximately 90 mA by

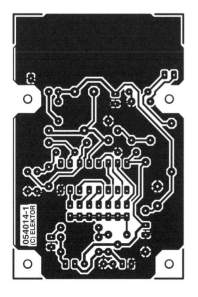

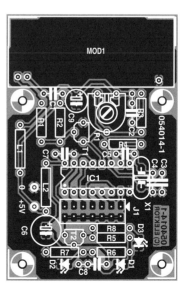

R18. R6 and R7 are connected in parallel with D1 and D2, respectively, to reduce the turn-off time. A normal LED (D3) is connected in parallel with the IR LEDs to provide a supplementary visible indication of a transmitted signal (IR or RF). The circuit needs nice stiff 5 V supply, which means that battery operation is not suitable (unless a hefty electrolytic capacitor is connected across the supply terminals). L1 and L2 ensure that the supply voltage for the receiver module is as clean as possible. The quiescent current consumption is approximately 1.3 mA, and it increases to around 8 mA on average with a good received signal level. Assembling the circuit board (see Figure 2) is straightforward. Naturally, a wire bridge can be used in place of J1 if you only want to use a single IR code. The receiver module is supplied with quite long connection leads and a prefitted antenna. It's a good idea to shorten the antenna slightly. With our modules, the length was nearly 18 cm. In practice, a length of 15.5 cm is more suitable at a frequency of 433.92 MHz. The circuit board is designed to allow the receiver module to be fitted to the board, but we chose to fit it next to the board (along its long axis). That puts the receiver a bit further away from the noise generated by the IR LEDs and the counter, which reduces the interference to signal reception. The module leads are long enough for this. You will also have to experiment with the orientation of the module and the antenna, since the arrangement proved to be critical in practice (the RF aspects, that is).

Ton Giesberts   (054014-1)

COMPONENTS LIST

Resistors:
R1,R2 = 10kΩ
R3 = 47Ω
R4 = 1MΩ
R5 = 18Ω
R6,R7 = 100Ω
R8 = 470Ω
P1 = 10kΩ preset

Capacitors:
C1, C7,C8 = 100nF ceramic
C2 = 100nF MKT
C3,C4 = 33pF
C5 = 15pF
C6 = 220µF 25V radial
C9 = 100µF 10V radial

Inductors:
L1 = 10µH
L2 = 47µH

Semiconductors:
D1,D2 = LD271
D3 = LED, red, high-efficiency
T1 = BC547B
T2 = BC639 IC1 = 74HC4060

Miscellaneous:
J1 = 14 (2x7) way pinheader + 1 jumper
X1 = 18.432MHz quartz crystal
MOD1 = 433.92MHz AM receiver (RX) module from set, Conrad Electronics # 130428
PCB, ref. 054014-1 from The PCBShop

221
FT639 One-Chip Servo Controller

To beginners in electronics, servos have a great appeal, probably because you can do lots of real-life 'things' with them. However, long faces soon appear and pages are turned when a servo control has to be designed and built. Even if that can be done with the good old 555 chip and a handful of parts, the next objection from the new generation is that 'it ain't computer-controlled'. The edeFT639 is an 8-pin chip that will control five servos through one 2400-baud serial line. As you can see from the schematic, the only external components required are a decoupling cap, plus two resistors and a diode to withhold the negative swing of the RS232 line the circuit is connected to.

In case TTL-swing (5 V) serial control is available, such as in PIC or Parallax Stamp systems, the control signal to the edeFT639 may be applied directly to pin 4. The serial data format is a dead simple 2400 bits/s, no parity, 1 stop bit so you can practice away using any terminal emulation or general purpose serial comms utility like HyperTerminal. The edeFT639 will always start in Setup mode, then switch to Active Mode. The chip can set a servo to one of 256 positions from 0 to 90 degrees using a 'short' pulse (1 ms) or from 0 to 180 degrees using a 'long' pulse length. The starting position of each individual servo can also be adjusted by using a different header length. Full details about bit-banging the device over a 2400-baud serial line are given in the 'FT639 Ferret' datasheet, which also contains a simple program (in QBASIC) showing the nitty-gritty of talking to the device using software and the PC's printer port (LPT:).

The 'get-to-know-u' program can be downloaded from the Elab Inc. website (see below) and should not be too difficult to convert into PIC or AVR code. On the same website we also found a Visual BASIC programming example complete with source & executable code to illustrate how the edeFT639 servo control-

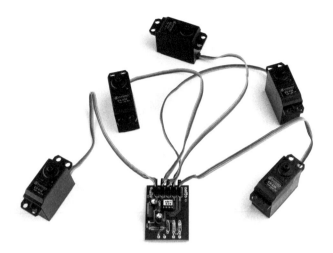

ler can be driven via Windows and RS232 (see screendump). Apart from the edeFT639 chip and its input protection network R1/R2/D1, the PCB shown here also accommodates a simple step-down voltage regulator.

RC Modellers wishing to use the circuit in a plane or boat may want to exchange the 7805 for a low-drop regulator and omit D1 in order to juice the craft battery. The servos are connected via the usual 3-pin con-

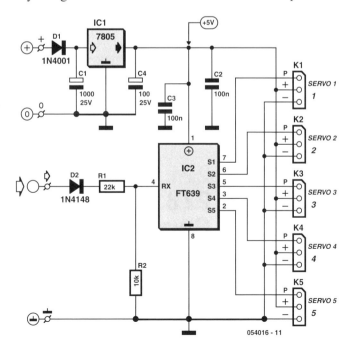

054016 - 11

COMPONENTS LIST

Resistors:
R1 = 22kΩ
R2 = 10kΩ

Capacitors:
C1,C4 = 100µF 25V radial
C2,C3 = 100nF

Semiconductors:
D1 = 1N4001

D2 = 1N4148
IC1 = 7805
IC2 = edeFT639 (Elab Inc.;
www.elabinc.com)

Miscellaneous:
5 off 3-way SIL pinheader
Evaluation software (VB application) from
Elab Inc. website
PCB, ref. 054016-1 from The PCBShop

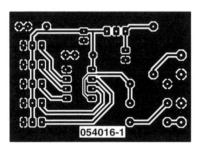

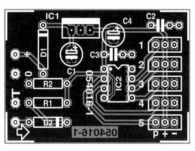

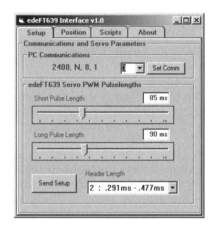

+5 V on their connectors, you should check the data available.

Source:
Elab Inc. (formerly FerreTronics) FT639 Ferret datasheet. www.elabinc.com

nectors. As servo manufacturers use different pinouts for pulse (p), ground and

Luc Lemmens   (054016–1)

222
Swapping Without a Buffer

Most programmers will have their own library of commonly used snippets of code. One task that appears very often is the exchange of the contents of two variables. The code for this usually looks as follows:

```
int c;
c=a;
a=b;
b=c;
```

There doesn't appear to be anything wrong with this, but it does make use of a third variable and this takes up more memory. In general, modern processors tend to have enough memory on board, but it never harms being economical with the available memory. Another way in which the variables can be exchanged is shown below:

```
a=a^b;
b=a^b;
a=a^b;
```

It isn't immediately obvious that the contents of the two variables are exchanged. However, the operation of this code is really quite simple.
We make use of the Boolean law that a^b^a = b, where the '^' symbol stands for a bitwise exclusive-or (XOR). One consequence of this law is that when we know that the content of register A is the XOR of two variables, where the value of one is known, we can recover the value of the unknown variable by XORing register A with the known value. It shouldn't come of much of a surprise that many encryption systems make use of this technique. We can imagine that it may still not be clear how the XOR routine works, so we've

	A	**B**
initial state	10101010	11001100
A = A ^B	01100110	11001100
B = A ^B	01100110	10101010
A = A ^B	11001100	10101010

XOR truth table		
In1	**In2**	**Output**
0	0	0
0	1	1
1	0	1
1	1	0

shown in the Table what each step of the program does. It should now be clear that at the end of the code the contents of variables a and b have been exchanged. You could try this yourself with pen and paper. You'll find that it works with any values for a and b.

Paul Goossens   (054019-1)

223
Carriage Detection for Model Railway

Model railway builders know all about this: it is a troublesome job to get a block system to function properly. We present here a simple, reliable and cheap solution on how to fit resistors between insulated wheels, as is used with the two-rail systems that operate a block detection system based on current consumption. A block, in this case, is an isolated section of rail. It is considered occupied if a load is detected. The locomotive usually has at least some cur-

rent consumption, even if it is just for the lights. Digital locomotives with a decoder always consume a few milliamps, which is also sufficient for detection. To be able to detect the rolling stock, a little more effort is required. When a single carriage is accidentally left behind in a block, the detector has to be able to sense this and indicate the block as occupied. In order to achieve this, all axles of all carriages have to be fitted with a small resistor, so that a small current

can flow. Carriages with internal lighting have additional sliding contacts (on the wheels) and a small lamp or LED as load. However, it is much too complicated to fit all carriages with additional sliding contacts. That is why it is usual for a resistor to be placed across the plastic insulating sleeve. One of the wheels is isolated with respect to the axle, otherwise the wheels would short circuit the rails. This is, of

course, not the intention with a two-rail system. Axles with resistors are available ready-made, but are somewhat expensive. Usually an SMD or 1/8 W type resistor is used, with a value ranging from about 4k7 to 10k. It is mechanically mounted with a little (epoxy) glue and the actual electrical connection is made with conductive glue (often containing silver particles). This will usually last for years, but regular inspection is required if carriages are subject to rough handling. In addition, conductive glue is expensive, not always available and dries out after a while. We consider the following a better method: grab the axle with small pliers and carefully pull of the isolated wheel from the axle. Also remove the insulating sleeve. Cut a two cm piece of thin hook-up wire and remove the individual strands. They have to be thin, no more than about 0.1 mm. Put the wheel back on, but now with a 0.1 mm wire between axle and sleeve and opposite that a 0.1 mm wire between sleeve and wheel. You now have created two connections. With a little bit of dexterity you can solder a resistor to those wires. First practise a little on a few old axles, the example in the pictures isn't the newest model either! Don't forget to glue the resistor down, otherwise the thin wires will certainly break after a short time. It is also possible to first solder the wires on an SMD resistor and then re-assemble the axle. Do it whichever way you personally find the easiest.

Karel Walraven (054017-1)

224
Noise Suppression for R/C Receivers

Receiver interference is hardly an unknown problem among model builders. Preventive measures in the form of ferrite beads fitted to servo cables are often seen in relatively large models and/or electrically driven models, to prevent the cables from acting as antennas and radiating interference to the receiver.

If miniature ferrite beads are used for this purpose, the connector must be first be taken apart, after which the lead must be threaded through the bead (perhaps making several turns around the core) and then soldered back onto the connector. An interference source can also cause problems in the receiver via the power supply connection. The battery is normally connected directly to the receiver, with the servos in turn being powered from the receiver. The servos can draw high currents when they operate, which means they can create a lot of noise on the supply line. This sort of interference can be kept under control by isolating the supply voltage for the receiver from the supply voltage for the servos.

All of these measures can easily be implemented 'loose' in the model, but it's a lot nicer to fit everything onto a single small circuit board. That makes everything look a lot ti-

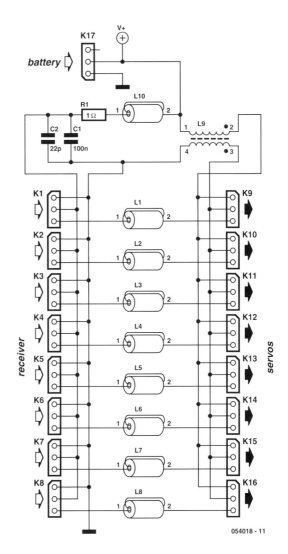

054018 - 11

Components list

Resistor:
R1 = 1Ω

Capacitors:
C1 = 100nF
C2 = 22pF

Miscellaneous:
L1-L8,L10 = ferrite inductor, SMD1206
(e.g., Digikey # PMC1206-202-ND)
K1-K8 = servo cable
K9-K17 = 3-way SIL pinheader
PCB, ref. 054018-1 from The PCBShop

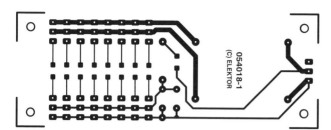

2

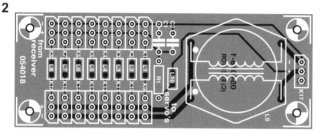

dier, and it takes up less space. The schematic diagram is shown in Figure 1. Connectors K1–K8 are located at the left. They are the inputs for the servo signals, which are connected to the receiver by the servo leads. The outputs (K9–K16) are located on the right. That is where the servos are connected. Finally, the battery is connected to K17. Interference on the supply voltage line due to the motors and servos is suppressed by a filter formed by L10, R1, C1 and C2. L10 is a ferrite-core coil with an impedance of 2000 ohms at 30 MHz. In combination with C1 and C2, it forms a substantial barrier to interference in the 35 MHz R/C band. Signals with frequencies close to the 10.4 MHz intermediate frequency (which is used in many receivers) are also effectively blocked by this filter. L9 filters out common-mode noise on the supply line for the servos, which effectively means that it prevents the supply servo signals are filtered out by ferritelines to the servos from acting as anten-beads in order to limit the antenna effectsnas. of these connection lines. Finally, high-frequency currents on the servo signals are filtered out by ferrite beads in order to limit the antenna effects of these connection lines.

Paul Goossens   (054018)

225
Dual Oscillator for µCs

The MAX7378 contains two oscillators and a power-on reset circuit for microprocessors. The Speed input selects either 32.768 kHz (LF) or a higher frequency, which is pre-programmed. The type number corresponds to the standard pre-programmed value and the value of the reset threshold. There is a choice of two threshold values: 2.56 V and 4.29 V. Both thresholds are available with all standard frequencies, which are 1 MHz, 1.8432 MHz, 3.39545 MHz, 3.6864 MHz, 4 MHz, 4.1943 MHz, and 8 MHz. However, any frequency between 600 kHz and 10 MHz is also possible. An internal synchronisation circuit ensures that no glitches occur when switching between the two oscillators. The Reset output of the MAX7378 is available in three different options. Two of the options are push-pull types, either active low or active high. The third option is open drain, which thus requires an external pull-up resistor. That is the only standard option (which is why a resistor in dashed outline is shown connected to the Reset output). The Reset signal remains active for 100 µs after

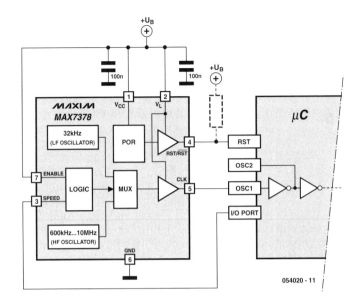

054020 - 11

The IC is housed in the tiny 8-pin µMAX package and has dimensions of only 3.05 × 5.03 mm including pins, with a pin pitch of only 0.65 mm. Unfortunately, the accuracy of the oscillators is not especially good. The HF oscillator has an error of ±2% at 25 °C with a 5 V supply voltage and a maximum temperature coefficient of +325 ppm, which doesn't exactly correspond to crystal accuracy, but it is certainly usable for most non-time-critical applications. The error over the full supply voltage range (2.7–5.5 V) is twice as large. The 32.768 kHz oscillator is more accurate, with an error of only 1% at 5 V and 25 °C, although this is still a bit too much for time measurements. The error can be as much as ±3% over the entire supply voltage range. The maximum current consumption is 5.5 mA, which is relatively low.

the supply voltage rises above the threshold voltage. The Reset signal becomes active immediately if the voltage drops below the threshold.

The IC is powered via two separate pins. The VL pin powers the reset and oscillator circuitry, while the VCC pin powers the remainder of the chip. The two pins must always have the same potential. Good decoupling in the form of two 100 nF ceramic capacitors (SMD types) is also necessary.

Ton Giesberts   (054020-1)

226
Video Sync Generator

All video signals include synchronisation signals, which help the television set keep the horizontal and vertical deflection synchronous with the picture. For experimental purposes, it can be handy to build a generator for these synchronisation signals. Synchronisation signals have a rather complicated structure. It's not easy to generate them accurately using analogue circuitry. By contrast, a design based on a CPLD is a

lot easier. This design uses the experimenter's board described in the May 2004 issue ('Design Your Own IC'). The hardware extension for this generator is shown in Figure 1. The extension could hardly be simpler. The 20-way connector must be connected to connector K3 of the experimenter's board by a flat cable. The synchronisation signals on pin 4 are routed to the output connector via voltage divider

1

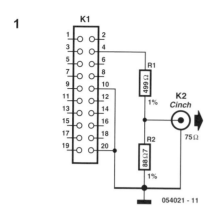

054021 - 11

R1/R2. R1 and R2 perform two tasks in this circuit. First, they provide the correct output impedance of 75 Ω. Second, they reduce the signal amplitude from around 4 V to 0.6 V. If the circuit is connected to a television set with an input impedance of 75 Ω, the maximum output voltage will be only 0.3 V, which complies with the specifications for the CVBS signal. As you've surely guessed by now, the majority of the design is contained in the CPLD. We generated this design in Quartus 4.2. The top-level schematic is shown in Figure 2. As can clearly be seen from that diagram, the design consists of three separate sections. In combination with the crystal on the circuit board, the inverter at the bottom provides the clock signal.

The crystal must generate a frequency of 10 MHz for this circuit. If a different crystal is already fitted, it must be replaced by a 10 MHz type. The clock signal finds its way back into the circuit via another input at the top left. It ensures that both parts of the circuit operate synchronously. The section named 'sync_state' keeps track of which part of the video signal is being generated. Here it should be remarked that each video line is divided into two parts in this circuit. That is done because the synchronisation signal occurs at twice the normal rate during the vertical retrace interval. The 'width' output indicated how long

the output signal should remain low. Assertion of the 'endline' input indicates to the subcircuit that the current portion of the video signal is ready and the next portion of the video signal can be processed. The 'sync_state' block then produces the associated value at the output.

The final block bears the revealing name 'pwm_10bit'. As its name suggests, this section (which is actually implemented as part of the program code) generates a pulse on the output whose width is equal to the value of 'width'. This PWM counts from 0 to 319, which corresponds to a total duration of 32 μs (half of one video line) at a clock frequency of 10 MHz. This period is divided into 320 intervals of 0.1 μs each. The PWM sets the 'endline' output high during the next-to-last interval. You might expect that this signal would be active during the final interval of the period, but making it active one interval earlier gives the 'sync_state' block at least one clock cycle to generate a new 'width' value. Naturally, the design files for the IC can be downloaded from the Elektor Electron-ics website (www.elektor-electronics.co.uk). The file number is 054021-11.zip. After compilation, the final result uses 65 macrocells. The number of macrocells can be reduced by optimising the overall code. For instance, there are several obvious places where the design can be made more efficient, but a few somewhat less optimisations are also possible. Try it for yourself, and see how much you can optimise the design. Your efforts might just spark an interesting discussion on our website Forum!

Paul Goossens   (054021-1)

2

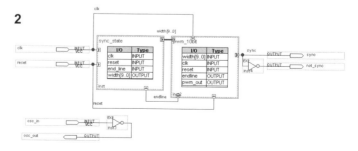

227
3V Headphone Amplifier

Many new devices require a headphone connection, but due to the high level of integration and miniaturisation there is usually little room left. The low supply voltage and/or battery voltage also causes problems. If no special techniques are used, the output power and headroom are severely limited. The MAX4410 made by Maxim overcomes these problems not just by virtue of its small size, but also by including an internal supply inverter (charge-pump). This requires only two small external ceramic SMD capacitors (C6 and C7). The supply voltage to the output stage is now symmetrical and the outputs are therefore relative to ground (no DC offset). This gets round the need for large output capacitors to stop a DC voltage from reaching the

headphones. A DC-coupled output can also be implemented using two bridge amplifiers, but virtually all plugs for stereo headphones are asymmetric and use 3-pole connectors (common ground), which can't be connected to a bridge output. Each channel can be individually turned off (SHDNL and SHDNR) by jumpers JP1 and JP2. During normal operation these two inputs should be connected to the positive supply. When both channels are turned off the charge pump is also switched off and the current consumption drops to about 6 µA. The IC also has thermal and short-circuit protection built in. The IC switches to standby mode when the supply voltage is too low and it has a circuit that prevents power-on and off plops at the out-

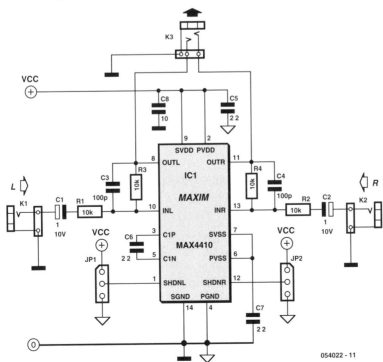

054022 - 11

puts. The recommended supply voltage is between 1.8 V and 3.6 V. The IC can deliver about 80 mW per channel into a 16 Ω load. The power supply should be able to output at least 200 mA. In practice this means that when you use a power supply that also powers other circuits, it should have at least 300 mA in reserve. The amplifiers are configured in inverting mode with a gain set by the ratio of two resistors (R3/R1 or R4/R2); the input impedance is determined by R1 and R2. C1 and C2 are required to decouple any possible DC-offset from the inputs. In the MAX4410 evaluation kit these are small tantalum capacitors, but we don't recommend these for use in audio applications. Plastic film types would be much better, although they take up much more room. HF decoupling is provided by 100 pF capacitors connected in parallel with R3 and R4. These set the bandwidth of the amplifiers to just over 150 kHz.

The typical distortion is 0.003%. For more details you should refer to the MAX4410 datasheet. It is also worth looking at the datasheet for the associated evaluation kit. The choice of capacitors for decoupling etc., their positioning on the board and the overall layout are very critical and demand a lot of attention. Furthermore, the 14-pin TSSOP package (with a pin spacing of 0.65 mm) and SMDs in 0402 packages make it very difficult to construct this circuit yourself. The IC is also available in a (much more difficult to solder) UCSP 16 package (ball grid array, only 2.02 by 2.02 mm).

Ton Giesberts   (054022-1)

228
Remote Control Blocker

This circuit was designed to block signals from infrared remote controls. This will prove very useful if your children have the tendency to switch channels all the time. It is also effective when your children aren't permitted to watch TV as a punishment. Putting the TV on standby and putting the remote control out of action can be enough in this case. The way in which we do this is very straightforward. Two IR LEDs continuously transmit infrared light with a frequency that can be set between 32 and 41 kHz. Most remote controls work at a frequency of 36 kHz or 38 kHz. The disruption of the remote control occurs as follows. The 'automatic gain' of the IR receiver in TVs, CD players, home cinema

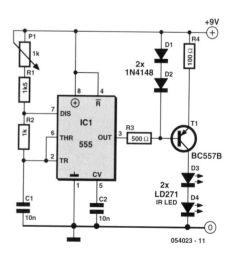

054023 - 11

systems, etc. reduces the gain of the receiver due to the strong signal from the IR LEDs. Any IR signals from a remote control are then too weak to be detected by the receiver. Hence the equipment no longer 'sees' the remote control! The oscillator is built around a standard NE555. This drives a buffer stage, which provides the current to the two LEDs.

Setting up this circuit is very easy. Point the IR LEDs towards the device that needs its remote control blocked. Then pick up the remote control and try it out. If it still functions you should adjust the frequency of the circuit until the remote control stops working. This circuit is obviously only effective against remote controls that use IR light!

Paul Goossens   (054023-1)

229
Phantom Supply from Batteries

Professional (directional) microphones often require a phantom supply of 48 V. This is fed via the signal lines to the microphone and has to be of a high quality. A portable supply can be made with 32 AA-cells in series, but that isn't very user friendly. This circuit requires just four AA-cells (or five rechargeable 1.2 V cells). We decided to use a standard push-pull converter, which is easy to drive and which has a predictable output voltage. Another advantage is that no complex feedback mechanism is required. For the design of the circuit we start with the assumption that we have a fresh set of batteries. We then induce a voltage in the secondary winding that is a bit higher than we need, so that we'll still have a high enough voltage to drive the linear voltage regulator when the battery voltage starts to drop (refer to the circuit in Figure 1). T1 are T2 are turned on and off by an astable multivibrator. We've used a 4047 low-power multivibrator for this, which has been configured to run in an astable

free-running mode. The complementary Q outputs have a guaranteed duty-cycle of 50%, thereby preventing a DC current from flowing through the transformer. The core could otherwise become saturated, which results in a short-circuit between 6V and ground. This could be fatal for the FETs. The oscillator is set by R1/C1 to run at a frequency of about 80 kHz. R2/R3 and D1/D2 make T1 and T2 conduct a little later and turn off a little faster, guaranteeing a dead-time and avoiding a short-circuit situation. We measured the on-resistance of the BS170 and found it was only 0.5 Ω, which isn't bad for this type of FET. You can of course use other FETs, as long as they have a low on-resistance. For the transformer we used a somewhat larger toroidal core with a high AL factor. This not only reduces the leakage inductance, but it also keeps the number of windings small. Our final choice was a TX25/15/10-3E5 made by Ferroxcube, which has dimensions of about 25 × 10 mm. This makes the construction of the transformer a lot easier. The secondary winding is wound first: 77 turns of a 0.5 mm dia. enamelled copper wire (ECW). If you wind this carefully you'll find that it fits on one layer and that 3 meters is more than enough. The best way to keep the two primary windings identical is to wind them at the same time. You should take two 30 cm lengths of 0.8 mm dia. ECW and wind these seven times round the core, on the opposite side to the secondary connections. The centre tap is made by connecting the inner two wires

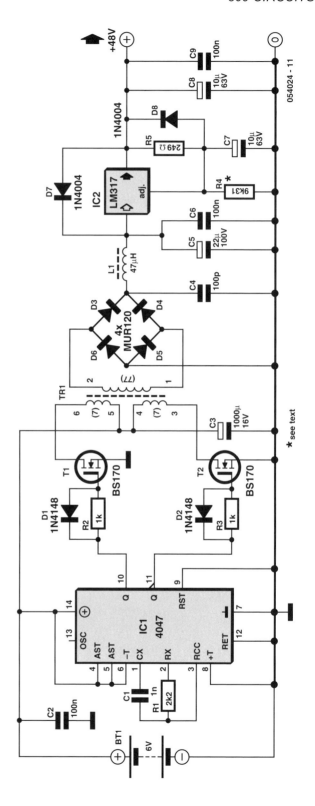

together. In this way we get two primary windings of seven turns each. The output voltage of TR1 is rectified by a full-wave rectifier, which is made with fast diodes due to the high frequency involved. C4 suppresses the worst of the RF noise and this is followed by an extra filter (L1/C5/C6) that reduces the remaining ripple. The output provides a clean voltage to regulator IC2. It is best to use an LM317HV for the regulator, since it has been designed to cope with a higher voltage between the input and output. The LM317 that we used in our prototype worked all right, but it wouldn't have been happy with a short at the output since the voltage drop would then be greater than the permitted 40 V.

If you ensure that a short cannot occur, through the use of the usual 6k81 resistors in the signal lines, then the current drawn per microphone will never exceed 14 mA and you can still use an ordinary LM317. D7 and D8 protect the LM317 from a short

at the input. There is virtually no ripple to speak of. Any remaining noise lies above 160 kHz, and this won't be a problem in most applications.

The circuit can provide enough current to power three microphones at the same time (although that may depend on the types used). When the input voltage dropped to 5.1 V the current consumption was about 270 mA. The reference voltage sometimes deviates a little from its correct value. In that case you should adjust R4 to make the output voltage equal to 48 V. The equation for this is:

$$R4 = \frac{48 - V_{ref}}{V_{ref} / (R5 + 50\mu A)}$$

To minimise interference (remember that we're dealing with a switched-mode sup-ply) this circuit should be housed in an earthed metal enclosure.

Ton Giesberts   (054024-1)

230
Balancing LiPo Cells

Things change fast in the electronics world, and that's also true for rechargeable batteries. The rate of development of new types of rechargeable batteries has been accelerated by the steadily increasing miniaturisation of electronic equipment. LiPo cells have conquered the market in a relatively short time. Their price and availability have now reached a level that makes them attractive for use in DIY circuits. Unlike its competitors Elektor Electronics has already published several articles about the advantages and disadvantages of LiPo batteries. One of the somewhat less well-known properties of this type of rechargeable battery is that the cells must be regularly 'balanced' if they are con-nected

in series. This is because no two cells are exactly the same, and they may not all have the same temperature. For instance, consider a battery consisting of a block of three cells. In this case the outer cells will cool faster than the cell in the middle. Over the long term, the net result is that the cells will have different charge states. It is thus certainly possible for an individual cell to be excessively discharged even when the total voltage gives the impression that the battery is not fully discharged. That requires action – if only to prolong the useful life of the battery, since LiPo batteries are still not all that inexpensive. One way to ensure that all of the cells have approximately the same charge state is limit the voltage of

1

IC2 = 74HC132

1 Lipo cell

IC1

MAX6385XS41D2-T

IC2.B

IC2.A

IC2.C

IC2.D

BAT85

MJD1274T4

054026 - 11

each cell to 4.1 V during charging. Most chargers switch over to a constant voltage when the voltage across the batter terminals is 4.2 V per cell. If we instead ensure that the maximum voltage of each cell is 4.1 V, the charger can always operate in constant-current mode. When the voltage of a particular cell reaches 4.1 V, that cell can be discharged until its voltage is a bit less than 4.1 V. After a short while, all of the cells will have a voltage of 4.1 V, with each cell thus having approximately the same amount of charge. That means that the battery pack has been rebalanced. The circuit (Figure 1) uses an IC that is actually designed for monitoring the supply voltage

of a microcontroller circuit. The IC (IC1) normally ensures that the microcontroller receives an active-high reset signal whenever the supply voltage drops below 4.1 V. By contrast, the out-put goes low when the voltage is 4.1 V or higher. In this circuit the output is used to discharge a LiPo cell as soon as the voltage rises above 4.1 V. When that happens, the push-pull output of IC1 goes low, which in turn causes transistor T1 to conduct. A current of approximately 1 A then flows via resistor R1. LED D2 will also shine as a sign that the cell has reached a voltage of 4.1 V. The function of IC2 requires a bit of explanation. The circuit built around the four NAND gates extends the 'low' interval of the signal generated by IC1. That acts as a sort of hysteresis, in order to prevent IC1 from immediately switching off again when the voltage drops due the internal resistance of the cell and the resistance of the wiring between the cell and the circuit. The circuitry around IC2 extends the duration of the discharge pulse to at least 1 s. Figure 2 shows how several circuits of this type can be connected to a LiPo battery. Such batteries usually have a connector for

2

Balancer 1

Balancer 2

Balancer 3

054026 - 11

a balancing device. If a suitable connector is not available, you will have to open the battery pack and make your own connections for it. The figure also clearly shows that a separate circuit is necessary for each cell.

Paul Goossens   (054026-1)

231
USB for the Xbox

The Xbox is the well-known Microsoft game computer. The fact that the Xbox is based on PC technology should hardly be surprising, since Microsoft specialises in this computer architecture. It's also not especially remarkable that if you open up an Xbox, you'll find several well-known ICs from the PC world. In fact, the Xbox is actually a PC. The major difference between the Xbox and a normal PC is that the operating system is stored entirely in Flash memory and users can-not add any functionality to the system. There is also a protection system that prevents any software from being accepted if it does not have a digital signature from Microsoft. At least, that was the intention. Naturally, there are people who have cracked the Xbox protection system and use the Xbox as a PC to run Linux. We plan to publish more information about this in Elektor Electronics in the near future.

The biggest problem with running Linux on an Xbox is that there is no keyboard for the Xbox. That's what we want to remedy here. As usual with game computers, the Xbox has connectors for controllers (enhanced joysticks) for playing the games. The Xbox has four such connecters on the front of the enclosure. They have a form that is totally new to us. After a bit of detective work, it turns out that the signals on these connectors are actually quite familiar: they're USB signals. Once you know that, it's easy to connect a keyboard to the Xbox. If you fit a standard USB connector somewhere on the enclosure and wire it to the signal lines for one of the connectors, you can then connect a USB keyboard. Another benefit of this is that most types of USB memory sticks can be used by the Xbox as Xbox memory cards. USB memory sticks are a lot less expensive than 'genuine' Xbox memory cards.

The wires to the connector at the front can be clearly seen on the photos. The yellow wire is not important for our purposes. The other wires are used for the actual USB signals. A nice detail is that Microsoft uses exactly the same colours for these wires as those specified in the USB standard, which

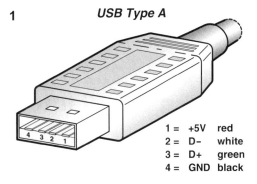

1   **USB Type A**

1 = +5V   red
2 = D–   white
3 = D+   green
4 = GND   black

054027 - 11

Table 1: USB Connections		
Pin 1	red	+5 V
Pin 2	white	D–
Pin 3	green	D+
Pin 4	black	GND

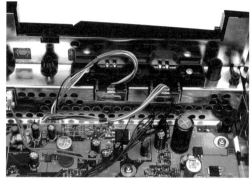

makes them much easier to recognise. For the installation, you will some wire and a USB connector (Type A). That's the same type of connector as the USB connectors in a PC. Figure 1 shows an example of such a connector, along with the proper pin numbers. Find a place on the Xbox where the connector can be fitted. It's easy to make a hole in the case with a small hand-held router or a drill and a small file. Next, make connections between the four wires leading to the Xbox connector and the four terminals of the USB connector, as shown in Table 1. It's a good idea to check the connections before using the modified unit, in order to ensure that there aren't any shorts between the lines. As a second test, we recommend checking whether a voltage can be measured between pin 1 of the USB connector (+3.6 to 5 V) and pin 4 (ground) after the Xbox power has been switched on. If the test results are all OK, the USB connector is ready for use. USB keyboards (and USB mice) are supported by Xbox Linux, and some Xbox games also use a USB keyboard to make it easier to control the game and/or chat with other players.

Paul Goossens   (054027-1)

232
DVI Interface

The PAL, NTSC and SECAM television standards are all several decades old. Even in this digital era, most people still have television sets that use complex analogue signals complying with these standards. Nevertheless, the end is nigh for these standards. The DVI standard is on its way to becoming the new standard

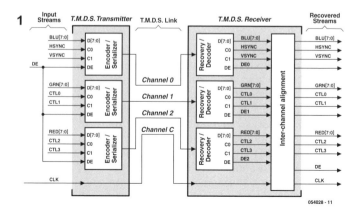

054028 - 11

for transmitting video information. The most important property of this interface is that the information is transferred in digital form (24-bit words) instead of in the form of analogue signals. Figure 1 shows how the required signals are divided into three data streams in a single-link interface, along with a clock signal. In DVI terminology, the clock signal is called 'Channel C'. A double-link interface with three additional information channels has also been specified to double the bandwidth. Whether these three additional channels are actually used depends on the selected resolution and repetition rate. Each channel consists of a differential signal pair. That makes this interface significantly less sensitive to interference. The technology behind these differential signal pairs is called 'transition-minimized

2

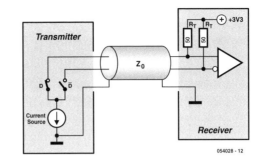

054028 - 12

3

Combined Analog and Digital Receptacle Connector ♀

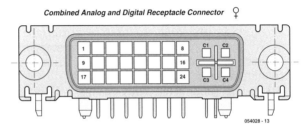

054028 - 13

differential signalling', or TMDS for short. Figure 2 shows how the signals are generated in electronic terms. The connector and associated signal names are shown in Figure 3 and Table 1. Besides the six TMDS signal pairs and clock signal (seven pairs in total), there are several other signals on the connector. DDC Clock and DDC Data al-

Table 1. DVI interface connector pin assignments							
Pin	**Signal**	**Pin**	**Signal**	**Pin**	**Signal**		
1	TMDS Data2 –	9	TMDS Data 1 –	17	TMDS Data 0 –		
2	TMDS Data2 +	10	TMDS Data 1 +	18	TMDS Data 0 +		
3	TMDS Data2/4 shield	11	TMDS Data 1/3 shield	19	TMDS Data 0/5 shield		
4	TMDS Data 4 –	12	TMDS Data 3 –	20	TMDS Data 5 –		
5	TMDS Data 4 +	13	TMDS Data 3 +	21	TMDS Data 5 +		
6	DDC clock	14	+5 V supply	22	TMDS clock shield		
7	DDC data	15	Ground	23	TMDS clock +		
8	Analog Vertical Sync	16	Hot Plug Detect	24	TMDS clock –		
C1	Analog Red	C2	Analog Green	C3	Analog Blue		
C4	Analog Horizontal Sync	C5	Analog ground				

Table 2. Resolutions supported by DVI-1.0	
VGA	640x400
SVGA	800x600
XGA	1024x768
SXGA	1280x1024
UXGA	1600x1200
HDTV	1920x1080
QXGA	2048x1536

low a connection to be established in accordance with the DDC protocol. That protocol allows a connected device to determine which resolutions and frequencies the monitor can handle.

The Hot Plug Detect signal makes it easy to recognise whether a monitor is connected. The +5 V line naturally does not need much explanation. Besides all these digital signals, the (old-fashioned) analogue RGB

signals and associated sync signals are still present. Unfortunately, the illustrated connector is not the only type that is available. The illustrated connector is a DVI-I connector. It can transport both analogue and digital signals. There is also a DVI-D (digital only) and DVI-A (analogue only) connector. With the digital version, when you acquire a monitor or playback equipment you have to check both types of equipment to see whether they use a double-link interface (six data channels) or a single-link interface (three data channels). The DVI 1.0 standard supports several different resolutions.

They are listed in Table 2. Each resolution can also be used with a variety of repetition rates, which are 60 Hz, 75 Hz and 85 Hz. That's a good selection of options, which (unfortunately) also makes for a good chance of confusion among consumers.

Paul Goossens   (054028-1)

233
Low-cost LiPo Charger

The charging of Lithium-Polymer (LiPo) cells takes place very differently to that of the well known NiCd and NiMH cells. This aspect has previously been covered in Elektor Electronics. And this isn't the first LiPo charger that we've published, but it is undoubtedly the smallest! Chip manufacturer Intersil has designed a LiPo charger IC that requires a minimum of external components. Since the IC itself is

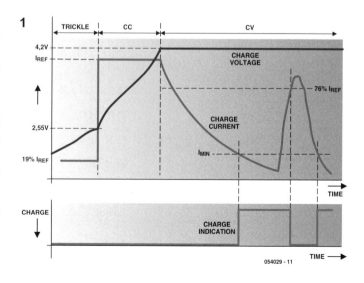

also extremely small (2 × 3 mm), the complete charger can be kept very small as well. This lets us design a charger that can easily be built into various pieces of equipment, especially when we use SMDs for all external components. For those of you who don't know how a LiPo cell should be charged, we'll give a short explanation. When the cell voltage is very low (<2.5 V), it should be charged using a small current (see Figure 1). This current is typically less than 0.1C (where C is the nominal battery capacity). When the voltage has risen sufficiently, but is still below 4.2 V, the cell is charged with a constant current. Most LiPo manufacturers specify a current of 1C for this stage. The voltage across the cell may not exceed 4.2 V, so the charger has to keep an eye on this as well. At this constant voltage the current through the cell will slowly reduce while the charge in the cell increases. At the point when the cell voltage is 4.2 V and the charging current has dropped to 0.1C, the cell is about 80-90% charged, depending on the manufacturer. Most chargers decide at this point that the cell is fully charged and switch to trickle charging the cell. Our charger works in exactly the same way. There are two parameters that can be adjusted in this charger, which are the normal charging current and the trickle charge that flows when the cell is 'full'. In the cir-

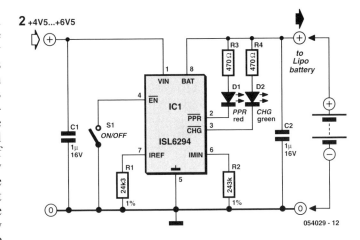

054029 - 12

cuit of Figure 2 resistor R1 sets the charging current to about 500 mA, and resistor R2 sets the trickle charge current to about 45 mA. R3, R4, D1 and D2 are optional in this design and provide the user with status information. D1 shows when the charging process is busy and D2 indicates that the correct input voltage is present. If you want to use different maximum and minimum charge currents you should use the following formulae for R1 and R2:

$$R1 = \frac{12 \cdot 10^3}{I_{ref}} \qquad R2 = \frac{11 \cdot 10^3}{I_{min}}$$

Keep in mind that the accuracy of the current source at 500 mA is about 10%; this drops to about 30% at 50 mA. You should therefore be conservative in your choice of charging current so that you keep below the manufacturer's maximum recommended charging current.

(054029-1)

234
High Voltage Regulator

The LR12 made by Supertex Inc. is a good choice for applications where a supply voltage of more than 35 to 40 V needs to be stabilised. This small regulator can cope

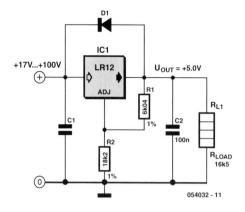

054032 - 11

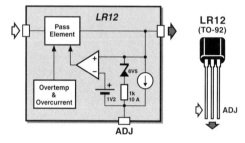

LR12
(TO-92)

tial divider the output voltage can be set using the following equation:

$$V_{out} = 1.2 \cdot \left(\frac{R2}{R1} + 1 \right) + I_{adj} \cdot R2$$

The circuit shows a standard application where the LR12 is used as a 5 V regulator. C1 decouples the input voltage. Its value and working voltage depend on the input voltage and the current consumption. The bypass capacitor (C2) is required to keep the LR12 stable. In cases where the voltage at the input may be smaller than at the output an extra protection diode is required, for example a 1N4004. The output current of the IC needs to be at least 0.5 mA. In the circuit shown here the potential divider made by R1/R2 already draws 0.2 mA. This means that with a 5 V output the load resistor needs to be less than 16k5. If the resistance is higher, the total output current drops below the required value of 0.5 mA. The output current of the LR12, with a 12 V difference between input and out-put, is limited to 100 mA (max. dissipation of a TO92 package: 0.6 W at 25 °C). The ripple suppression is at least 50 dB. The current consumption of the IC itself is very low at only 5 to 15 µA.

Ton Giesberts   (054032-1)

with input voltages of up to 100 V, when the output voltage can be adjusted between 1.2 and 88 V. A small disadvantage is that the input voltage needs to be at least 12 V more than the output voltage. The regulator keeps the voltage between the output and adjust pin constant at 1.2 V. With a poten-

235
MP3 Adapter for TV

Nowadays there are many ways in which you can listen to music. A portable MP3 player with headphones is often used while on the move. But when you need to stay in a hotel and would like to listen without headphones to your favourite music in your hotel room, things become more difficult. Most hotel rooms have a TV, but rarely a music centre. We have designed a

fairly compact adapter that lets you connect an MP3 player (or any other portable device) to the SCART socket of a TV. It is obvious that this will only work if the TV has a SCART socket! The board directly feeds the headphone signals to the correct SCART pins. If the TV only has mono sound, then the left input signal is probably used. The connection between the adapter

and TV is made with a SCART cable. It is best to make your own lead for the connection of the headphone signals. There are three pins on the board for the signal con-

nections (L, R and ground). Unfortunately, this alone is not enough in practice. Most TVs expect a correct video signal on an extern connection; otherwise the input is turned off (muted). To get round this problem we used a programmable logic device (IC1, an EPM7064 made by Altera) to generate a video sync signal. More detailed information can be found in the sync generator article elsewhere in this issue. It is sufficient to state that this IC generates a sync signal according to the PAL/SECAM specification. Several outputs are connected in parallel to give a stronger output signal. Potential divider R5/R6 produces a signal that is slightly bigger than a normal video signal (normally 30% of 1 Vpp into

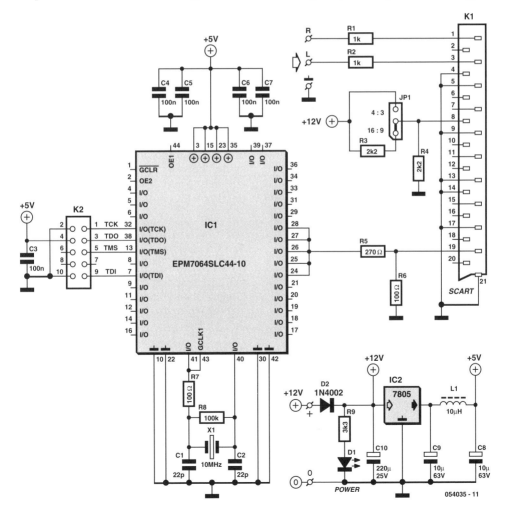

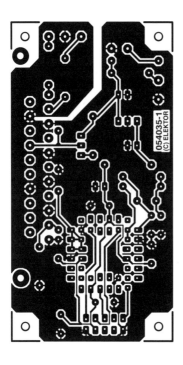

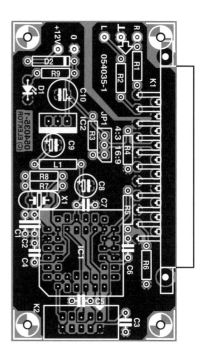

75 Ω). The signal is now a bit under 0.5 Vpp into 75 Ω. The output impedance of R5/R6 is 75 Ù, which reduces reflections should the cable be incorrectly terminated. ISP connector K2 has been included for those of you who would like to experiment with this circuit (see

http://www.altera.com). The TV can be automatically switched to an external source using the video-status signal (pin 8). Modern widescreen TVs use this signal to switch between two display modes. A voltage between 9.5 and 12 V results in a 4:3 display and a voltage between 4.5 and 7 V

Components list

Resistors:
R1,R2 = 1kΩ
R3, R4 = 2kΩ2
R5 = 270Ω
R6,R7 = 100Ω
R8 = 100kΩ
R9 = 3kΩ3

Capacitors:
C1,C2 = 22pF
C3-C7 = 100nF ceramic, lead pitch 5mm
C8,C9 = 10ìF 63V radial
C10 = 220 ìF 25V radial

Inductor:
L1 = 10 ìH

Semiconductors:
D1 = LED, low-current
D2 = 1N4002
IC1 = EPM7064SLC44-10 PLC44 with socket (programmed, Publisher's order code 054035-31 *)
IC2 = 7805

Miscellaneous:
JP1 = 3-way SIL header with jumper
K1 = SCART socket (female), PCB mount, angled
K2 = 10-way boxheader
X1 = 10MHz quartz crystal
PCB, Publisher's order code 054035-1 *
Disk, EPM7064 code, Publisher's order code 054035-11 * or Free Download

* See Elektor SHOP pages or
http://www.elektor-electronics.co.uk

switches to a widescreen format. Jumper JP1 can be used to control this function of the TV. For older TVs the jumper should be in the 4:3 position. In all cases the display remains dark though.

A supply voltage is required for IC1 as well as the video-status signal. This is provided by a mains adapter with a stabilised 12 V output, preferably a modern switch-mode supply. Voltage regulator IC2 provides a stable 5 V to IC1 and K2. LED D1 indicates that the circuit is powered up. The current consumption is mainly determined by IC1 and is about 80 mA.

(054035-1)

236
Stable USB Power Supply

A common problem when an AC mains adapter is used to power a USB device is that the voltage does not match the nominal 5 V specified by the USB standard. The circuit shown here accepts an input voltage in the range of 4–9 V and converts it into a 6 V output voltage, which is then stabilised to a clean 5 V level by a series regulator. The combined boost/buck converter used here operates on the SEPIC principle. That principle is quite similar to the operating principle of the Cuk converter (see the Jan-

uary 2005 issue of Elektor Electronics), but without the disadvantage of a negative output voltage. The circuit is built around a MAX668, which is intended to be used as a controller for boost converters. The difference between a SEPIC converter and a standard boost (step-up) converter is that the former type has an additional capacitor (in this case C2) and a second inductor (in this case, the secondary winding of transformer L1). If C2 is replaced by a wire bridge and the secondary winding of L1 is

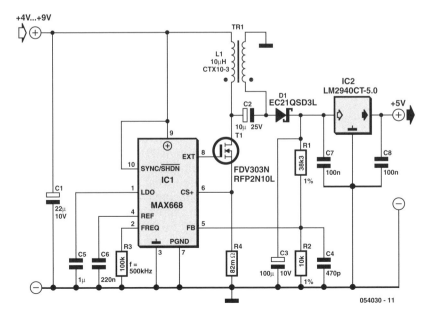

054030 - 11

left open, the result is a normal boost converter. In that case, a current can always flow from the input to the output via L1 and D1, even when the FET is not driven by IC1. Under these conditions, the output voltage can never be less than the input voltage less the voltage drop across the diode. The operation of a SEPIC converter can be explained in simple terms by saying that C2 prevents any DC voltage on the input from appearing at the output, so the output voltage can easily be made lower than the input voltage. The second coil causes a defined voltage to be present at the anode of D1. It is also possible to replace the transformer by two separate coils that are not magnetically coupled. However, the efficiency of the circuit is somewhat higher if coupled coils are used as shown here. The value of resistor R4 is chosen to limit the maximum current to 500 mA, which is also the maximum current that a USB bus can provide according to the specifications. Resistors R1 and R2 cause the voltage across C3 and C7 to be regulated at a value of around 6 V. A low-drop regulator (LM2940) is used to generate a stabilised 5 V from the 6 V output (with ripple voltage). The efficiency should be somewhere between 60 % and 80 %.

(054030-1)

237
Protection for Voltage Regulators

People often forget that many voltage regulator ICs have an upper limit (usually 35 V) on the input voltage they can handle. That applies primarily to types with a fixed output voltage. Adjustable voltage regulators also have a maximum voltage specification, in that case between the input and output (commonly 40 V). The input voltage must thus be limited to that level in a fault situation in which the output is shorted. This circuit shows a way to allow such regulators to be used in situations with higher input voltages. Although the solution consists of an additional three components, it is simple and can be built using commonly available components. The voltage across the regulator is limited by the combination of T1 and zener diode D1 to a value that allows the regulator to work properly with loads up to the maximum rated load. R1 provides an adequate operating current for D1 and the bias current for T1. It's a good idea to use a Darlington type for T1 in order to keep the value of R1 reasonably high. The current through D1 is only 10 mA with an input voltage of 60 V. Naturally, we also measured what the circuit does when no load is connected. Surprisingly enough, the nominal output voltage of 5.02 V increased to only 5.10 V (with a 60 V input voltage). In our experiments, we used a BDV65B for T1 and a value of 4.7 kΩ for R1. If you want to ensure that the circuit is truly short-circuit proof with an input voltage of 60 V, you must use a transistor that remains

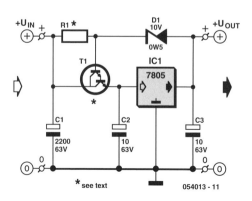

within its safe operating area at the maximum input voltage with the short-circuit current of the regulator (which can exceed 2 A). The BDV65B and TIP142 do not meet this requirement. The maximum voltage for the BDV65B is actually 40 V, and for the TIP142 is 50 V. If the transistor breaks down, the regulator will also break down. We verified that experimentally. One possibility is to add SOA protection for T1, but that amounts to protecting the protection. Another option is to relax the requirements. For that purpose, R1 must provide enough current to ensure that T1 receives sufficient current in the event of a short circuit to keep the voltage across T1 lower, but that doesn't make a lot of difference in practice, and it also increases the minimum load. Besides that, it should be

evident that adequate cooling for T1 and IC1 must be provided according to the load. Ripple suppression is only marginally affected by the protection circuit, since the input is already well stabilised by T1, but the current through D1 does flow through the output. The presence of C2 must also be taken into account. In this circuit, with an adjustable voltage regulator such as the LM317 and an output voltage greater than 40 V, C2 will cause the voltage to be briefly higher than 40 V in the event of a short circuit, which can also cause the IC to be damaged. In that case, it will be necessary to find a different solution or use a different type of voltage regulator.

Ton Giesberts   (054031-1)

238
PCI Express

Many new PCs now come equipped with one or more PCI Express slots on the motherboard. Eventually, PCI Express will make the old PCI standard obsolete, but for now most expansion cards still use the (old) PCI standard. The old standard (PCI 1.0) had a maximum bandwidth of 133 MB per second. In the meantime there have been many developments in the computer industry. We can now watch videos over the Internet, as well as listen to the radio. MP3s (also unheard of then) are decoded in real-time, while we might also watch a DVD in another window. All these applications place a big demand on the PC hardware. To process these data streams efficiently, contemporary PCs have a separate memory bus, another bus for the graphics card (AGP) and yet another bus (PCI) that lets the processor communicate with expansion cards. There have been several enhancements to the PCI bus (e.g. the

66 MHz PCI bus, 64 bit versions, etc.), but the time has come for a complete overhaul. The result is PCI Express. The x1 version has a bandwidth of 250 MB/s, but other versions are waiting in the wings (including a x16 version) to provide even greater bandwidths. One way in which the data rate can be increased is to increase the number of bits that are moved at the same time. This technique has already been used

PCI-Express 1x

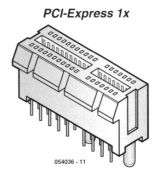

054036 - 11

in processors to increase their speed. For PCI Express however, a serial transport mechanism was chosen (as with SATA, USB, firewire, etc.). The slowest version of PCI Express uses a connector with only 26 connections. The connection details for the PCI Express connector are shown in the Table. There are only four signals on this connector that take care of the actual data transmission via the PCI Express protocol.

These are the signal pairs PETn0/PETp0 and PERn0/PERp0. Signal pair PETx0 (PCI Express Transmit 0) moves data from the host (PC) to the slave (slot), while PERx0 (PCI Express Receive 0) moves the data in the opposite direction. The small letter p or n denotes the polarity (positive or negative). The data has to be in step with a clock signal, provided by the signal pair CLK+ and CLK-. The rest of the connec-

Pin	Name	From To Description
1	GND	N/A Ground
2	USBD–	bidirectional USB signals
3	USBD+	
4	CPUSB	slave host Detection of USB device
5	reserved	
6	reserved	
7	SMBCLK	bidirectional SMBus clock
8	SMBDATA	SMBus data
9	+1.5V	N/A +1.5V supply voltage
10	+1.5V	
11	WAKE slave host	Wake-up signal for host
12	+3.3V	N/A +3.3V supply voltage
13	PERST	host slave Reset signal
14	+3.3V	N/A +3.3V supply voltage
15	+3.3V	
16	CLKREQ	slave host Clock request
17	CPPE	slave host PCI Express detection
18	REFCLK– host slave	Differential clock signal pair synchronous with data
19	REFCLK+ host slave	
20	GND	N/A Ground
21	PERn0 slave host	Differential data signal pair from slave to host
22	PERp0 slave host	
23	GND	N/A Ground
24	PETn0 host slave	Differential data signal pair from host to slave
25	PETp0 host slave	
26	GND	N/A Ground

tions are for the supply, plus a few more signals for housekeeping tasks. It is noteworthy that the connector also has a USB port and an SMBus (a type of I_C bus). This bus is used in PCs for power management and system monitoring.

PCI Express cards could, for example, return measurements of the supply voltage and temperature. The PC could also put the

expansion card into standby mode while it wasn't needed. Faster versions of PCI Express make use of multiple transmit and receive channels, (hence the '0' in, for example, PERn0). In this way a form of parallelism is still used to provide a speed increase.

Paul Goossens   (054036-1)

239
Plant Growth Corrector

House plants can make things more pleasant and cheerful. However, they have the drawback that they require a fair amount of care, since otherwise their life expectancy is usually quite short. This care is not limited to watering them and occasionally adding a bit of fertilizer to the soil. The problem is that sun-loving plants always try to grow toward the source of sunlight. Regular rotation of these plants can prevent them from growing crooked. That's not especially difficult with small plants, but it can be a highly unpleasant (and difficult) task with large types, which often have large pots as well. It's even worse if the plants are located in the garden. These plants always try to grow toward the south (or for readers located south of the equator, toward the north; readers located close to the equator don't have to worry about this!). One solution is to place a light source on the side of the plant that receives the least amount of sunlight. As the selected light source must be energy-efficient, a standard grow lamp is out of the ques-

tion. LEDs, by contrast, are a very good choice for this application. Plants contain three different substances that can extract energy from incident light. From Figure 1, it can be seen that blue light can be used as a source of energy by all three substances. Red light can only be used by chlorophyll a and chlorophyll b. From this, it can be deduced that blue light is the most suitable source of light for plants, but red light is also quite usable. That's quite fortunate, because red LEDs are a lot cheaper than blue LEDs. The circuit shown in Figure 2 can be used as a guide for building your own plant lamp. The AC voltage at the

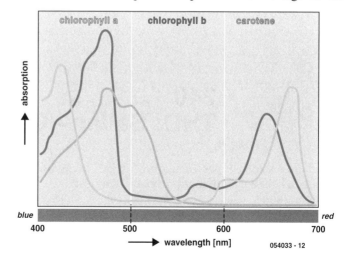

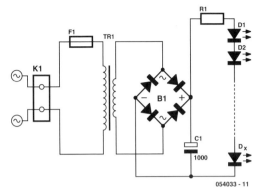

054033 - 11

transformer output is converted into a DC voltage by bridge rectifier B1. C1 smoothes this voltage to produce a more constant voltage across the resistor and LEDs. R1 ensures that the proper current flows through the LEDs. These LEDs emit energy in the form of red light. No component values are given here, since suitable val-ues must be selected according to your particular wishes.

The first thing you need to know is the nominal current rating of the LEDs and the voltage across each LED at the nominal current. The total voltage across the LEDs is then:

$$V_{LEDtotal} = V_{LED} \times \text{(number of LEDs)}$$

You should ensure that at least 5% of this voltage appears across R1, which means that:

$$V_{R1min} = 0.05 \times V_{LEDtotal}$$

The voltage across C1 is approximately 1.4 times the output voltage of the transformer, so the secondary voltage of the transformer must be

$$V_{xfmr} = (V_{R1min} + V_{LEDtotal}) \div 1.4$$

Select a transformer that can provide at least this voltage at the desired nominal current.

Now you can calculate the actual voltage across R1:

$$V_{R1} = (1.4 \times V_{xfmr}) - V_{LEDtotal}$$

Finally, calculate the value of R1:

$$R1 = V_{R1} \div I_{nominal}$$

R1 also has to dissipate a certain amount of power, so you have to first calculate this in order to determine what type of resistor to use for R1:

$$P_{R1} = V_{R1} \times I_{nominal}$$

The value if C1 is not especially critical. Something between 100 µF and 1000 µF is a good guideline value. Ensure that the capacitor is suitable for use with the selected supply voltage.

Paul Goossens (054033-1)

240
THD: Sallen–Key versus MFB

There are various types of active filters, and the Sallen–Key version is probably the most commonly used type. A voltage follower is usually used for such filters, although gain can also be realised using two additional resistors. A disadvantage of this type of filter is its relatively high sensitivity to component tolerances. Measurements made on such filters have shown that component variations affect not only the filter characteristic but also the amount of distortion. However, an advan-

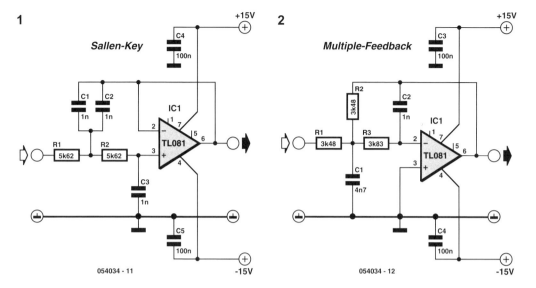

1 Sallen-Key
054034 - 11

2 Multiple-Feedback
054034 - 12

tage is that filters more complex than third-order types can also be realised using a single amplifier stage, although severe requirements are placed on the component values in such cases. One of the alternatives to the Sallen–Key filter is the 'multiple feedback' (MFB) filter. It owes its name to the fact that the feedback occurs via two paths. The inverting architecture can perhaps be regarded as a slight disadvantage, but that is offset by the fact that non-unity gain can be obtained without using extra components. In addition, the filter is less sensitive to component tolerances.

Another drawback is that the implementation is restricted to third-order filters, so additional stages (and thus opamps) are necessary for higher-order filters.

That's all very nice, you might think, but how can multiple-feedback filters be calculated? That's practically impossible to do by hand. Fortunately, various software programs have been developed to do this for you, such as the quite usable FilterPro program from Texas Instruments, which can even calculate component values that exactly match the various E series. For both types of filter, we designed a 20 kHz low-pass Butterworth bandpass filter using a standard TL081 IC (Figures 1 and 2) and then measured the distortion in the output signal for an input signal of 5Vrms. Standard polyester (MKT) capacitors were used in the circuits. To make the ultimate result more distinct, we intentionally used a simple opamp (TL081) and avoided using expensive polypropylene, polystyrene or silver-mica capacitors. The results of the mea-

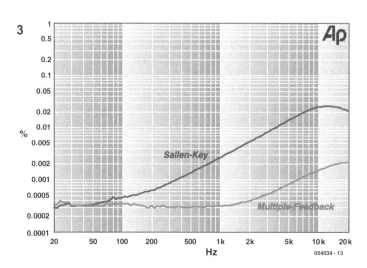

3
Sallen-Key
Multiple-Feedback
054034 - 13

surements can be characterised as astonishing. The multiple-feedback filter proved to generate considerably less distortion than the Sallen–Key architecture. Figure 3 shows the measurements for the two filters, which speak for themselves. The amplitude curves were the same within a few tenths of a dB. The Sallen–Key filter clearly generates up to more than ten times as much distortion at certain frequencies. With the Sallen–Key architecture, better

results can be obtained by using better capacitors and opamps (such as an OPA627). From the results, it is clear the multiple-feedback architecture is less sensitive to the compo-nents used in the filter.

Ton Giesberts (054034-1)

FilterPro:
http://focus.ti.com/docs/toolsw/ folders/print/filterpro.html

241
Bridge-Rectifier LED Indicator

Using a few diodes and a LED, you can make a nice indicator as shown in associated schematic diagram that can be used for a lot of applications (with a bit of luck). It's quite suitable for use in series with a doorbell or thermostat (but don't try to use it with an electronically controlled central-heating boiler!). This approach allows you to make an attractive indicator for just a few pennies. The AC or DC current through the circuit causes a voltage drop across the diodes that is just enough to light the LED. As the voltage is a bit on the low side, old-fashioned red LEDs are the most suitable for this purpose. Yellow and green LEDs require a somewhat higher forward voltage, so you'll have to first check whether it works with them. Blue and white LEDs are not suitable. You also

don't have to use modern high-efficiency types (sometimes called '2 mA LEDs' or '3 mA LEDs'). If a DC current flows through the circuit and the LED doesn't light up, reverse the plus and minus leads. When building the circuit, you'll notice that despite its simplicity it involves fitting quite a few components to a small printed circuit board or a bit of prototyping board.

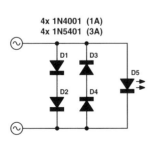

4x 1N4001 (1A)
4x 1N5401 (3A)

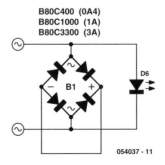

B80C400 (0A4)
B80C1000 (1A)
B80C3300 (3A)

054037 - 11

D5, D6

That's why we'd like to give you the tip of using a bridge rectifier, since that allows everything to be made much more compact, smaller and more tidy, and it eliminates the need for a circuit board to hold the components. Besides that, you can surprise friend and foe alike, because even an old hand in the trade won't understand the trick at first glance and will likely mumble something like "Huh? That's impossible." A bridge rectifier contains four diodes, which is exactly what you need. If you

short the + and − terminals of the bridge, you create a circuit with two pairs of diodes connected in parallel with opposite polarity. Select a bridge rectifier that can handle the current that will flow through it. In the case of a doorbell, for example, that can easily be 1 A. Select a voltage of 40 or 80 V. Never use this circuit in combination with mains voltage, due to the risk of contact with a live lead.

Karel Walraven (054037-1)

242
Searching for Components

This is a tip we think you'll appreciate: the large mail-order firms have a lot of information on their sites that can be quite handy even if you never order anything from them. As an example we can mention www.Farnell.com. If you're looking for a certain component, you can simply use the search menu to retrieve its most important characteristics and immediately see whether it is still generally

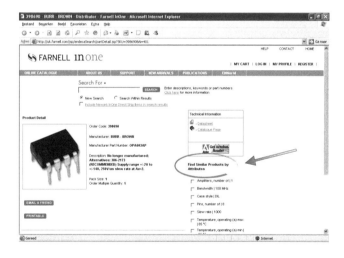

available and approximately what it costs. The full data sheet is also usually available. If the component is no longer available or you're looking for an alternative for some reason, you can use the 'find similar products by attributes' function to find several similar products. You can refine your choice in several steps until you finally end up with the components with the desired properties. You can't get the same type of results on a manufacturer's site, because such sites are usually limited to the manu-

facturer's own products, while a dealer also shows competitive products and you can directly compare their availability and prices. That means you can make things a lot easier for yourself, since it takes time to check out all the manufacturer sites, with the added risk that even if you find a suitable component it may turn out to not be available unless you plan to buy 100,000 at the same time.

Karel Walraven (054039-1)

243
PC IrDA Port

Many PC motherboards (perhaps most of them, in fact) are fitted with a connector for an infrared communication port. That connector is generally not used, since IrDA never became truly successful and has probably already seen its best days. Still, quite a few modern devices that can use this link to communicate with a PC are available, including printers, PDAs, mobile phones, and laptops. However, the link between the above-mentioned connector on the motherboard and the outside world, the IrDA interface, is not supplied with any motherboard and is usually not available in computer shops as an accessory. Fortunately, the necessary hardware is quite simple, and most solder artists should find it dead easy to build it as a 'point-to-point' construction. Selection of the IrDA IC is not critical, and just any about any type is suitable. However, you should check the size of the module when making your purchase, because quite a few types are so small that they're difficult to solder. Naturally, the interface must be able to see outside, and it should preferably be fitted at the front of the computer. A cover plate for a free floppy-disk or CD drive bay is quite suitable for this purpose. It's easy to remove from the case, and it should be easy to make a neat opening in it for the interface. The connection to the motherboard is another story. Unfortunately, the motherboard manufacturers never agreed on a standard for this. That means you'll have to consult your user's

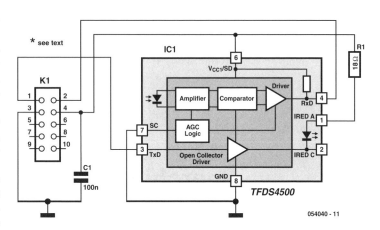

manual, and if that documentation is no longer to be found, the manufacturer's website can remedy the situation in 99% of the cases, since most manuals are available as downloads. Be careful: a mistake in the connection can cause serious damage to the motherboard, so check everything thoroughly and carefully before switching on the PC. The next step is to check the BIOS settings of the PC (to verify that the IrDA port is enabled) and install the drivers. What this involves varies from one PC to the next, and it also depends on the operating system, so there's no single recipe for it. If you aren't keen on figuring all this out for yourself, you can try searching the Internet. A highly suitable site is http://www.infrarotport.de, which clearly explains in German and English what you have to do and has all sorts of settings, drivers and patches, including supplementary explanations and experience with various types of motherboards. That's often what takes the most time in getting this serial interface to start talking.

Luc Lemmens   (054040-1)

244
Luxeon LEDs

It won't be very long before your new sitting room lamp will last forever. While our parents had to replace the lamp at least once a year, these days the energy-efficient lamps last five to ten times as long. Our children will buy an LED lamp once, which will keep working until their grave. Granted, what is written here is possibly a slight exaggeration, but that the development is in the direction of increased lifespan and higher efficiency is beyond dispute. The company Lumileds (from Agilent and Philips) has as its goal the development of LEDs that are suitable for lighting. That is why they are available in various shades of white: 3200 K (warm white), 4100 K (commercial white) and 5500 K (cool white). By combining multiple LEDs of different brightness in one fitting, the different fittings will have equal brightness, even if you buy another new one years later. With respect to power output, there are models rated 1 W and 5 W, and when you read this, the range is quite likely to have grown. Because the light output and life expectancy are strongly related to temperature, not only unmounted LEDs are available but also types integrated with a heatsink. This series has the

beautiful name of Luxeon Star LED, because the heatsink has somewhat of a star shape. Because of the heatsink it is possible to use the Luxeon Star LED at the rated current without any additional measures. For the 1 watt type that is 350 mA DC. The current may be as high as 500 mA of the LED is multiplexed, however the average value may not be higher than 350 mA. Don't use a switching frequency of less than 1 kHz, otherwise the temperature of the chip varies too much. The 3 watt type can be driven with a maximum of 1 A, also when multiplexing. If you are going to use the LEDs at these limits it is advisable to drive them with an electronically con-

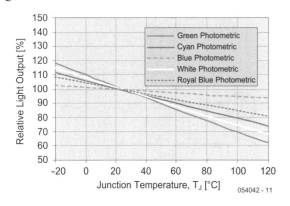

054042 - 11

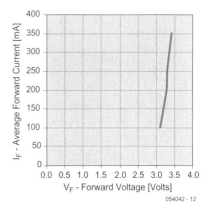

054042 - 12

trolled current source, so that you can be sure that the limits are not exceeded. In general this is not necessary, since a slightly lower current does not reduce the brightness much. This is because the brightness reduces significantly when the temperature of the chip increases. This can be as much as 10 % per 20 degrees of junction temperature! It is therefore always a good idea to provide the LED with as much additional cooling as is practicable, for example by mounting the heatsink on a thermally conducting part of the fitting. We recommend that you choose a current which is a little less than the maximum value. A simple current limiting resistor is then all you need and additional electronics is not required. For examples and calculations see:
http://www.luxeonstar.com/resistor-calculator.pdf
Note that the LED, in contrast to a halogen lamp, requires DC. So in the event of an AC power supply, apart from the resistor, it is also necessary to add a bridge rectifier between transformer and LED! More information can be found in the Custom Luxeon Design Guide, which can be downloaded from:
http://www.lumileds.com/[dfs/AB12.pdf

Karel Walraven   (054042-1)

245
Bidirectional I2C Level Shifter

In some cases I2C signals need to be level-converted if they are exchanged between (sections of) logic systems operating at different supply voltages. For example, one section of a circuit may work at 5 V while a newly added I2C device is happy at just 3.3 V. Without a suitable bidirectional level converter, signals from the 5 V system may disrupt or even damage the SDA/SCL inputs of the 3.3 V device, while the other way around signals emitted by the lower voltage device may not be properly detected. As shown in the diagram, two n-channel enhancement MOSFETs inserted in the SDA and SCL lines do the trick.
Note that each voltage section has its own pull-up resistors Rp connected between the respective supply rail (+3.3 V or +5 V) and the MOSFET source or drain terminals.

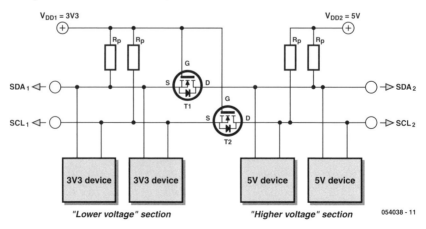

"Lower voltage" section   *"Higher voltage" section*   054038 - 11

Both gates (g) are connected to the lowest supply voltage, the sources (s) to the bus lines of the lower voltage section, and the drains (d) to the same of the higher voltage section.

The MOSFETs should have their source connected to the substrate – when in doubt check the datasheet. Other supply voltages than 3.3 V and 5 V may be applied, for example, 2 V and 10 V is perfectly feasible. In normal operation VDD2 must be higher than or equal to VDD1.

Luc Lemmens   (054038-1)

Source:
Philips Semiconductors
Application Note AN97055.

246
IrDA Interface

Many modern motherboards are equipped with an infrared data interface compliant with the IrDA standard, but this interface not very often used. However, it is not difficult to build a data transmission module and connect it to the corresponding header. As can readily be seen from the schematic diagram, this doesn't exactly involve a large array of ICs. This is because transceiver ICs are available for the IrDA standard, so only a few passive components have to be added to obtain an operational circuit. The author has successfully built this circuit many times using the TFDU5102 from Vishay Semiconductors (formerly Telefunken). If this IrDA transceiver is no longer available (it has been officially discontinued), the largely pin- and function-compatible TFDU6102 can be used without any problems. This IC is faster and meets the latest IrDA specification. The TFDU6102 low-power receiver IC supports IrDA at data rates up to 4 Mbit/s (FIR), HP-SIR, Sharp ASK, and carrierbased remote control modes up to 2 MHz. The IC contains a photodiode, an infrared emitter and CMOS control logic. The IC also has internal protection against electromagnetic immissions and emissions, so no external screening is necessary. The IC works with a supply voltage of 2.7–5.5 V, so it is suitable for use in

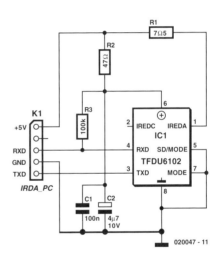

desktop PCs, notebooks, palmtops, and PDAs. It is also used in digital still and video cameras, printers, fax machines, copiers, projectors, and many other types of equipment. The author has designed a printed circuit board for the IrDA module that is only 20 × 20 mm square. Of course, this means that all of the components are SMD types. The TFDU6102 in the 'babyface' package is available in upright and flat versions. Here the upright version

COMPONENTS LIST

Resistors:
R1 = 7Ω5 (shape 1210)
R2 = 47 Ω (shape 1206)
R3 = 100 kΩ (shape 1206)

Capacitors:
C1 = 100nF (shape 1206)
C2 = 4µF7 (shape 1210)

Semiconductors:
IC1 = TFDU6102TR3 (Vishay) (Farnell)

Miscellaneous:
X1 = 5-way SIL pinheader

(suffix 'TR3') is used. Thanks to its small size, the assembled circuit board can easily be placed behind a drive bay cover or the like. It is connected to the motherboard by a five-way flat cable. The pin assignments for header X1 must match the mating connector on the motherboard. After you have fitted the module, you may have to edit the BIOS settings to activate the UART for IrDA operation. These settings enable the (Windows) operating system to boot the new device and automatically install it. You may have to briefly insert the Windows CD to modify the settings. There is an abundance of free programs on the Internet that use the IrDA interface.

A. Bitzer   (020047-1)

247
Constant Current Sink

The precision current sink shown in Figure 1, which uses a Si4936 DY dual MOSFET (IC1) as a controlled load resistance, is

1

014015 - 4

very energy-efficient and thus particularly suitable for use in battery-powered equipment. The voltage drop across current-sense resistor R4 should be approximately 85 mV. IC2 and R4 form the actual current sink, which has a very low dropout voltage. The supply voltage can be reduced to as low as the operating voltage of the load. If the supply voltage is greater than this, transistor T1 in IC2 regulates the supply current. Each of the two MOSFETS T1 and T2 can switch 4.6 A. IC2 is only available in an SO-8 SMD package. Figure 2 shows the voltages across T1, T2 and R4 for various currents. IC1, which is a MAX951, includes a 1.200 V reference potential source that is brought out to pin 6. R1 and

2

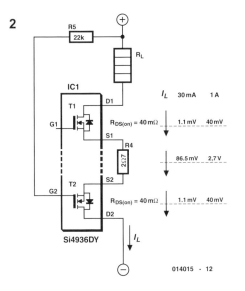

R5 22k

R_L

IC1

T1

G1

D1

$R_{DS(on)} = 40\,m\Omega$

S1

R4

$2\,\Omega\,7$

T2

G2

S2

$R_{DS(on)} = 40\,m\Omega$

D2

Si4936DY

I_L

I_L	30 mA	1 A
	1.1 mV	40 mV
	86.5 mV	2,7 V
	1.1 mV	40 mV

014015 - 12

R2 determine the voltage present at the non-inverting input of opamp IC1 (approximately 90 mV). The sense voltage across R4 is applied to the inverting input, where it is compared with the voltage on pin 3. T2 is only included in the circuit to provide reverse-polarity protection. The maximum allowable voltage between the source and gate also determines the maximum allowable supply voltage for this circuit. The values of resistors R4 and R2 must be adjusted to match the desired constant current value. The circuit is primarily intended to be used with small currents. If the load current is greater than 200 mA, the value of shunt resistor R4 should be reduced.

K. Thiesler (014015-1)

248
Automatic Switch for Voltage Converters

New applications for DC voltage converters, such as the 'workhorse' LT1070, arise every day. These converters can be adapted to nearly every imaginable ratio of input and output voltages. However, all of these circuits and devices have the same shortcoming, which is that they lack an on/off switch. Especially when they are used as a source of 6 V / 12 V power for a car radio, this is highly impractical. The circuit described here adds automatic load detection to the converter. For use in a car, the additional circuitry must be small and fit into a compact enclosure together with the converter. Since the battery voltage and ambient temperature vary over wide ranges, a simple form of load detection must be used. Besides this, the voltage drop across the load sensing circuitry must naturally be as small as possible. This can be achieved by using 'ultra-modern' SiGe technology. The 6 V from the battery and the 12 V from the converter are combined in the MB

R2545 dual diode. Consequently, a voltage of at least 6 V is always applied to the radio (for memory retention). If the radio is switched on, it draws a current from the 6 V battery, which may be around 100 mA. This current produces a voltage across R1. If this voltage is 75 mV or greater, the AC128 germanium transistor starts conducting and charges electrolytic capacitor C1, which is connected to the gate of the BUZ10. The MOSFET energises RE1 and thus connects the supply voltage to the converter. As a result, 12 V power is connected to the radio. The resulting increased current causes the voltage drop across R1 to increase, which is undesirable, so a 10 A Schottky diode is connected in parallel. The total voltage drop is thus approximately 0.6 V. The RC network connected to the BUZ10 ensures that the transistor always remains switched on for at least several seconds, to prevent the circuit from 'chattering' with varying current con-

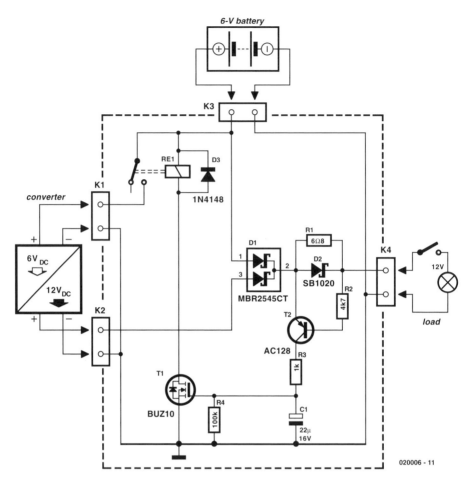

020006 - 11

sumption. If the load is switched off, the AC128 cuts off, the electrolytic capacitor discharges and the relay again disconnects the voltage converter. The residual current consumption is so small that the circuit can also be connected ahead of the ignition switch. The Schottky diodes need only be rated for the necessary voltages and currents, and above all, they should have the lowest possible saturation voltage. The exact type is not critical. Two separate diodes can also be used. A small heat sink for the MBR diode won't hurt, but this is normally not essential. Practically any type of PNP germanium transistor that is still available or on hand can be used (AC125, AC126 and AC128 work perfectly). It may be necessary to modify the value of R1. In combination with the germanium transistor, R1 determines which level of current will be ignored (for memory retention) and which level of current will cause the converter to be switched on. With the component values shown in Figure 1, this level is between 10 mA and 25 mA. It is recommended to measure the quiescent current (at 6 V) and switch-on current of the load and then simulate the switching process using dummy load resistors. When selecting the 6 V relay, ensure that its contacts have an adequate current rating. The actual value can be significantly greater than the nominal output current. With a load of 5 A at 12 V and a converter efficiency of 70 percent, the current through the relay contacts rises to 14.3 A.

C. Wolff (02006-1)

249
Card Radio

Among some of our modern contemporaries, 'musical' postcards evoke strong reactions of astonishment about hypermodern microcontroller technology. However, such flat melody memories would only have elicited a weary smile from our forefathers. As early as 1928, there are reports that radio cards with the dimensions of a regular postcard and a thickness of only a few millimetres were being made. These cards concealed a basketwork coil with a sliding tap for tuning the frequency of the received signal, a fixed capacitor and a miniscule detector device consisting of a small crystal with a 'whisker' contact. A similarly simple circuit can also be implemented using current resources. For this, you will need an interesting local medium-wave transmitter and a high-impedance headphone (1–2 kΩ), as well as a good aerial (such as a metal downpipe or an earthed radiator). The aerial is connected to an LC resonant circuit tuned to the frequency of the local transmitter, and a diode provides the demodulation. The necessary capacitance following the diode is provided by the cable to the headphone or amplifier. The coil can be made using a circular piece of stiff cardboard with a diameter of a couple of centimetres. Cut an odd number of slots into the cardboard disc. Then wind enamelled copper wire (diameter 0.15–0.2 mm) back and forth through the slots. Forty turns will give an inductance of around 80 µH. The coil looks like the bottom of a reed basket, which explains its cryptic name in RF jargon. To tune the coil to the frequency of the local transmitter and determine the required frequency of the resonant circuit, connect a dual-gang or multiple-gang variable capacitor (500–1000 pF) to the coil, with the stator sections (the fixed portion of the capacitor plates) connected in parallel. The rotor sections, which are connected to the shaft of the rotary capacitor, must without fail be connected to ground in order to prevent a 'hand effect' while tuning. Incidentally, the resonant circuit formula cannot be used to determine the tuning capacitance, since it ignores the effect of the aerial. After the capacitor has been adjusted, estimate the value of the capacitance (or even better, measure it), dig out a suitable

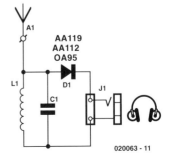

020063 - 11

fixed capacitor from your parts box and solder it to the coil at the centre of the cardboard disc, along with a general-purpose germanium diode (AA119, AA112, OA95, etc.). Secure the capacitor and diode with glue. For terminals, you can use 4 mm tubular rivets for miniature plugs, as shown in the photo. A suitable 'enclosure' can be made from 'customer discount' cards in credit-card format (you probably already have more than you really need). Use one card as the 'circuit board' for the receiver, and cut an opening in a second card to receive the circuitry. Ideally, this card should thick enough to fit the full height of the receiver. The cover is formed by a third card. After a final check, glue or rivet the cards together, and your card radio is finished. It's not high-end, but it has astonishingly good performance for such a simple circuit. One final glimpse into the past: already in the 1930s, such fixed-tuned detector receivers were available in the form of 'Berlin plugs', 'Hamburg plugs', and so on, for receiving local transmitter signals in various locations.

G. Stabe   (020063-1)

250
Model Train Identification Circuit

This circuit is intended to be used to detect trains on a model railway track. It is partially based on designs described in past issues of Elektor Electronics. As can be seen from Figure 1, it consists of a simple infrared transceiver built around an LM358 and an NE567. With the component values shown in the schematic, it has a working range of around 40 to 50 cm. When several detectors are used, crosstalk among the detectors can be avoided by setting the frequency of the NE567 to a different value for each detector. The LED and series resistor shown at the far right can be added to

1

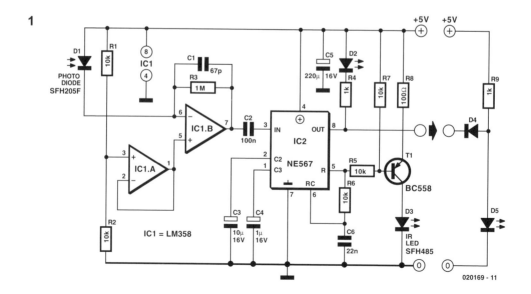

020169 - 11

2

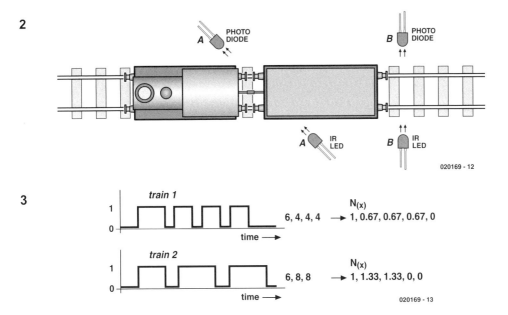

020169 - 12

3

020169 - 13

indicate that a track section is occupied. The most interesting aspect of this circuit is how it is used. For this, refer to Figure 2. If the transmitter and receiver are placed in the 'A' positions, the circuit acts as a 'normal' train detector. The diagonal placement makes it very reliable as a train detector. If instead the transmitter and receiver are placed in the 'B' positions, the circuit can be used for train identification.

When a train passes through the gate formed by the diodes in the B positions, the light beam is interrupted at certain intervals, which generates a sort of 'bar code' for the train configuration in question. Using a PC or microcontroller, it is then possible to distinguish various trains from each other using this bar code. This in turn makes it possible to have different trains travel differently on the track, or to handle them differently at the mimic station. Figure 3 shows two examples of such bar codes. Here train 1 consists of a locomotive with three short wagons, while train 2 has two long wagons.

Measuring the lengths of the individual pulses yields the series '6-4-4-4' for train 1 and the series '6-8-8' for train 2. Train speed information can be eliminated by normalising the values to the first number

of the series. This yields a series that indicates how long the wagons are relative to the length of the locomotive, regardless of the speed of the train. If zeroes are added as necessary to achieve a certain train length and the detected values are compared with all previously stored train configurations, surprisingly reliable train identification is possible.

The comparison can be performed in a variety of manners. One way that works well is to determine the quadratic error between the measured value and all stored 'reference trains'. However, this method is sensitive to differences in train length, so a reference train with one less wagon will not be recognised. The error estimate can be strongly reduced by taking train length into account (by reducing the reference trains to the same length as the measured train). The numerical value determined using quadratic error estimation is also a measure of the reliability of the estimate. An error threshold can be used to distinguish between trains that have been recognised and trains that are uncertain. In the author's system, the light gate operates at a frequency of approximately 20 Hz, which yields good train recognition. A final tip: if you place the transmitter and receiver diodes at the level of the floor of the wagons, you can prevent the light from passing through carriage windows to the receiver. This prevents measurement errors, and it allows flat and open wagons to be properly sensed.

M. Hajer

(020169-1)

251
Secure On/Off Pushbutton

The author used to employ a switching central that allows mains appliances among his PC equipment to be individually switched on and off by pressing pushbuttons. One of these pushbuttons switches the PC on and off. Good economics, but not perfect, because awkward situations can arise when a wrong pushbutton is pressed, causing the PC to be switched off instead of, say, powering up the printer. A solution was found in the use of relay which is switched on and off only if the relevant pushbutton is pressed twice within a certain perie first press causes a LED inside the pushbutton to flash for about 7 seconds. If you press again within this period, the re-

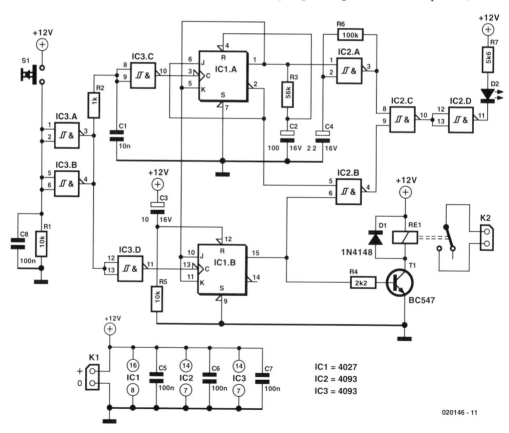

020146 - 11

lay is energised and the LED lights permanently. The switch-off procedure is identical. In this way, the switch is given a level of security. After the first press on S1, IC1.A is switched on and the resulting state is indicated by D2 starting to flash in the rhythm defined by oscillator IC2.A. The flash rate is determined by components R6 and C4. At the same time, the second flip-flop IC1.B is enabled via its J/K inputs. Capacitor C2 is slowly charged via R3. Once the switching threshold is reached, flip-flop IC1.A is reset again. Within this period, the pushbutton has to be pressed again. IC1.B, now operating as a data flip-flop (toggling on the positive pulse edge), toggles and causes the relay to be energised. Also, IC2.B then causes D2 to light constantly (or remain off when the relay is switched off). Capacitor C3 ensures that the relay contact cannot

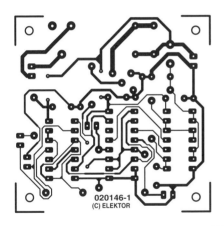

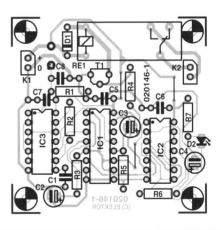

COMPONENTS LIST

Resistors:
R1,R5 = 10kΩ
R2 = 1kΩ
R3 = 56kΩ
R4 = 2kΩ2
R6 = 100kΩ
R7 = 5kΩ6

Capacitors:
C1 = 10nF 5mm lead pitch
C2 = 1000µF 16V radial
C2 = 100µF 16V radial
C2 = 2µF2 16V radial

C5-C8 = 100nF

Semiconductors:
D1 = 1N4148
D2 = LED, red, low current
T1 = BC547
IC1 = 4027
IC2,IC3 = 4093

Miscellaneous:
K1, K2 = solder pin
Re1 = relay, 12V (e.g. Siemens V23057-12V)
S1 = pushbutton, 1 make contact

close of its own should the mains voltage disappear. The circuit may be built on a PCB of which the layout is shown here. As usual the artwork file may be obtained from the Free Downloads section of the Publishers' website.

L. Libertin

(020146-1)

252
Alternating Blinker for Model Vehicles

This type of blinker is primarily used in model vehicles, where it can be used to achieve a realistic imitation of the dual flasher bar on an ambulance or fire engine. The LEDs, which are connected to the two open collectors, blink twice on the right and then twice on the left. The visual effect is simply fantastic. The electronics consists of a simple 555-based clock generator and a 4017 decimal counter.

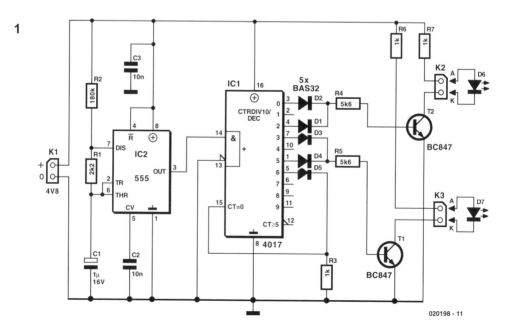

1

020198 - 11

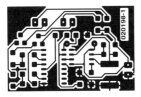

2

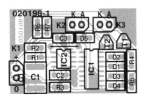

COMPONENTS LIST

(all parts SMD)
Resistors:
R1 = 2kΩ2
R2 = 180kΩ
R3,R6,R7 = 1kΩ
R4,R5 = 5kΩ6

Capacitors:
C1 = 1µF 16V
C2 = 10nF
C3 = 100nF

Semiconductors:
D1-D5 = BAS32
D6,D7 = LED, low current*
T1,T2 = BC847
IC1 = 4017T
IC2 = NE555

Miscellaneous:
K1 = supply connections
* not SMD, see text

The 555 is wired as a free-running oscillator. The timer output signal on pin 3 is fed to the Count input of the 4017, whose outputs go high sequentially.
Each time an output goes high, the previous output goes low, as it should be with a decimal counter. The 'program' runs in six steps. In the first and third steps, T2 conducts, while in the fourth and sixth steps T1 conducts. Each time one of the transistors conducts, the connected LED flashes. On the second and fifth clock pulse, both transistors are switched off and the LEDs are dark. The leading edge of the seventh clock pulse resets the counter via Q6, and the cycle starts again from the beginning. Other blinking patterns can also be implemented using different logical-OR combinations of the IC outputs. The open-collector outputs allow other loads, such as relays or incandescent lamps, to be connected in place of the LEDs. If this is done, R6 and R7 should be replaced by 0 Ω resistors or wire bridges.

However, the load current should not exceed 300 mA. Circuit construction using the printed circuit board shown in Figure 2 is not difficult, despite the use of SMD components, so it should work properly right from the start.

U. Neubert   (020198-1)

253
Solar Relay

With extended periods of bright sunshine and warm weather, even relatively large storage batteries in solarpower systems can become rather warm. Consequently, a circuit is usually connected in parallel with the storage battery to either connect a high-power shunt (in order to dissipate the excess solar power in the form of heat) or switch on a ventilation fan via a power FET, whenever the voltage rises above approximately 14.4 V. However, the latter option tends to oscillate, since switching on a powerful 12 V fan motor causes the voltage to drop below 14.4 V, causing the fan to be switched off. In the absence of an external load, the battery voltage recovers quickly, the terminal voltage rises above 14.4 V again and the switching process starts once again, despite the built-in hysteresis. A solution to this problem is provided by the circuit shown here, which switches on the fan in response to the sweltering heat produced by the solar irradiation instead of an excessively high voltage at the battery terminals. Based on experience, the risk of battery overheating is only present in the summer between 2 and 6 pm. The intensity of the sunlight falling within the viewing angle of a suitably configured 'sun probe' is especially high precisely during this interval. This is the operating principle of the solar relay. The trick to this apparently rather simple circuit consists of using a suitable combination of components. Instead of a power FET, it employs a special 12 V relay that can handle a large load in spite of its small size. This relay must have a coil resistance of at least 600 Ω, rather than the usual value of 100-200 Ω. This requirement can be met by several Schrack Components relays (available from, among others, Conrad Electronics). Here we have used the least expensive model, a type RYII 8-A printed circuit board relay. The light probe is connected in series with the relay. It consists of two BPW40 phototransistors wired in par-

allel. The type number refers to the 40-degree acceptance angle for incident light. In bright sunlight, the combined current generated by the two phototransistors is sufficient to cause the relay to engage, in this case without twitching. Every relay has a large hysteresis, so the fan connected via the a/b contacts will run for many minutes, or even until the probe no longer receives sufficient light. The NTC thermistor connected in series performs two functions. First, it compensates for changes in the resistance of the copper wire in the coil, which increases by approximately 4 percent for every 10 °C increase in temperature, and second, it causes the relay to drop out earlier than it otherwise would (the relay only drops out at a coil voltage of 4 V). Depending on the intended use, the 220 Ω resistance of the thermistor can be modified by connecting a 100 Ω resistor in series or a 470 Ω resistor in parallel. If the phototransistors are fastened with the axes of their incident-angle cones in parallel, the 40-degree incident angle corresponds to 2 pm with suitable solar orientation. If they are bent at a slight angle to each other, their incident angles overlap to cover a wider angle, such as 70 degrees. With the tested prototype circuit, the axes were oriented nearly parallel, and this fully met our demands. The automatic switch-off occurs quite abruptly, just like the switch-on, with

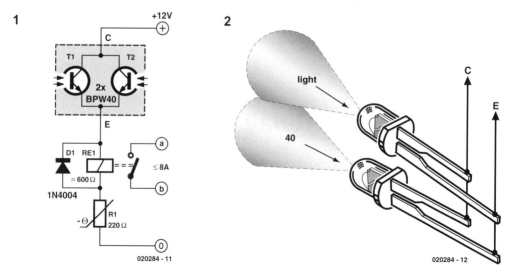

020284 - 11

020284 - 12

no contact jitter. This behaviour is also promoted by the NTC thermistor, since its temperature coefficient is opposite to that of the 'PTC' relay coil and approximately five times as large. This yields exactly the desired effect for energising and de-energising the relay: a large relay current for engagement and a small relay current for disengagement. Building the circuit is actually straightforward, but you must pay attention to one thing. The phototransistors resemble colourless LEDs, so

there is a tendency to think that their 'pinning' is the same as that of LEDs, with the long lead being positive and the short lead negative. However, with the BPW40 the situation is exactly the opposite; the short lead is the collector lead. Naturally, the back-emf diode for the relay must also be connected with the right polarity. The residual current on cloudy days and at night is negligibly small.

W. Zeiller (020284-1)

254
Step-Up Booster Powers Eight White LEDs

Tiny white LEDs are capable of delivering ample white light without the fragility problems and costs associated with fluorescent backlights. They do pose a problem however in that their forward voltage can be as high as 4 V, precluding them being from powered directly from a single Li-Ion cell. Applications requiring more white LEDs or higher efficiency can use an LT1615 boost converter to drive a series connected array of LEDs. The high efficiency circuit (about 80%) shown here can provide a constant-current drive for up to eight LEDs. Driving eight white LEDs in series requires at least 29 V at the output and this is possible thanks to the internal 36 V, 350 mA switch in the LT1615. The constant-current design of the circuit guarantees a steady current through all LEDs, regardless of the forward voltage differences between them. Although this circuit was designed to operate from a single Li-Ion battery (2.5V to 4.5V), the LT1615 is

also capable of operating from inputs as low as 1 V with relevant output power reductions. The Motorola MBR0520 surface mount Schottky diode (0.5 A 20 V) is a good choice for D1 if the output voltage does not exceed 20 V. In this application however, it is better to use a diode that can withstand higher voltages like the

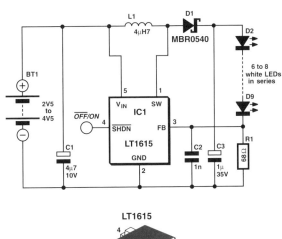

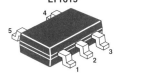

MBR0540 (0.5 A, 40 V). Schottky diodes, with their low forward voltage drop and fast switching speed, are the best match. Many different manufacturers make equivalent parts, but make sure that the component is rated to handle at least 0.35 A. Inductor L1, a 4.7 µH choke, is available from Murata, Sumida, Coilcraft, etc. In order to maintain the constant off-time (0.4 ms) control scheme of the LT1615, the on-chip power switch is turned off only after the 350 mA (or 100 mA for the LT1615-1) current limit is reached. There is a 100 ns delay between the time when the current limit is reached and when the switch actually turns off. During this delay, the inductor current exceeds the current limit by a small amount.

This current overshoot can be beneficial as it helps increase the amount of available output current for smaller inductor values. This will be the peak current passed by the inductor (and the diode) during normal operation. Although it is internally current-limited to 350 mA, the power switch of the LT1615 can handle larger currents

without problems, but the overall efficiency will suffer. Best results will be obtained when IPEAK is kept well below 700 mA for the LT1615. The LT1615 uses a constant off-time control scheme to provide high efficiencies over a wide range of output current. The LT1615 also contains circuitry to provide protection during start-up and under short-circuit conditions. When the FB pin voltage is at less than approximately 600 mV, the switch offtime is increased to 1.5 ms and the current limit is reduced to around 250 mA (i.e., 70% of its normal value). This reduces the average inductor current and helps minimize the power dissipation in the LT1615 power switch and in the external inductor L1 and diode D1.

The output current is determined by $V_{ref}/R1$, in this case, $1.23V/68 = 18$ mA. Further information on the LT1615 may be found in the device datasheets which may be downloaded from: www.linear-tech.com/pdf/16151fa.pdf.

D. Prabakaran   (020349-1)

255
Telephone Free Indicator

Depending on local regulations and the telephone company you happen to be connected to, the voltage on a free telephone line can be anything between 42 and 60 volts. As it happens, that's sufficient to make a diac conduct and act like a kind of zener diode maintaining a voltage of 38 V or so. The current required for this action causes the green high-efficiency LED in the circuit to light. Line voltages higher than about 50 V may require R1 to be changed from 10 k to a slightly higher value. When the receiver is

lifted, the line voltage drops to less than 15 V (typically 12 V) causing the diac to

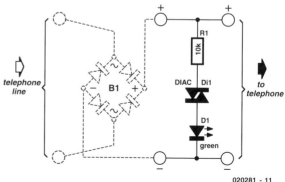

020281 - 11

block and the LED to go out. The circuit diagram indicates + and − with the phone lines. However, in a number of countries the line polarity is reversed when a call is established.

To make sure the circuit can still function under these circumstances, a bridge rectifier may be added as indicated by the dashed outlines. The bridge will make the circuit independent of any polarity changes on the phone line and may consist of four discrete diodes, say, 1N4002's or similar. Finally, note that this circuit is not BABT approved for connection to the public switched telephone network (PSTN) in the UK.

R. J. Gorkhali   (020281-1)

256
Universal Clock Generator

This universal clock generator is implemented using an Atmel AT90S2313 microcontroller, so it does not require very many external components. It is a versatile

Divider (S3)	Frequency (S2) (frequency in Hz)							
	0	1	2	3	4	5	6	7
0	1 MHz mode (activated by Reset)							
1	stopped	0.1	1	10	100	1,000	10,000	100,000
2	stopped	0.05	0.5	5	50	500	5,000	25,000
3	stopped	0.0333	0.333	3.333	33.33	333.3	3,333	16,666
4	stopped	0.025	0.25	2.5	25	250	2,500	12,500
5	stopped	0.02	0.2	2	20	200	2,000	10,000
6	stopped	0.0166	0.1666	1.666	16.66	166.6	1,666	8,333
7	stopped	0.0143	0.143	1.43	14.3	143	1,430	7,143
8	stopped	0.0125	0.125	1.25	12.5	125	1,250	6,250
9	stopped	0.0111	0.1111	1.111	11.11	111.1	1,111	5,555
A	stopped	0.01	0.1	1	10	100	1,000	5,000
B	stopped	0.0091	0.091	0.91	9.1	91	910	4,545
C	stopped	0.00833	0.0833	0.833	8.33	83.3	833	4,166
D	stopped	0.0077	0.077	0.77	7.7	77	770	3,846
E	stopped	0.00714	0.0714	0.714	7.14	71.4	714	3,571
F	stopped	0.00666	0.0666	0.666	6.66	66.6	666	3,333

Table 1: Fixed frequencies

(Values rounded to 1% as necessary)

clock generator for use in pulse generator and timer circuits in a lab setting. It provides approximately 100 reference frequencies with 'crystal accuracy', and it can be used to implement dimmer circuits, generate arbitrary pulse waveforms for simulations, as a freely settable frequency generator and much more. Operation and adjustment are simple and easy to understand. As can be seen, all of the

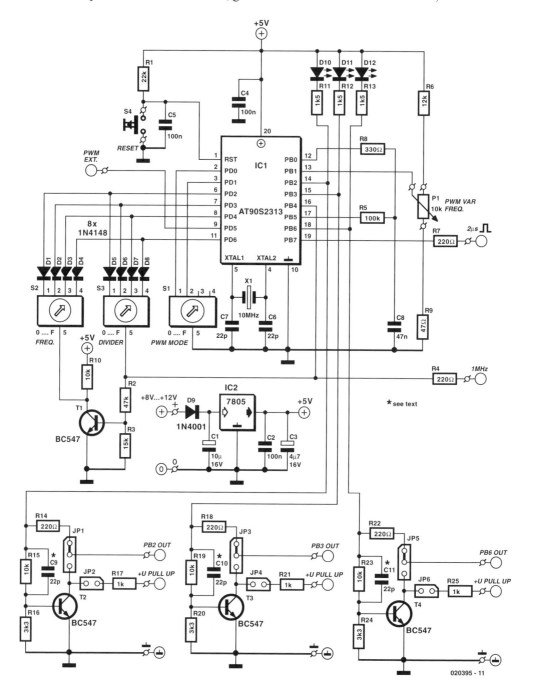

020395 - 11

microcontroller ports are fully used. Port lines PC2–PD4, PD6 and PB4 have dual functions. This is not a problem for a microcontroller as long as the software is suitably designed. In this case, two BCD switches are read using the same inputs. The switches are selected using PB4 and inverter T1, and they are isolated from each other by diodes D1–D8. The 1 MHz clock signal is also output via PB4. During normal operation, the 1 MHz output is inactive; only when it is in the 1 MHz mode does the microcontroller enter a permanent loop, in which the other functions are anyhow not used.

Each of the outputs PB2, PB3 and PB6 is connected to a transistor stage, which can be configured using jumpers according to the intended application. The microcontroller outputs can be fed out either directly or inverted via the transistor stages. Another jumper can be used to connect a pull-up resistor, which can be connected

externally to any desired voltage below the maximum specified value for the transistor. With the given component values, the output circuits are short-circuit proof for transient shorts. For some applications, it may be necessary to alter the component values. If necessary, suitable external circuitry can also be added.

Capacitors C9–C11 are only necessary if it is important to have extremely steep pulse edges. These capacitors accelerate the switch-on response of the transistors and reduce the delay time when the transistors switch off. The frequency or pulse duty cycle is set using PB0, PB1 and PB5. Via PB1, the microcontroller measures the time required for capacitor C8 to charge to a threshold level of 2.2 V. Within this range, the charging curve of the capacitor is still relatively linear. The maximum value is set by R6, and the minimum value by R9; these can be modified if desired. For good longterm stability, C8 should be a polystyrene type if possible. The microcontroller inputs are configured to use internal pull-up resistors.

Table 2: Variable frequencies

FREQUENCY (S2)	Frequency range (Hz)
8	stopped
9	196 – 50,000
A	39 – 10,000
B	3.9 – 1,000
C	0.39 – 100
D	0.039 – 10
E	0.0039 – 1
F	stopped (free)

Frequency range resolution: 8 bits (255 steps)

Table 3: Pulse width modulator (PWM)

PWM MODE (S1)	PB1	PB0	PWM clock
0	1	1	external (PD5)
1	1	0	19.6 kHz
2	0	1	2.45 kHz
3	0	0	306.4 Hz

The following is a summary of the utilisation of the I/O lines:

PB0, PB1 & PB5
Analogue signal processing using a simple RC network. The comparator in the microcontroller switches when the voltage on PB0 exceeds the value on PB1. The capacitor is charged and discharged under program control via PB5.

PB2 & PB6
Clock frequency on complementary outputs, switchable during operation. Approximately 100 fixed frequencies and six variable ranges can be selected. The outputs can be used directly or buffered and inverted by open-collector drivers.

PB4
1 MHz clock output when the fixed-frequency divider switch is set to '0' and Reset is pressed. This mode can only be exited by pressing Reset, which causes the

output to go high and instigate the change-over via BCD switch S3. In normal operation, the output is low.

PB7
Pulse edge marker for each switching transition on PB2 and PB6. Pulse width 2 µs. This output is active for both fixed and variable frequencies.

PB3
Pulse width modulator (PWM) output; operates in parallel with the frequency outputs (8-bit resolution). One of four PWM clock frequencies can be selected using PD0 & PD1 (19.6 kHz, 2.45 kHz, 306.4 Hz or an external clock on PD5). The PWM can be adjusted over a range of approximately 0–99.5 percent of the supply voltage using an analogue voltage (P1). In the 1 MHz mode, the PWM continues to operate with its most recent setting. Same output options as PB2 and PB6.

PD0 & PD1
Configuration inputs for selecting the PWM clock frequency using DIP switches or a rotary BCD switch. The setting is captured on Reset (see Table 1).

PD2, PD3, PD4 & PD6
Configuration inputs for fixed-frequency and divider settings using two BCD switches. The frequency setting is captured immediately following any changes to the input values; the divider setting is only captured on Reset, at which time S3 is evaluated, following which PB4 switches back to S2.

PD5
External PWM clock frequency. If Reset is pressed when S1 = 0, the PWM is switched to the external clock frequency.

The circuit can be powered by a small (mains adapter) power supply providing an output voltage of 8–12 V. On the circuit board, a fixed voltage regulator converts the supply voltage to a stabilised 5 V level. D1 protects the circuit against reverse-polarity connection of the power supply. Decoupling capacitors should be soldered to the circuit board as close as possible to the supply pins of the microcontroller. The reset switch is debounced using an RC network. The crystal must have a frequency of 10 MHz, since all of the calculations in the software are based on this value. It should be located as close as possible to the microcontroller, along with the associated capacitors.

The pushbutton switch and BCD switches are also fitted to the PCB. It is a practical idea to use IC sockets to fit the BCD switches. This allows the switches to be fitted external to the PCB if necessary and connected using cables and plugs. The provisions of the EMC Directive should generally be observed in the construction of the generator. After all, here you are working with a fast microcontroller with a 10 MHz clock frequency. The decoupling capacitors directly connected to the

---

Controls

FREQUENCY (S2, BCD): selects the basic frequency
Fixed: 0.1, 1, 10, 100, 1000, 10 000, 50 000 Hz (settings 1–7)
Variable: 50 000, 10 000, 100, 100, 10, 1 Hz (settings 9–14)
The clock generator is stopped (disabled) for settings 0, 8, and 15.

DIVIDER (S3, BCD): divides the basic frequency in 15 steps (1:1 to 1:15). The selected division factor only becomes effective after Reset or when the FREQUENCY setting is changed. If step 0 is selected and Reset is pressed, the 1 MHz mode is activated.

PWM MODE (S1, BCD): selects one of four clock frequencies. The PWM operates in parallel to the frequency outputs. The pulse width can be adjusted using a potentiometer or an analogue voltage.

microcontroller crystal should be located as close as possible to the microcontroller, and the pertinent capacitors should be connected directly to ground. It is recommended to have the largest possible ground plane or a gridded ground reference plane, use short or screened wiring and use freewheeling diodes with inductive loads.

R. Zenzinger (020395-1)

257
Small RC5 Transmitter

We have published many articles in past issues of *Elektor Electronics* about IR data transmission using the Philips RC5 code. A prerequisite for these using these circuits has always been an RC5 remote control made by Philips or Loewe, or else a 'home-brew' RC5 transmitter based on the SAA3010 encoder IC. Philips has now officially discontinued its RC5 encoder, effective the end of last year. Nevertheless, it does not appear to be difficult to obtain small quantities of this IC from existing stocks – and that could remain true for some time to come. It is thus certainly worthwhile to design a simple RC5 remote control unit, especially given the availability of the PT2211 from Princeton in Taiwan and the HT6230 from Holtek, which are nearly identical alternatives. Our objective is to have a versatile transmitter that can communicate with several different devices. As this can only be achieved by using a user-modifiable address code, an 8-section DIP switch is used to allow the system address to be selected within the range of 0–7. According to the RC5 protocol, this makes it possible to address the devices listed in table 1 – and for our pur-

1

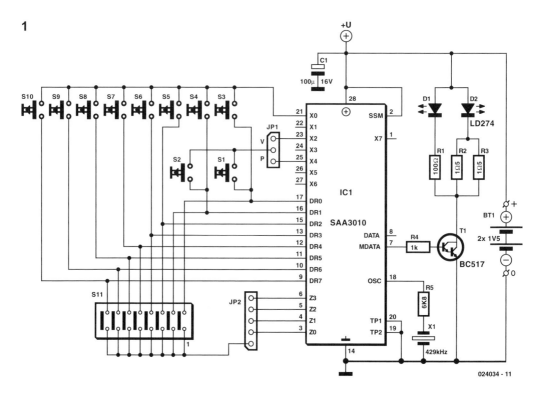

024034 - 11

Table 1		
S11	**Address**	**Device**
1	0	TV1
2	1	TV2
3	2	Videotext
4	3	Extension TV1/TV2
5	4	Laser Vision Player
6	5	VCR1
7	6	VCR2
8	7	Reserved

Table 2	
Address range	**Jumper JP2**
0–7	1–2
8–15	1–3
16–23	1–4
24–31	1–5

poses, the 'DIY system address' (7) is naturally especially interesting. A jumper (wire bridge) allows other address ranges to be alternatively selected, as shown in table 2. This allows the RC5 transmitter to work with all possible RC5 receivers. Using the ten pushbutton switches, ten different commands can be sent to the device selected by JP2 and S11.

Buttons S1 and S2 have a special role here: depending on the setting of JP1 (which may be a jumper, a wire bridge or a small slide switch), S1 and S2 control the volume (audio) or the preset values, as shown in table 3. This makes our simple RC5 transmitter compatible with the Model Remote Control circuit design published in the May 2001 issue of Elektor Electronics. In other respects, the circuit shown in Figure 1 corresponds to the specifications in the data sheet. The SSM pin is connected to Vpp, since address selection is hard-wired using the DIP switch. Philips prescribes a resonator with a 429 kHz clock frequency for compliance with the timing specified

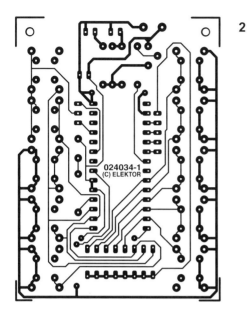

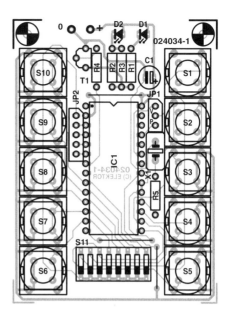

Table 3			
Button	**Command**	**JP1**	**Command**
S3	0	–	0
S4	1	–	1
S5	2	–	2
S6	3	–	3
S7	4	–	4
S8	5	–	5
S9	6	–	6
S10	7	–	7
S1	16	1–2	+ Volume
S2	17	1–2	– Volume
S1	32	2–3	+ Preset
S2	33	2–3	– Preset

COMPONENTS LIST

Resistors:
R1 = 100Ω
R2,R3 = 1Ω5
R4 = 1kΩ
R5 = 6kΩ8

Capacitor:
C1 = 100µF 16V radial (low profile)

Semiconductors:
D1 = LED, red, 3 mm
D2 = LD274
T1 = BC517
IC1 = SAA3010 (see text)

Miscellaneous:
BT1 = two 1.5V batteries with holder
S1-S10 = pushbutton, PCB mount (e.g. ITT/Shadow D6)
X1 = 429kHz 2-pin ceramic resonator
JP1 = 3-way pinheader with jumper, or wire link
JP2 = wire link
S11 = 8-way SIP switch

for the RC5 protocol. Using a 455 kHz resonator (which is easier to obtain) is not allowed, since it would cause the bit intervals to be shortened by around 40 µs. However, this could be taken into account in a DIY receiver design. A BC517 Darlington transistor is used as a driver for the IR transmitter diode (D2), which emits in the 950 nm region, because of its high current gain – which is so great that it can also drive a normal LED 'on the side' as an operational indicator.

The current level is set to give the remote control transmitter a range of around eight metres. The circuit works with a supply voltage of 3 V. Two AAA or AA cells can be used as a power source. In the quiescent state, the load on the batteries is less than 10 µA. We have designed a circuit board for this simple IR transmitter (Figure 2).

The layout can be downloaded from the *Elektor Electronics* website (Free Downloads, item 024034-11).

F. Wohlrabe (024034-1)

258
Boolean Expression Parser

Given a Boolean expression or its minterms, the program described here prints the simplified ('reduced' or 'parsed') expression. Anyone who's ever designed a fairly complex digital circuit will know that what looks like a tangled network of logic gates and inverters may often be simplified by writing down everything that happens in the circuit in the form of Boolean expressions and then discovering that these can often be condensed in a way that allows your to actually cut down on the number of ICs to be used!

The program is also useful if you need to design logic with nothing more front of you than a large truth table describing the functionality. The program was written to run under DOS which means that it is not particularly user friendly but then again it's free! Boolean equations may be entered in two ways.

One is to simply enter the equation, the other, to define the minterms. The first method is fairly straightforward. The use of brackets is not allowed. An inverted signal is entered by adding an apostrophe (') behind it, for example, AB becomes AB', and AB is entered as A'B'.

The minus sign (–) is used to terminate the entry. The use of minterms is practical if the Boolean equation is not available, but you do have a truth table. Using this method you mark those positions in the truth able that should use a output to become '1'. The value –1 is used to terminate the entry. Next, the program will ask you how many

variables should be used. An example may help to explain the method. The table below has to be entered into the program by typing 1[Enter]3[Enter]–1[Enter] and then replying '3' to the number of variables prompt (we need A, B and C).

A	B	C	OUT
0	0	0	0
0	0	1	1
0	1	0	0
0	1	1	1
1	0	0	0
1	0	1	0
1	1	0	0
1	1	1	0

After a valid entry the program will print the simplified Boolean expression on the screen.

The program executable code as well as the source code written in C++ may be obtained from the Free Downloads section at the publishers' website, www.elektor-electronics.co.uk.

The file number is 020391-11, see month of publication.

R. Sridhar (020391-1)

259
Motorcycle Battery Monitor

A circuit for monitoring the status of the battery and generator is undoubtedly a good idea for motorcyclists, as for other motorists. However, not every biker is willing to drill the necessary holes in the cockpit for the usual LED lamps, or to screw on an analogue accessory instrument. The circuit shown here manages to do its job with a single 5 mm LED, which can indicate a total of six different conditions of the onboard electrical system. This is done using a dual LED that can be operated in pulsed or continuous mode (even in daylight). Built on a small piece of prototyping board and fitted in a mini-enclosure, the complete circuit can be tucked inside the headlamp housing or hidden underneath the tank.

The heart of the circuit is IC2, a dual comparator. The comparator circuit is built without using any feedback resistors, with the indication being stabilised by capaci-

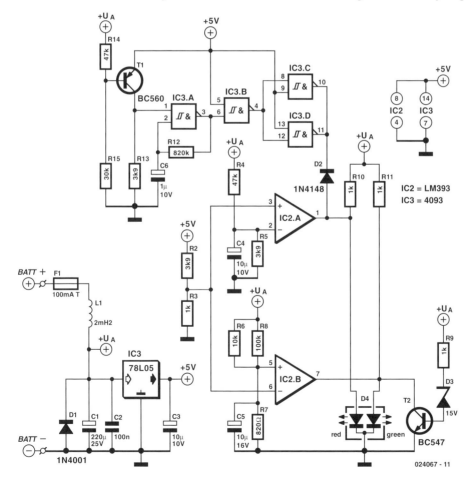

024067 - 11

tors C4 and C5 instead of hysteresis. Small 10 μF tantalum capacitors work well here; 220 μF 'standard' electrolytic capacitors are only necessary with poorly regulated generators. Voltage regulator IC1 provides the reference voltage for IC2 via voltage divider R2/R3. The onboard voltage is compared with the reference voltage via voltage dividers R4 /R5 and R6/R7, which are connected to the inverting and non-inverting comparator sections, respectively. Using separate dividers allows the threshold levels to be easily modified by adjusting the values of the lower resistors. IC2a drives the anode of the red diode of LED D4 via pull-up resistor R10. The anode of the green diode is driven by IC2b and R11. T2 pulls R11 to ground, thereby diverting the operating current of the green diode of the LED, if the voltage of the electrical system exceeds a threshold level of 15 V (provided by Zener diode D3). The paralleled gate outputs on pins 10 and 11 of IC3 perform a similar task. However, these gates have internal current limiting, so they can only divert a portion of the current from the red diode of the LED.

The amount of current diverted depends on the battery voltage. The two gates are driven by an oscillator built around IC3a,

Indication	Ignition on, lights off	Motor rpm > 2000
Blinking red	Battery deeply discharged ($U_{BATT} < 7.1$ V)	Electrical system Defective
Pulsing red	Battery empty (7.1 V < U_{BATT} < 11.3 V)	Generator not Charging
Continuous red	Battery half full (11.3 V < U_{BATT} <12.1 V)	Generator barely charging
Continuous yellow	Battery full (12.1 V < U_{BATT} <13.0 V)	Generator overloaded
Continuous green	13.0 V U_{BATT} <16.0 V	Normal state
Dark	U_{BATT} = 16.0 V	Regulator defective

which is enabled via voltage divider R14/R15 and transistor T1 when the battery voltage is sufficiently high. Depending on the state of IC3a, the red diode of the LED blinks or pulses. The circuit is connected to the electrical system via fuse F1 and a low-pass filter formed by L1 and C1. If you cannot obtain a low-resistance choke, a 1 Ω resistor can be used instead. In this case, the values of C3, C4 and C5 should be increased somewhat, in order to help stabilise the indication. D1 protects the circuit against negative voltage spikes, as well as offering protection against reverse-polarity connection. Due to its low current consumption (less than 30 mA), the circuit could be connected directly to the battery, but it is better to power it from the switched positive voltage.

A. Eschhold (024067-1)

260
Power Buzzer

How often on average do you have to call members of your family each day to tell them that dinner is ready, it's time to leave, and the like? The person you want is usually in a different room, such as the hobby room or bedroom. A powerful buzzer in the room, combined with a pushbutton at the bottom of the stairs or in the kitchen, could

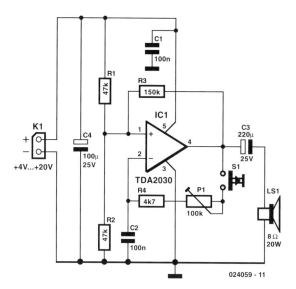

024059 - 11

where you want to have the pushbutton. The loudspeaker can then be placed in a strategic location, such as in the bedroom or wherever is appropriate. Use speaker cable to connect the loudspeaker. Normal bell wire can cause a significant power loss if the loudspeaker is relatively far away. The loudspeaker must be able to handle a continuous power of at least 6W (with a 20 V supply voltage). The power quickly drops as the supply voltage decreases

$$P = \frac{U_{rms}^2}{R_L}.$$

be very handy in such situations. The heart of this circuit is formed by IC1, a TDA2030. This IC has built-in thermal protection, so it's not likely to quickly give up the ghost. R1 and R2 apply a voltage equal to half the supply voltage to the plus input of the opamp. R3 provides positive feedback. Finally, the combination of C2, R4 and trimmer P12 determines the oscillation frequency of the circuit. The frequency of the tone can also be adjusted using P1. There is no volume control, since you always want to get attention when you press pushbutton S1. Fit the entire circuit

The power supply for this circuit is not particularly critical. However, it must be able to provide sufficient current. A good nominal value is around 400 mA at 20 V. At 4 V, it will be approximately 25 mA. Most likely, you can find a suitable power supply somewhere in your hobby room. Otherwise, you can certainly find a low-cost power supply design in your *Elektor Electronics* archive that will fill the bill!

G. Baars   (024059-1)

261
Cheap and Cheerful Transistor Tester

By using a simple visual indicating system, this small transistor tester allows you to run a quick 'go/non-go' check on NPN as well as PNP transistors. If the device under test is a working NPN then the green LED (D1) will flash, while the red counterpart will flash for a functional PNP device. However if the transistor is shorted, both LEDs

will flash, and an open-circuit device will cause the LEDs to remain off. The circuit is based on just one CD4011B quad NAND gate IC, four passive parts and two LEDs. The fourth gate in the IC is not used and its inputs should be grounded. Alternatively, you may want to connect its inputs and output in parallel with IC1.C to increase its

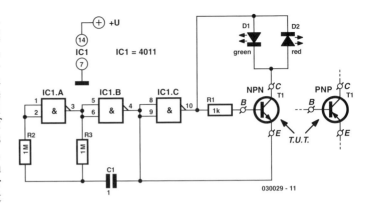

drive power to the transistor test circuit. IC1.A and IC1.B together with R2, R3 and C1 form an oscillator circuit that generates a low-frequency square wave at pin 4. This signal is applied to the emitter of the transistor under test as well as to inverter IC1.C. The inverted signal from IC1.C and the oscillator output then drive the test circuit (LEDs, device under test, R1) in such a away that the voltage across that part of the circuit is effectively reversed all the time. For example, with an NPN transistor under test, when pin 10 is High and pin 4, Low, current flows through LED D1 and the forward biased transistor. However, no current will flow when pins 10 and 4 change states, since the transistor is then reverse-biased. The green LED, D1, will therefore flash at the rate determined by the oscillator. As you would expect to happen, a PNP transistor will be forward biased when pin 10 is Low and 4, High, en-

abling current to flow through the red LED in that case. A supply rail of around 3 V (two series connected 1.5 V batteries) should be adequate. To prevent damage to the transistor under test, supply voltages higher than 4.5 V should not be used. Because the LED currents are effectively limited to a few mA by the output of IC1.C (also slightly dependent on the supply voltage), it is recommended to use high-efficiency devices for D1 and D2.

R. J. Gorkhali   (030029-1)

262
555 DC/DC Converter

It is all too often necessary to augment the power supply of an existing electronic circuit because exactly the voltage that you need is missing. The circuit presented here may provide a solution in a number of cases, since it can be used to convert a single-ended supply voltage into a balanced set of supply voltages. That's not so remarkable by itself, but the special feature of this circuit is that this is accomplished without using difficult to obtain, exotic ICs. All of the components used in the circuit are ones that every electronics hobbyist is likely to have in a drawer somewhere.

The heart of the circuit is formed by an 'old reliable' 555 timer, which is wired here as a free-running oscillator with a frequency of approximately 160 kHz. The oscillator is followed by two voltage-doubling rectifiers, consisting of C1, D1, D2, C3 and C7, D3, D4, C5. They are followed in turn by two voltage regulators to stabilise the positive and negative voltages generated in this manner. The duty cycle of the 555 is set to approximately 50 percent using R1 and R2. The square-wave signal at the output of the timer IC has a DC offset, which is eliminated by C4 and R3.

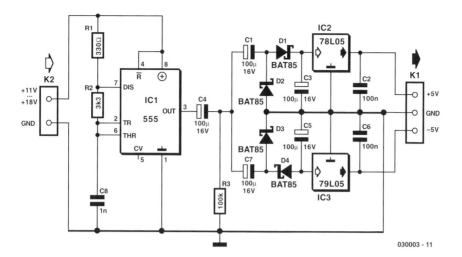

030003 - 11

The amplitude of the output signal from the 555 is approximately equal to the supply voltage less 1.5 V, so with a 12 V input voltage, there will be a square-wave signal on pin 3 with an amplitude of approximately 10.5 Vpp. With respect to ground (across R3), this is this +5 V / –5 V. Although this yields a symmetric voltage, its positive and negative amplitudes are somewhat too small and it is not stabilised. In order to split the square-wave signal into sufficiently large positive and negative amplitudes, C1/D2 are added for the positive voltage, causing the positive half to be doubled in amplitude. For the negative half, the same effect is achieved using C7/D3. Following this, the two signals are

smoothed by D1/C3 and D4/C5, respectively. Both voltages are now high enough to be input to normal 5 V voltage regulators, yielding symmetric +5 V and –5 V supply voltages at the output. The input voltage does not have to be regulated, although it must lie between +11 V and +18 V. The maximum output current is ±50 mA with an input voltage of 12 V. This circuit is an excellent choice for generating auxiliary voltages, such as supply voltages for low-power opamps. Naturally, the fact that the converter can be powered from the in-vehicle voltage of a car is a rather attractive feature.

L. de Hoo   (030003-1)

263
Poor Man's MW/LW Wideband Noise Reduction

Many radio signals in the medium and long wave bands (MW/LW) but also shortwave (SW) are infested by noise of wide variety and of such levels that weak stations are virtually obliterated. The worst noise is the wideband variety which stretches across several hundred kilohertz across the band.

However the very fact that the noise is wideband in nature allows it to be suppressed using a little known and inexpensive method of using a second receiver tuned just beside the frequency you want to listen to. The principle is illustrated in the drawing. The second receiver effectively isolates the noise by adding it in anti-phase to the wanted signal. As shown the loudspeaker outputs of the two radios are connected in phase while the loudspeaker in the second radio is disconnected or removed. With the second

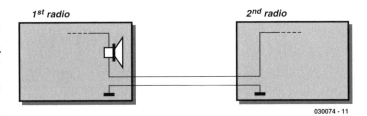

030074 - 11

receiver tuned a little higher or lower than the main one, the loudspeaker in the main receiver is fed with the difference between the two signals. Using the volume adjustment on the second radio, a setting can be found that should result in considerable reduction of wideband noise. If an external antenna is used, it may be connected in parallel to the receiver inputs. Ferrite rod antennas need to turned in the same direction. Best results are obtained when two identical radios are used.

G. Baars   (030074-1)

264
EMC Improver

All electronic equipment these days must conform to the requirements of the electromagnetic compliance (EMC) regulations. A particular problem is the radiation of interference, for example the clock signal, from digital circuits.

The problem is conventionally attacked using shielding and complex smoothing circuits. A simpler and cheaper approach is offered by the company PulseCore in the form of a crystal oscillator IC. It distributes the energy in the interference signal over a band of frequencies, rather than concentrating it at one frequency.

The energy at any one particular frequency is therefore reduced. In practice reductions in interference of 10 dB to 20 dB can be achieved. The technique used is known as 'Spread Spectrum'. The P2010 device includes a crystal oscillator, suitable for frequencies between 10 MHz and 35 MHz, and the spread spectrum jitter circuit. ICs designed to work at higher frequencies are also available. Crystal X1 is designed for fundamental frequency oscillation; for overtone crystals coil L1 ensures that the crystal oscillates at the correct (third or fifth) harmonic. The frequency range is se-

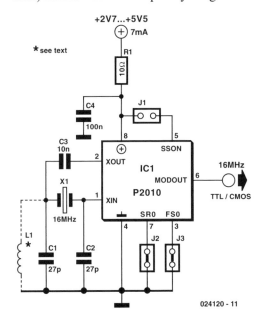

024120 - 11

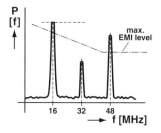

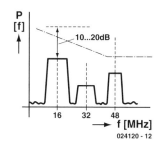

024120 - 12

FS0	SR0	Frequency range	Spreading range
1	0	10 to 20 MHz	± 1.50 %
1	1	10 to 20 MHz	± 2.50 %
0	0	20 to 35 MHz	± 1.25 %
0	1	20 to 35 MHz	± 2.00 %

lected using pin FS0, while SR0 allows one of two different spreading ranges to be selected (see table). In the table a '0' indicates that Br 2 (respectively Br 3) bridged while a '1' indicates that Br 2 (respectively Br 3) is open.

The device draws a current of about 7 mA and operates with 3.3 V and 5 V logic. Br 1 allows the clock jittering to be disabled for test purposes.

G. Kleine　　　　　(024120-1)

265
RS232 Voltage Regulator

There are many small applications where it would be preferable to power a device directly from an RS232 (V.24) interface, avoiding a mains power supply. Most ICs require 5 V, and the interface can provide a current of around 8 mA, almost all of which would be consumed by a readily-available voltage regulator, leaving nothing for the actual circuit. Using just four transistors we can construct a voltage regulator with current limiting which will allow us to draw more than the permitted 8 mA from an RS232 interface without damaging it. The example circuit in is configured for an output voltage of 5 V from an input voltage of at least 8 V, and a short-circuit current of 19 mA. The current drawn by the regulator itself is only 0.2 mA. The circuit appears very simple, but it is more cunning than it looks. Few people appreciate what a handy device the transistor is. To meet the requirements for the circuit, the gains of the transistors need to be controlled carefully. Here only Bclass

devices are used, which have a gain of between about 220 and 280. Diodes D1 to D3 extract the positive voltage from the serial interface. Current limiting is achieved via resistor R1 and transistor T1. As soon as the voltage across the resistor reaches 0.7 V (at 18 mA with R1 = 39 Ω) the tran-

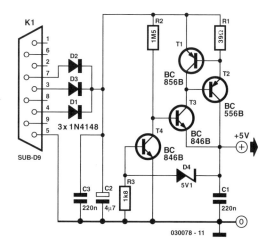

030078 - 11

sistor turns on and thus turns off the output voltage by turning off T2. The output voltage of 5 V is set by Zener diode D4. Note that the output voltage is only approximate: beware when using components which have narrow supply voltage tolerances. When the Zener diode voltage and the voltage across transistor T4 are added together, the total is 5.8 V. However, because of T3, the diode is operating at a low current and the actual threshold for T4 is 4.9 V. The main regulation loop is built around R2 and T2. The high value of R2 (1.5 MΩ) is important, since this limits the maximum current through T2. At the output we would like to be able to draw a maximum current of 19 mA. The base of T2 must therefore be supplied with exactly 1/220 (the gain of the transistor) of 19 mA, and likewise the current into the base of T3 should be just 1/220 of 80 µA. With an input voltage of 9 V the voltage drop across R2 will be 3.3 V, and so a current of 2.2 µA will flow. Transistor T3 multiplies this current by 220 to 0.5 mA, which is also the minimum quiescent current of the circuit.

M. Müller   (030078-1)

266
Cat and Dog Repeller

Nowadays, just about every house has an outside lamp with a motion sensor. Such a device eliminates the need to feel your way to the front door, and it apparently also scares away intruders. The only problem is that free-running dogs and cats in the neighbourhood have little regard for such lamps and continue to deposit their excrement in the garden, once they have found a habitual location there for this purpose. This gave rise to the idea of connecting a sort of siren in parallel with the outside lamp to clearly advise dogs and cats that they are not welcome. Naturally, it would be nice to avoid startling the entire neighbourhood with this alarm signal. Here we can take advantage of the fact that dogs and cats have a significantly better sense of hearing than people. Not only are their ears more sensitive, they can also perceive significantly higher frequencies. With people, the upper limit is around 18 kHz, but dogs and cats can hear frequencies in excess of 20 kHz. We can take advantage of this by building a siren that emits a frequency just above 20 kHz. This will scare off dogs and cats, but people will simply not hear it. All we need for this is an oscillator with an amplifier stage and a tweeter that can reproduce such high frequencies, such as a piezoelectric tweeter. The schematic diagram shows how easily this can be implemented. The power supply for the entire circuit is formed by the components up to and including C2. The 230 V leads are connected in parallel with the motion-sensor lamp. C1 and R1 provide capacitive coupling to reduce the 230 V to an acceptable voltage.

A DC voltage of approximately 9.1 V is generated from this voltage using a bridge rectifier and D1, filtered and buffered by C2. The oscillator is built around R3, C3 and IC1a. The frequency of this oscillator is rather dependent on the specific characteristics of IC1, so the values shown here should be regarded as guidelines. If the oscillator frequency is too high, it can be reduced by increasing the value of R3 and/or C3. If the frequency is too low (which means that the siren tone it is audible), the value of R3 and/or C3 should be increased. The square-wave signal from the oscillator is applied to the input of an H bridge com-

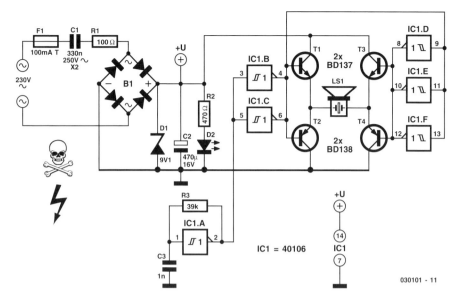

030101 - 11

posed of several Schmitt triggers in combination with the final output stages (T1–T4). This approach causes the peak-to-peak value of the square wave signal to be twice the supply voltage. As a result, a respectable 18 V is obtained across the piezoelectric tweeter, which is sufficient to produce a quite loud whistle tone. When building the circuit, you should bear in mind that it is directly powered from 230 V and not electrically isolated from the mains network. It is thus necessary to avoid

contact with all of the components when the circuit is in use. In practice, this means that the circuit must be fitted into a well-insulated, waterproof box. If you want to test the circuit, it is a good idea to first discharge C1 using a resistor, since it can hold a dangerous charge. You must also ensure that components F1, C1, R1 and B1 all have a mutual insulation separation of at least 6 mm!

I. Fietz   (030101-1)

267
Voltage Inverter Using Switch-Mode Regulator

This circuit uses a step-up switch-mode regulator, which is usually used to produce a positive supply, to generate a regulated negative output voltage. The device used here is the MIC4680 from Micrel (www.micrel.com), but the idea would of course work with similar regulators from other manufacturers. Because of coil L1, which performs the voltage conversion by

the intermediate storage of energy in the form of a magnetic field, the output is effectively isolated from the input. We can therefore connect the right-hand side of L1 to ground rather than to the positive output without causing a large current to flow. Then we connect the ground pin of the regulator IC and all the components connected to it as the negative voltage output, isolated

U_{OUT}	I_{OUT}	C3	D1	L1	R1	R2
-5 V	0.5 A max.	10 nF	SS14 (1 A, 40 V)	15 μH	3 kΩ	1 kΩ
-12 V	0.15 A max.	22 nF	ES1B (1 A, 100 V)	33 μH	8.87 kΩ	1 kΩ

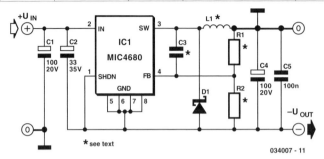

034007 - 11

from ground. The components on the output side of the regulator are connected as usual: flywheel diode D1, coil L1 and the voltage divider formed by R1 and R2. These last two components set the output voltage, according to a formula given in the data sheet. Example component values for the MIC4680 used here are given in the table. The input voltage should lie within the permitted range for the regulator used, and must in any case be at least as great in magnitude as the desired output voltage (here +5 V or +12 V), so that the step-down regulation technique can work. It is important to take care when building this circuit to mount the regulator using an insulator, since generally the GND pin of the device is connected to the heatsink tab. Also, the ON/OFF control input cannot be driven using a normal logic signal, since the regulator's ground reference is the output voltage rather than ground itself. If the ON/OFF function is required, a level shifter or optocoupler must be used.

G. Kleine (034007-1)

268
Electronic Telephone Ringer

This circuit produces a ringing sound similar to that made by more recent telephones. It consists of three almost identical oscillators connected in a chain, each generating a squarewave signal.

The frequency of each oscillator depends on the RC combination: R4 and C1 around IC1.A, R8 and C2 around IC1.B and R12 and C3 around IC3.C. The pairs of 100 kΩ resistors divide the asymmetric power supply voltage (between 5 V and 30 V) so that, in conjunction with the 100 kΩ feedback resistors (R3, R7 and R11) either one third or two thirds of the supply voltage will be present at the non-inverting inputs to the opamps. The voltage across the capacitor therefore oscillates in a triangle wave between these two values. The first oscillator is free-running at a frequency of approximately 1/3 Hz. Only when its output is high, and D1 stops conducting, can the second oscillator run.

The frequency of the second oscillator is about 13 Hz, and optional LED D3 flashes when it is running. When the output of the second oscillator is low, the third is allowed to run. The frequency of the third oscillator is around 1 kHz, and this is the tone that is produced. The second oscillator is not absolutely necessary: its function is

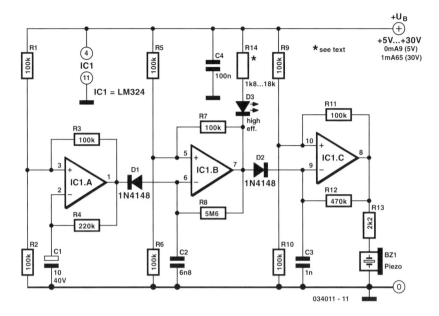

034011 - 11

just to add a little modulation to the 1 kHz tone. A piezo sounder is connected to the output of the third oscillator to convert the electrical signal into an acoustic one. The current consumption of the circuit is just under 1 mA with a 5 V power supply, rising to about 1.65 mA with a supply voltage of 15 V.

L. Libertin   (034011-1)

269
Reducing Relay Power Consumption

Relays are often used as electrically controlled switches. Unlike transistors, their switch contacts are electrically isolated from the control input. On the other hand, the power dissipation in a relay coil may be unattractive for battery-operated applications. Adding an analogue switch lowers the dissipation, allowing the relay to operate at a lower voltage. The circuit diagram shows the principle. Power consumed by the relay coil equals V^2/R_{COIL}. The circuit lowers this dissipation (after actuation) by applying less than the normal operating voltage of 5 V. Note that the voltage required to turn a relay on (pickup voltage)is usually greater than that to keep it on (dropout voltage). In this respect the relay shown has specifications of 3.5 and 1.5 V respectively, yet the circuit allows it to operate from an intermediate supply voltage of 2.5 V. Table 1 compares the relay's power dissipation with fixed operating voltages across it, and with the circuit shown here in place. The power savings are significant. When SW1 is closed, current flows through the relay coil, and C1 and C2 begin to charge. The relay remains inactive because the supply voltage is less than its pickup voltage. The RC time constants are such that C1 charges almost completely be-

Table 1

Voltage (V)	Current (mA)	Total Power Dissipation (mW)
5 (normal operating voltage)	90	450
3.5 (pickup voltage)	63	221
2.5 (application circuit)	45	250

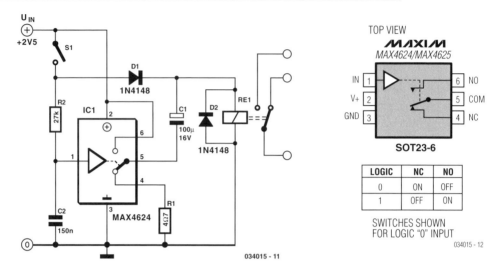

034015 - 11

fore the voltage across C2 reaches the logic threshold of the analogue switch inside the MAX4624 IC. When C2 reaches that threshold, the on-chip switch connects C1 in series with the 2.5 V supply and the relay coil.

This action causes the relay to be turned on because its coil voltage is then raised to 5 V, i.e., twice the supply voltage. As C1 discharges through the coil, the coil voltage drops back to 2.5 V minus the drop across D1. However, the relay remains on because the resultant voltage is still above the dropout level (1.5 V). Component values for this circuit depend on the relay characteristics and the supply voltage.

The value of R1, which protects the analogue switch from the initial current surge through C1, should be sufficiently small to allow C1 to charge rapidly, but large enough to prevent the surge current from exceeding the specified peak current for the analogue switch. The switch's peak current (U1) is 400 mA, and the peak surge current is

$$I_{PEAK} = \frac{V_{IN} - V_{D1}}{R1 + R_{ON}}$$

where RON is the on-resistance of the analogue switch (typically 1.2 Ω). The value of C1 will depend on the relay characteristics and on the difference between V_{IN} and the pickup voltage. Relays that need more turn-on time requires larger values for C1. The values for R2 and C2 are selected to allow C1 to charge almost completely before C2's voltage reaches the logic threshold of the analogue switch. In this case, the time constant R2C2 is about seven times C1(R1 + R_{ON}). Larger time constants increase the delay between switch closure and relay activation. The switches in the MAX4624 are described as 'guaranteed break before make'. The opposite function, 'make-before break' is available from the MAX4625. The full datasheets of these interesting ICs may be found at: http://pdfserv.maxim-ic.com/en/ds/MAX4624-MAX4625.pdf

Source: Maxim Design Showcase   (034015-1)

270
Vehicle Interior Lighting with Switch-Off Delay

Nowadays, a switch-off delay for the vehicle interior lighting is a naturally a standard feature. However, with certain models having only spartan fittings, or older-model vehicles, you're left sitting in the dark as soon as you climb in and close the door. That calls for an aftermarket accessory! The author built this circuit using 'normal' components (with leads), but in the SMD manner, which means fitting the components on the copper side.

The only holes drilled in the circuit board were the four fixing holes, and the entire assembly was firmly attached to the surface of the heatsink for power transistor T3 (the author used a finned heat sink rated at 7.2 °C/W). The heatsink is at ground potential.

A value of 1 Ω was used for R3 with satisfactory operation of the darlington. The light goes on when the door is opened. After the door is closed, it continues to illuminate the interior of the car at full brightness for around 30 seconds, after which it

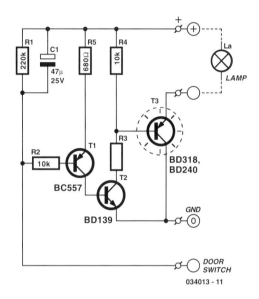

slowly dims. Approximately 1 minute after the door is closed, the quiescent current drops to zero.

L. Libertin   (034013-1)

271
Fan Controller using just Two Components

The Maxim MAX 6665 (www.maxim-ic.com) provides a complete temperature-dependent fan controller. It can switch fans operating at voltages of up to 24 V and currents of up to 250 mA. The IC is available from the manufacturer in versions with preset threshold temperatures be-

tween +40 °C (MAX6665 ASA 40) and +70 °C (MAX6665 ASA 70). The device's hysteresis can be set by the user via the HYST input, which can be connected to +3.3 V, connected to ground, or left open. The following table shows the hysteresis values available:

HYST	Hysteresis
open	1 °C
ground	4 °C
+3.3 V	8 °C

The other pins of the SO8 package are the $\overline{\text{FORCEON}}$ input and the status outputs $\overline{\text{WARN}}$, $\overline{\text{OT}}$ and FANON. The test input $\overline{\text{FORCEON}}$ allows the fan to be run even below the threshold temperature.

The open-drain output $\overline{\text{WARN}}$ goes low when the temperature rises more than 15 °C above the threshold temperature, while the open-drain output $\overline{\text{OT}}$ indicates when the temperature is more than 30 °C above the threshold. The pushpull output FANON can be used to indicate to a con-

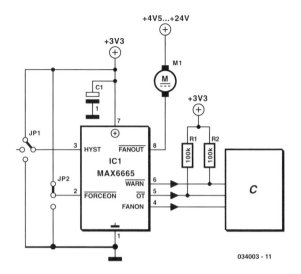

034003 - 11

nected microcontroller that the fan is turned on.

G. Kleine

(034003-1)

272
2708 Replacement

At our editorial offices we are surprised (and delighted) how often we are being asked about relatively ancient circuits. In not a few cases, these circuits contain components that are currently no longer available, such as, for example, the 1 K size EPROM 2708, which required three(!) power supply voltages. The 2716 and 2732

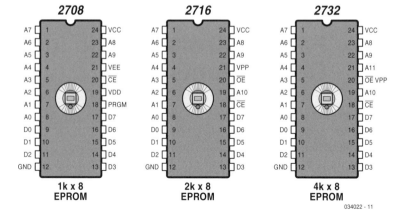

034022 - 11

are parts that are still reasonably easily obtained and programmed, and they also have 24 pins. These parts require only a single 5 V power supply. Because the two additional power supplies of +12 V and –5 V are no longer required, these two pins have become available on the 2732 and 2716. But the 2716, being 2 K in size, needs an additional address line, of course, and the 2732 (4 K) needs two more. These extra address lines are simply connected to ground permanently, so that both the 2716 and 2732 behave lke a 1 K EPROM, just like the 2708. Pin 19 (VDD on the 2708) became address line A10 on the 2716 and 2732. This line is connected straight to ground. Pin 21 (VBB on the 2708) became the programming pin on the 2716. In order to be able to read the EPROM, this pin must be connected to the +5 V power supply. On the 2732, this pin became address line A11, and therefore has to be connected to ground. Unfortunately there have been additional changes to pins 20 (CE, chip enable) and 18 (programming voltage). Pin 18 on the 2708 is the programming voltage but is the CE (chip-enable) on the 2716 and 2732. Pin 20 used to be CE on the 2708 but is OE (output enable) on the 2716 and 2732. The 2708 did not have an OE pin. The simplest solution is as follows: continue to use pin 20 to enable the data output

from the chip. This is possible because CE for the 2708 is now used as OE for the 2716 and 2732, provided pin 18 (CE) is permanently connected to ground. This way, the chip is continuously selected, but the data only appears on the outputs when OE becomes active. The only disadvantage is that there is no reduction in current consumption. We can imagine that you may have become a little confused after this long-winded story. Hence a set of foolproof instructions:

To replace a 2708 with a 2716:
1. Break the track to pin 18 and connect pin 18 to ground.
2. Break the track to pin 19 and connect pin 19 to ground.
3. Keep the connection to pin 20.
4. Break the track to pin 21 and connect pin 21 to +5 V.

To replace a 2708 with a 2732:
1. Break the track to pin 18 and connect pin 18 to ground.
2. Break the track to pin 19 and connect pin 19 to ground.
3. Keep the connection to pin 20.
4. Break the track to pin 21 and connect pin 21 to ground.

K. Walraven   (034022-1)

273
IR Remote Control Receiver

With many audio systems consisting of separate units, you'll often find that due to economic reasons only the amplifier has a remote control receiver module. The control signals are then sent to the other units using patch cables. The tuner and CD player, for example, won't have a built-in receiver module. When the tuner from such a system is bought separately it can there-

fore not be used directly with a remote control, which is a big disadvantage in practice. The only way in which this can be accomplished is to connect an IR receiver to the input used by the patch cable. And that is exactly what this circuit is for. In practice it is not always clear which signal should be used and what its polarity should be. However, it will most likely be a de-

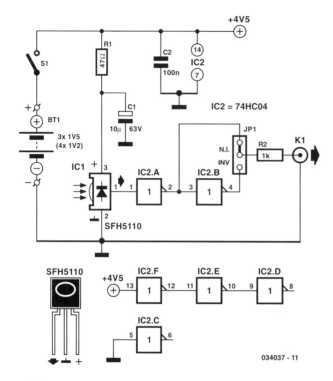

034037 - 11

JP1 is used to select which of the signals is presented at the output. R2 protects the output from short circuits or possible overloading of the electronics in the equipment it's driving (for example when the input circuit uses 3 V logic). R1/C1 suppress any possible supply spikes. Batteries are suitable for the power supply, because the circuit only takes about 1 mA.

With a set of four rechargeable batteries with a capacity of 1800 mAh the circuit can function continuously for 2.5 months. Four NiMH cells and a charger are therefore perfect for the power supply. If you can be sure that the circuit will always be switched off when not in use, you could also use three ordinary alkaline batteries (AA cells). Because of their slightly larger capacity they will probably last for about half a year. When making your choice you should of course keep in mind that rechargeables are better for the environment.

modulated signal. For these reasons we've combined a standard IR receiver module and two inverters.

The first inverter also functions as a buffer, since the output of the module has a high impedance. The output of the receiver module is active low, so the first inverter outputs a non-inverting signal. The second inverter inverts this signal again. Jumper

(034037-1)

274
AVR Dongle

This circuit is intended to program AVR controllers such as the AT90S1200 via the parallel port. The circuit is extremely simple. IC1 provides buffering for the signals that travel from the parallel port to the microcontroller and vice versa. This is essentially everything that can be said about

the circuit. The two boxheaders (K2 and K3) have the 'standard' ISP (in system programming) pinout for the AVR controllers. The manufacturer recommends these two pinouts in an attempt to create a kind of standard for the in-circuit programming of AVR-controllers. These connections can be found on many development boards for these controllers. The software carries out the actual programming task. It is therefore necessary to have a program (ATMEL AVR ISP), which is available as a free download from http://www.atmel.com/. The construction of the circuit will have to made on standard prototype board, since we didn't design a PCB for this circuit. This should not present any difficulties considering the small number of parts involved.

We recommend that inexperienced builders first make a copy of the circuit and cross off each connection on the schematic once it has been made on the board. This makes it easy to check afterwards whether all connections have been made or not.

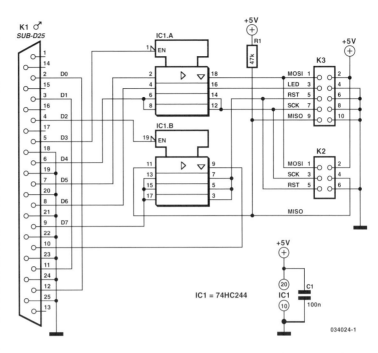

P. Goossens   (034024-1)

275
LED Flasher

The idea behind this economical flashing LED indicator is very simple: it uses a very low frequency oscillator to drive two LEDs. As can be seen in the circuit diagram, the LEDs are connected in anti-parallel. At every change in level there will only be a current flow for a short time, due to the large capacitor (C3) that is connected in series with D1 and D2.

When C3 is charging D1 lights up and when it is discharging D2 lights up. The current through the LEDs is limited by R4. When the circuit is switched on the peak current is determined by R4 and for a red LED will be just over 7 mA:

$$\frac{5-1.6}{470} = 7.2 \ \text{mA}$$

This value has been chosen because it is the maximum safe current for certain types of low-current LEDs. When the circuit is in operation the (dis)charging of C1 determines the maximum current.

The electrolytic has a minimum voltage remaining which is equal to the LED voltage, causing the pulses to be smaller than the very first pulse:

$$\frac{5-(2\times 1.6)}{470}=3.8 \text{ mA peak.}$$

If one LED is sufficient then you can replace either D1 or D2 with a Schottky diode; the pulse through the remaining LED then becomes slightly larger. The use of two LEDs has the advantage that the discharge current of C3 is also used by the indicator, which makes the circuit more efficient. The oscillator can of course be replaced by a different type, for example the classic Schmitt-trigger/inverter type using a 74HC14 and RC network. The version used here has the advantage of offering a better stability. C1 makes IC1f switch faster. The resistors have been made as large as possible, so the adjustment of the flash frequency is best done by varying the value of C2. The period with the component values given is about 6 seconds, so a LED will flash every 3 seconds. The current consumption is determined mainly by IC1f, since this is al-

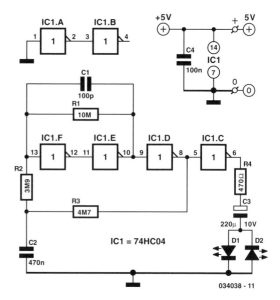

most working as a linear amplifier. The circuit doesn't draw more than about 1.3 mA in total. The LEDs used should preferably be red 'low-current' or 'high-efficiency' types. The current pulses will then be slightly larger and the efficiency of red LEDs is still better than that of green or yellow ones.

T. Giesberts   (034038-1)

276
LCD module in 4-bit Mode

In many projects use is made of alphanumeric LCDs that are driven internally by Hitachi's industry-standard HD44780 controller. These displays can be driven either in 4-bit or 8-bit mode. In the first case only the high nibble (D4 to D7) of the display's data bus is used. The four unused connections still deserve some closer attention. The data lines can be used as either inputs or outputs for the display. It is well known that an unloaded output is fine, but that a

floating high-impedance input can cause problems. So what should you do with the four unused data lines when the display is used in 4-bit mode? This question arose when a circuit was submitted to us where D0-D3 where tied directly to GND (the same applies if it was to +5 V) to stop the problem of floating inputs. The LCD module was driven directly by a microcontroller, which was on a development board for testing various programs

and I/O functions. There was a switch present for turning off the enable of the display when it wasn't being used, but this could be forgotten during some experiments. When the R/W line of the display is permanently tied to GND (data only goes from the microcontroller to the display) then the remaining lines can safely be connected to the supply (+ve or GND). In this application however, the R/W line was also controlled by the microcontroller. When the display is initialised correctly then nothing much should go wrong. The data sheet for the HD44780 is not very clear as to what happens with the low nibble during initialisation. After the power-on reset the display will always be in 8-bit mode. A simple experiment (see the accompanying circuit) reveals that it is safer to use pull-down resistors to GND for the four low data lines. The data lines of the display are configured as outputs in this circuit (R/W is high) and the 'enable' is toggled (which can still happen, even though it is not the intention to communicate with the display). Note that in practice the RS line will also be driven

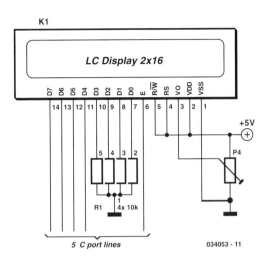

by an I/O pin, and in our circuit the R/W line as well. All data lines become high and it's not certain if (and if so, for how long) the display can survive with four shorted data lines. The moral of the story is: in 4-bit mode you should always tie D0-D3 via resistors to ground or positive.

L. Lemmens   (034053-1)

277
Battery Saver Switch

There are circuits that consume only a little power or require power occasionally for short periods of time. An example of this last type is the 'FM Remote Control Transmitter'. This type of circuit can easily be powered by a small lithium button cell. The CR2032 immediately comes to mind, since it is a popular battery used in many applications and is widely available. The circuit shown here was designed to protect these button cells from short circuits or extended current drain. Its primary function is to restrict the time that power can be drawn from the cell after a button has been pressed. In order to keep the construction

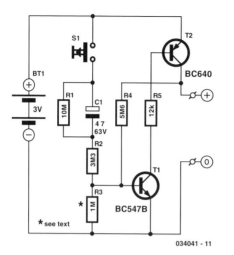

034041 - 11

COMPONENTS LIST

Resistors:
R1 = 10MΩ
R2 = 3MΩ3
R3 = 1MΩ*
R4 = 5MΩ6
R5 = 12kΩ

Capacitors:
C1 = 4µF7 63V axial

Semiconductors:
T1 = BC547B
T2 = BC640

Miscellaneous:
S1 = 6 mm 'tactile switch', e.g.,
MCDTS6-5R (Farnell # 312-1033)
BT1 = 3 V CR2032 with PCB mount holder
(dia. 22.75 or 27.76 mm)*
* see text

as easy as possible we have designed two small PCBs that can accommodate the three most popular types of battery holders for the CR2032, including those with solder pins. There are at least two types of battery holder available for single cells: a small and wider version. Their diameters are 22.75 and 27.76 mm respectively. The PCB has been made a bit larger for use with the wider type. When such a battery holder is ordered it is not always clear if it will be a small or wider type, hence the two PCB layouts. The circuit consists of no more than a simple comparator using two tran-

sistors. Because of its simplicity it is likely that the value of R3 has to be adjusted to obtain the required switching characteristics. In one of our prototypes the best value for R3 was 1 MΩ and in another it was 2.2 MΩ. It should be chosen such that the circuit can switch on properly with a supply voltage of about 2 V. Determining the correct value for R3 is very easy. Using a variable power supply, the voltage should be decreased slowly (starting at about 3 V),

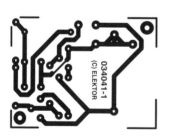

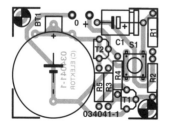

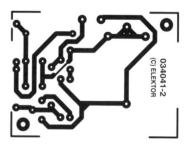

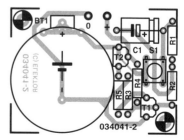

and C1 should be discharged every time before pressing S1 again. At 4 V or just above the circuit should still switch off (keep pressing S1 and slowly increase the supply voltage). The switching characteristics also depend upon the current gain of both transistors and the voltage drop across the base/emitter junctions. When there is a short circuit the current is limited to a few mA's because in that case R4 won't provide extra bias to T1. The circuit is so simple that it should be easy to adapt it for your own needs. There is another standard battery holder (also with a diameter of 27.76 mm) that takes two CR2032 cells on top of each other, giving a nominal supply voltage of 6 V. In that case you will have to determine the component values yourself. R1 discharges C1 when the switch is open.

This time has been made fairly large on purpose, about 50 seconds! The advantage of this circuit is that as the battery voltage becomes lower, T1 stops conducting earlier and hence saves the battery even more. With C1 fully discharged and a fresh battery the on-time is between about 15 to 20 seconds. We have assumed that the circuit will be used in applications where S1 will only be switched on momentarily. When S1 is accidentally switched on for longer then C1 will prevent the battery from discharging quickly. The current drain will then be below 0.3 µA. A construction tip: if the circuit is housed in a small case it may be easier to mount the switch on the solder side of the PCB.

T. Giesberts   (034041-1)

278
Replacement for Standard LCDs

Many circuits published in Elektor Electronics use LC displays. These displays are usually only intended to show text. Such displays are much less expensive than LCDs that can display complex graphic images. The legibility of most LC displays is without question good with normal lighting and proper contrast adjustment. Unfortunately, it degrades with reduced ambient light. Some LCDs have a 'backlight', which is a light source located behind the actual screen.

When a backlight is used, the entire screen lights up, except where pixels are to be made visible. These locations are dark. The presentation of the display can be improved by having the letters light up while the rest of the screen remains dark. This could be achieved using software, but due to the way the displays are made, it's not possible to darken the entire screen. There is a type of display that does not have this

problem, which is called a 'VFD'. Such displays show bright pixels against a dark background, instead of making the pixels dark. However, this type of display has the disadvantage that it requires a high voltage to illuminate the pixels. Fortunately, the Japanese firm Noritake has recently launched a series of VFD modules (the CU series) that is compatible with standard LC displays. These displays have converters to generate the high voltage needed to illuminate the pixels, and the programming interface and connector are exactly the same as those of standard LCD modules. This means that in any project that uses a standard LCD, the display can be replaced by a CU-series VFD from this manufacturer. Of course, you must take into account that these modules draw more current than standard LCD modules.

This is hardly surprising, since they generate their own light. For a standard display with 16 × 2 characters, the maximum current consumption is approximately 150 mA, while the current consumption of a 4 × 40 type can be as much as 550 mA! If you are interested in these new displays, you might want to take a look at the manufacturer's website at http://www.noritake-

itron.com/. These displays may not be readily available, but you can certainly enquire at your local electronics shop. In Europe, Noritake has offices in the UK and Germany.

P. Goossens (034065-1)

279
Spike Detector for Oscilloscopes

Dynamic flip-flops ignore pulses at their inputs that are shorter than 40 ns or do not have TTL levels. This means that TTL flip-flops are poorly suited to capturing noise pulses having unknown durations and magnitudes. Anyone who has ever tried to observe very short laser pulses (15–25 ns) is familiar with this problem. By contrast, this circuit can detect impulses with widths less than 8 ns and amplitudes between +100 mV and +5 V. The heart of this circuit is formed by a MAX903, a very fast comparator with internal memory. The IC has separate supply pins for its analogue and digital portions. The analogue portion is powered by a symmetrical ±5 V supply. This allows the detector to also handle input voltages that are negative relative to ground. The internal memory and output stage operate from a single-ended +5 V supply, so the output signal has proper TTL levels. The MAX903 (IC1) has a special internal memory circuit (latch). The latch either connects the output of the internal comparator directly to the signal output or stores the most recent TTL level and blocks the output of the internal comparator, causing the most recent TTL level appears at the output. This allows short input pulses to be stretched to any desired length. Despite its extremely short switching times, the MAX903 consumes only a modest 18 mW. In the quiescent state, the voltage on the Latch input (pin 5) is at 1.75 V.

This reference voltage is provided by LED D1, which draws its current via R2. In this state the latch is transparent, and a positive edge at the input appears will appear as a negative transition on the output after a propagation delay of 8 ns (tPD). This only happens if the peak voltage on the input is more positive than ground potential. C1 passes this change in the output voltage level to the Latch input (pin 5). As soon as the voltage on the Latch input drops below 1.4 V, the internal latch switches to the Hold state. In this state, the output is no longer connected to the comparator, and the output remains low for the duration of the latch hold time, regardless of what happens with the input signal. The latch hold time is determined by the time constant of the C3/R1 network; it has an adjustment range of 100–500 ns. Pulses of this length can be readily observed using practically any oscilloscope. This latch function in this circuit is only triggered if the input signal has a rising edge that crosses the zero-voltage level.

The internal latch remains transparent for signals in the range of –5 V to 0 V, so such pulses will not be stretched. If only positive input voltages are anticipated, the negative supply voltage is not necessary and the circuit can be powered from a single +5 V supply. A fast circuit such as this requires a carefully designed circuit board layout. All connections to the IC must be

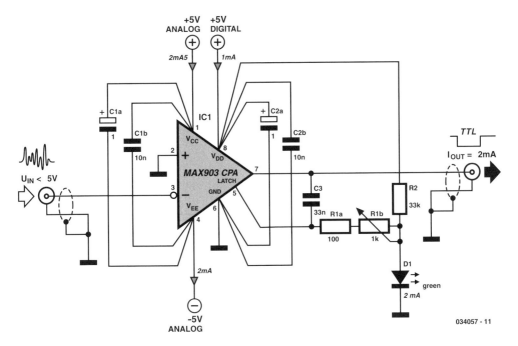

034057 - 11

kept very short. Decoupling capacitors C1 and C2 should preferably be placed immediately adjacent to the supply pins. Pin 3 of the IC can be bent upward and soldered directly to a length of coax or twisted-pair cable (air is still the best insulator). If a coax cable is used, the unbraided screen must not be formed into a long pigtail. It's better to peel back a short length of the screen, wrap a length of bare wire around it and

solder it directly to the ground plane. The supply traces for the analogue and digital portions must be well separated from each other, and each supply must be well decoupled, even if only a single supply voltage (+5 V) is used. The preferred solution is to use two independent voltage regulators.

K. J. Thiesler   (034057-1)

280
IR–S/PDIF Transmitter

The best-known ways to transmit a digital audio signal (S/PDIF) are to use a standard 75 Ω coaxial cable or Toslink optical modules with matching optical cable. Naturally, it can happen that for whatever reason, you cannot (or don't wish to) run a cable between the equipment items in question. With a wireless solution, you

have the choice of a wideband RF transmitter or an optical variant. Here we describe a simple optical transmitter. The matching IR-S/PDIF receiver is described on page 378. Although designing such an IR transmitter/receiver system does not have to be particularly difficult, in practice there are still several obstacles to be overcome. For one thing, the LEDs must have sufficient optical switching speed to properly pass the high frequencies of the S/PDIF signal, and they must also produce sufficient light intensity to deliver a noise-free signal at the receiver over a reasonable distance. At a sampling frequency of 48 kHz, it's necessary to be able to transfer pulses only 163 ns wide! The LEDs selected here (Agilent HSDL-4230) have optical rise and fall times of 40 ns, which proved to be fast enough in practice. With a beam angle of only 17°, they can also provide high light intensity. The downside is that the combination of transmitter and receiver is highly directional, but the small beam angle also has its advantages. It means that fewer LEDs are necessary, and there is less risk of continuously looking into an intense in-frared source. The circuit is essentially built according to a standard design. The S/PDIF signal received on K1 is amplified by IC1a to a level that is adequate for further use. JP1 allows you to use a Toslink module as the signal source if desired. JP1 is followed by a voltage divider, which biases IC1b at just below half of the supply voltage. This causes the output level of the buffer stage driving switching transistor T1 to be low in the absence of a signal, which in turn causes IR LEDs D1 and D2 to remain off. The buffer stage is formed by the remaining gates of IC1. This has primarily been done with an eye to elevated capacitive loading, in the unlikely event that you decide to use more LEDs. A small DMOS transistor (BS170) is used for T1; it is highly suitable for fast switching appli-

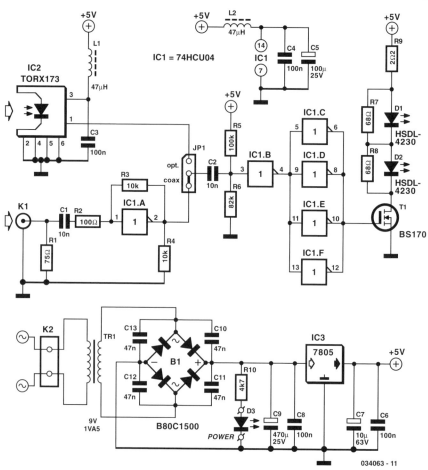

034063 - 11

cations. Its maximum switching time is only 10 ns (typically 4 ns). Getting D1 and D2 to conduct is not a problem. However, stopping D1 and D2 from conducting requires a small addition to what is otherwise a rather standard IR transmitter stage, due to the presence of parasitic capacitances. This consists of R7 and R8, which are connected in parallel with the LEDs to quickly discharge the parasitic capacitors. The drawback of this addition is naturally that it somewhat increases the current consumption, but with the prototype this proved to be only around 10 percent. With no signal, the circuit consumes only 25 mA. With a signal, the output stage is responsible for nearly all of the current consumption, which rises to approximately 170 mA. In order to prevent possible interference at such high currents and avoid degrading the signal handling of the input stage, everything must be well decoupled. For instance, the combination of L2, C4 and C5

is used to decouple IC1. The circuit around T1 must be kept as compact as possible and placed as close as possible to the voltage regulator, in order to prevent the generation of external interference or input interference.

If necessary, place a noise-suppression choke (with a decoupling capacitor to ground) in series with R9. Note that this choke must be able to handle 0.3 A, and if you use additional stages, this rating must be increased proportionally. The circuit should preferably be fitted into a well-screened enclosure, and it is recommended to provide a mains filter for the 230 V input of the power supply. For the sake of completeness, we have included a standard power supply in the schematic diagram, but any other stabilised 5 V supply could be used as well. LED D3 serves as the obligatory mains power indicator.

T. Giesberts (034063-1)

281
On-Train Radio Camera

A radio camera on a model railway should transmit constantly while the train is moving and continue transmitting for a few minutes after the train stops. But if the train starts up again after a relatively long halt, imagery should be transmitted immediately. Consequently, the power source for the camera cannot be rechargeable batteries (since they take too long to charge), nor can it be primary batteries (for environmental reasons). Instead, GoldCaps provide a good alternative. They can be charged in no time flat, and they assure sufficient reserve power for operating the radio camera for a few minutes. Coming from the left in the schematic shown in Figure 1, the dc voltage arrives at the supply circuit and is buffered by capacitor C1,

which bridges brief power interruptions. The actual reserve power source consists of four GoldCaps connected in series, each rated at 22 F / 2.3 V, which yields a net capacitance of 5.5 F / 9.2 V. The maximum charging voltage must never exceed 9.2 V. This is ensured by a modern adjustable low-drop voltage regulator (LT1086), which is set to a nominal output voltage of 9.57 V by resistors R2 and R3 (since there is an 0.6 V voltage drop across D5). The LT1086 can handle a current of 1.5 A (with current limiting), so even completely empty GoldCaps can be charged in a few seconds. Whenever the dc voltage is present, the GoldCaps are charged via D2. When the dc voltage is present, the camera is not powered from the Gold-Caps, but in-

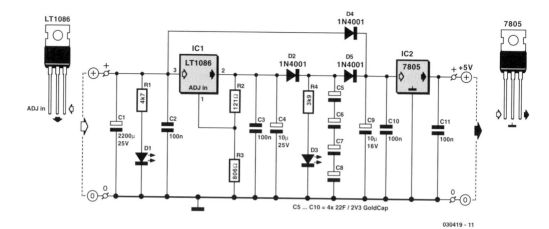

030419 - 11

stead directly from the track via D4. Diode D5 prevents this voltage from reaching the bank of capacitors, and D4 prevents the GoldCaps from discharging via the track when no voltage is present on it. D4 and D5 thus form a sort of OR gate. The radio camera used by the author requires 5 V and draws a current of approximately 70 mA. This means the circuit must have an output stage consisting of a 'normal' 7805 fixed

voltage regulator and the usual capacitors (C9, C10, C11). The two low-current LEDs respectively indicate whether voltage is present on the track and whether the storage capacitors are charged. They can also be omitted. The 100 nF capacitors must be placed as close as possible to the voltage regulators.

Bernd Oehlerking   (030419-1)

282
Balanced Microphone Amplifier

A number of years ago (November 1997), we published a design for a stereo microphone preamplifier with balanced inputs and a phantom power supply. The heart of this circuit was a special Analog Devices IC, the SSM2017. Unfortunately, this IC has been discontinued. In its place, the company recommends using the pin-compatible AMP02 from its current product line. However, and again unfortunately, the specifications of this opamp make it considerably less suitable for use as a microphone amplifier. By contrast, Texas Instruments (in their Burr Brown product line) offer an integrated instrumentation

amplifier (type 1NA217) that has better specifications for this purpose. Incidentally, this IC is also recommended as a replacement for the SSM2017. It features internal current feedback, which ensures low distortion (THD + noise is 0.004 % at a gain of 100), low input-stage noise (1.3 nV/vHz) and wide bandwidth (800 kHz at a gain of 100). The supply voltage range is ±4.5 V to ±18 V. The maximum current consumption of the 1NA217 is ±12 mA. The gain is determined by only one resistance, which is the resistance between pins 1 and 8 of the IC. The circuit shown here is a standard application circuit

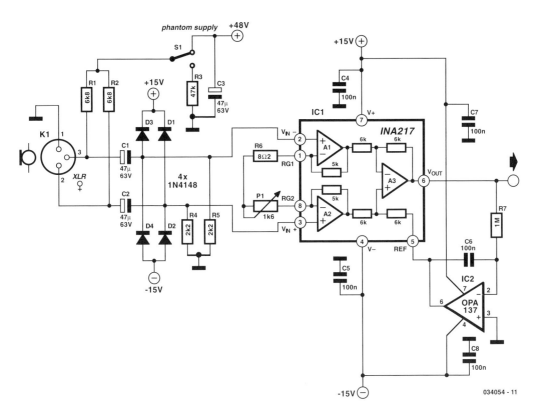

034054 - 11

for this instrumentation amplifier. R1 and R2 provide a separate phantom supply for the microphone connected to the amplifier (this is primarily used with professional equipment). This supply can be enabled or disabled using S1. C1 and C2 prevent the phantom voltage from appearing at the inputs of the amplifier. If a phantom supply is not used, R1 and R2 can be omitted, and it is then better to use MKT types for C1 and C2. Diodes D1–D4 are included to protect the inputs of the 1NA217 against high input voltages (such as may occur when the phantom supply is switched on). R4 and R5 hold the bias voltage of the input stage at ground potential. The gain is made variable by including potentiometer P1 in series with R6. A special reverse logtaper audio potentiometer is recommended for P1 to allow the volume adjustment to follow a linear dB scale. The input bias currents (12 μA maximum!) produce an offset voltage across the input resistors

(R4 and R5). Depending on the gain, this can lead to a rather large offset voltage at the output (several volts). If you want to avoid using a decoupling capacitor at the output, an active offset compensation circuit provides a solution. In this circuit, a FET-input opamp with a low input offset (an OPA137) is used for this purpose. It acts as an integrator that provides reverse feedback to pin 5, so the DC output level is always held to 0 V. This opamp is not in the audio signal path, so it does not affect signal quality. Naturally, other types of low-offset opamps could also be used for this purpose. The current consumption of the circuit is primarily determined by the quiescent current of IC1, since the OPA137 consumes only 0.22 mA.

(From a Texas Instruments application note)

T. Giesberts (034054-1)

283
IR–S/PDIF Receiver

This simple circuit proves to achieve surprisingly good results when used with the IR–S/PDIF transmitter described on page 373. The IR receiver consists of nothing more than a photodiode, a FET and three inverter gates used as amplifiers. The FET is used as an input amplifier and filter, due to its low parasitic capacitance. This allows R1 to have a relatively high resistance, which increases the sensitivity of the receiver. The bandwidth is primarily determined by photodiode D1, and with a value of 2k2 for R1, it is always greater than 20 MHz. The operating current of the FET is intentionally set rather high (around 10 mA) using R2, which also serves to ensure adequate bandwidth. The voltage across R2 is approximately 0.28–0.29 V. The combination of L1 and R3 forms a high-pass filter that allows signals above 1 MHz to pass. L1 is a standard noise-suppression choke. From this filter, the signal is fed to two inverters configured as amplifiers. The third and final inverter (IC1c) generates a logic-level signal. This 74HCU04 provides so much gain that there

is a large risk of oscillation, particularly when the final stage is loaded with a 75 Ω coaxial cable. In case of problems (which will depend heavily on the construction), it may be beneficial to add a separate, decoupled buffer stage for the output, which will also allow the proper output impedance (75 Ω) to be maintained in order to prevent any reflections. When building the circuit, make sure that the currents from IC1 do not flow through the ground path for T1. If necessary, use two separate ground planes and local decoupling. Furthermore, the circuit must be regarded as a high-frequency design, so it's a good idea to provide the best possible screening between the input and the output. With the component values shown in the schematic, the range is around 1.2 metres without anything extra, which is not especially large. However, the range can easily be extended by using a small positive lens (as is commonly done with standard IRDA modules). In our experiments, we used an inexpensive magnifying glass, and once we got the photodiode positioned at the focus after a

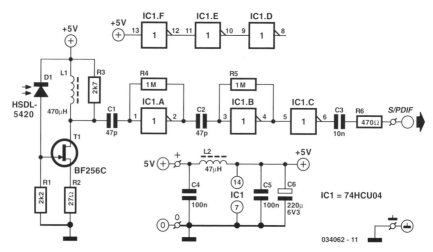

034062 - 11

bit of adjustment, we were able to achieve a range of 9 metres using the same transmitter (with a sampling frequency of 44.1 kHz). This does require the transmitter and receiver to be physically well aligned to each other. As you can see, a bit of experimenting certainly pays off here! It may also be possible to try other types of photodiode. The HDSL-5420 indicated in the schematic has a dome lens, but there is a similar model with a flat-top case (HDSL-5400). It has an acceptance angle of 110°, and with the same level of illumination, it generates nearly four times as much current. The current consumption of the circuit is 43 mA with no signal and approximately 26 mA with a signal (fs =

44.1kHz) That is rather high for battery operation, but it can handled quite readily using a pair of rechargeable NiMH cells. Incidentally, the circuit will also work at 4.5 V and even 3 V. If a logic-level output is needed, C3 at the output can be replaced by a jumper. Finally, there is one other thing worth mentioning. With the HSDL-5400 that we had to play with, the cathode marking (a dark-blue line on the side below one lead) was on the wrong side (!). So if you want to be sure that the diode is fitted properly, it's a good idea to measure the DC voltage across R1, which should be practically zero.

T. Giesberts   (034062-1)

284
Low-Drop Constant Current Source

for Ultrabright LEDs

Ultrabright LEDs are becoming increasingly attractive for use in lighting and warning-signal applications. LEDs must be operated at a constant current to ensure that they continue to emit light at the same brightness. The usual approach is to use a series resistor, but in order to prevent the non-linear voltage/current characteristic, the NTC property of the LED and variations in the supply voltage from affecting brightness, we'd like to have something better. The circuit described here is a low-drop constant current source for ultrabright LEDs with blinking capability, for use in headlights, taillights, dog blinkers, light chains, car alarms and the like. It provides a constant current of 20 mA with a supply

voltage of 4.5–30 V, or 50 mA with a 4.5–12 V supply. The voltage drop of the circuit is only 0.6 V, so practically the entire supply voltage can be used for the LEDs. When the supply voltage is switched on, D5 generates a start-up voltage if the circuit is exposed to light (an

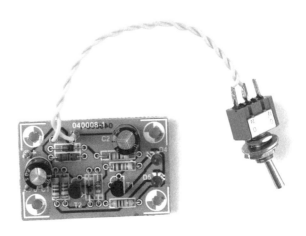

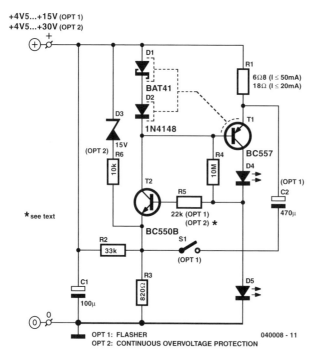

+4V5...+15V (OPT 1)
+4V5...+30V (OPT 2)

D1
BAT41
D2
1N4148
D3
15V
(OPT 2) R6
10k

R1
6Ω8 (I ≤ 50mA)
18Ω (I ≤ 20mA)

T1
BC557

R4
10M

D4
(OPT 1)
C2
470μ

★ see text

T2
R5
22k (OPT 1)
(OPT 2) ★
BC550B

S1
(OPT 1)

R2
33k

C1
100μ

R3
820Ω

D5

OPT 1: FLASHER
OPT 2: CONTINUOUS OVERVOLTAGE PROTECTION

040008 - 11

duced proportional to the supply voltage, which further stabilises the current. R2 is thus configured such that the current actually decreases slightly as the battery voltage increases. This causes T1 to be driven into full saturation at low voltages. D2 is intended to compensate for the base–emitter voltage of T1, while Schottky diode D1 provides the 'miserly' voltage drop across R1. D1 and D2 are thermally coupled to T1. As a result, with increasing temperature the current is more likely to decrease than increase. LED D4 is included in the circuit to prevent T2 from having to go into saturation at low operating voltages. That wrings out an extra 0.2 V or so. In this way, the battery is mercilessly sucked dry when it becomes weak. It's totally defenceless. However, rechargeable batteries should never be deep-discharged! Overvoltage protection should be included if the circuit is intended to be used for general experimentation or with LED chains. In case of overvoltage, D3 starts conducting and takes current away from T2 (which is essentially what R2 does as well). To prevent overheating of T1, particularly with a current of 50 mA, the current is sharply reduced for voltages greater than approximately 15 V. The upper voltage limit is then practically determined solely by the maximum collector voltage (U_{CE}) ratings of the two transistors. An LED chain can best be inserted in series with the supply voltage, in order to avoid making any changes to the circuit. Naturally, at least D5 must still remain in the circuit. R5 and C2 form part of the blinking circuit. R5 and C1 smooth out large current spikes and RF oscillations during blinking operation. Low-impedance feedback is provided by C2. The resulting pulses are long enough to

LED can generate a photoelectric current). To enable the constant-current source to also reliably start up in the dark, R4 provides an initial base current to T1 and T2. In any case, both transistors initially only pass a small amount of current. But since each transistor provides the base current for the other one, the current rises to its setpoint value. The setpoint is stably maintained as follows. The voltage across D5, less the base–emitter voltage of T2, is also present across R3 (R5 has no effect at the beginning). A stabilised current thus flows through R3. Most of the constant current flowing through T2 comes from D1 and D2, with only a secondary contribution from the base current of T1. The voltage across D1 and D2 is also stable, due to the stabilised current. This voltage, less the base–emitter voltage of T1, is present across R1, where it causes a constant current to flow through T1 and D5 (and D4 if present). This closes the loop. The circuit thus consists of two constant-current sources that stabilise each other. That's the basic principle. R2 increases the voltage across R3. This causes the current to be re-

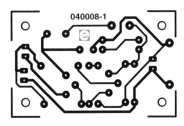

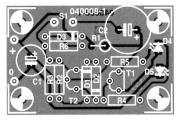

COMPONENTS LIST

Resistors:
R1 = 6Ω8 or 18Ω(see text)
R2 = 33kΩ
R3 = 820Ω
R4 = 10MΩ
R5 = 22kΩor 0 Ω(see text)
R6 = 10kΩ

Capacitors:
C1 = 100µF 40V radial
C2 = 470µF 40V radial

Semiconductors:
D1,D2 = BAT41 or similar Schottky diode
with IF>80mA
D3 = zener diode 15V, 0.5W
D4,D5 = LED (see text)
T1 = BC557B
T2 = BC550B

Miscellaneous:
S1 = switch (see text)
PCB, order code 040008-1 from the
PCBshop

be clearly visible and short enough to use as little energy as possible. The duty cycle is only 10 %. However, at base voltages less than 5 V (with two LEDs fitted) it increases to 50 %, with an accompanying decrease in current. Below 4.6 V, shortly before the circuit 'runs dry', the pulse heads toward to zero. The 10-% pulses achieve the rated pulse current level of 100 mA for

50 mA LEDs, and they have such steep edges that two blinkers connected to a single power source blink in unison if they are not decoupled by at least 1 Ω. The voltage fluctuations on C2 are so small (approximately 0.6 V) that hardly any energy is lost. We have designed a small printed circuit board for the constant-current source, which could hardly be easier to build and does not require any comments.

Diodes D1 and D2 are placed immediately next to T1 and thus adequately thermally coupled. For a maximum constant current of 50 mA, the value of R1 should be 6.8 Ω, while for 20 mA a value of 18 Ω should be used. Naturally, the value of R1 can be increased even more to reduce the value of the constant current any the desired level. If T1 does not have the anticipated current gain of 140, R3 should be reduced to 680 Ω. The current flowing through diodes D1 and D2 should be at least three times the base current of T1. Naturally, the base current flowing through T2 is not multiplied. The value of IR3 is thus $4 \times I_{B,T1}$ (since $I_{B,T2}$ can be neglected). As D5 determines the voltage across R3, we thus have the formula:

$$R3 \leq B_{T1} \cdot \left(\frac{U_{D4} - 0.65 \ V}{4 \cdot I_{const}} \right)$$

The maximum permissible value of UR1 is 340 mV. From the author's experience, when setting the level of the constant current it helps to try several diodes for D2 with different tolerance values. In stubborn cases of excessively high current levels (or if you want to be on the safe side but don't want to or can't measure, adjust or whatever), you can simply connect two 1N4148s in parallel. This will cause the operating point to lie somewhat lower on the characteristic curve. Another important tip for avoiding eye injury (retina damage): never look directly at an ultrabright LED, especially in the dark!

Olaf König (040008-1)

285
Amplifier with Squelch

This addition to the receiver part of the 'FM Remote Control' is mainly intended for use as a simple intercom or P.A. system. With a few changes this circuit could possibly also be used with different receiver modules. In the previously mentioned project the decoder and dual D-type flip-flop (IC2 and IC3) are not required for use with this circuit.The main feature of this circuit is that the supply voltage to the power amplifier is switched on or off automatically. Implementing a mute state this way is very simple and also saves on current consumption (especially important in battery powered applications). Opamp IC1d, which is configured as a comparator, compares the signal strength of the RX5 receiver with a level set by P2. The range of P2 is a little bit larger than the voltage provided by the RSSI output of the receiver module, and are 1.23-2.74 V and 1.27-2.63 V respectively. IC1d switches T1, which then provides a supply voltage to the power amplifier. Potential divider R13/R14 makes T1 switch cleanly and the voltage across R14 is used via D1

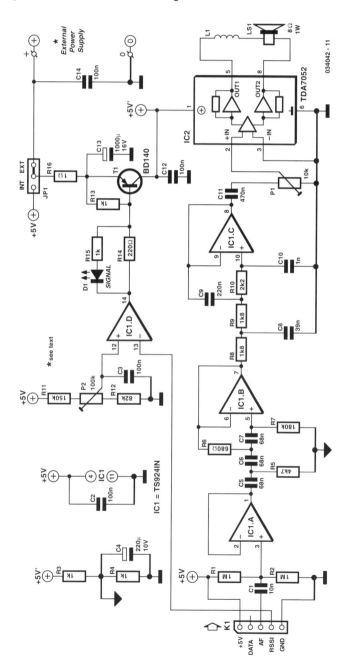

to indicate when a signal is received. For the power amplifier a small bridge amplifier is used, which comes in an 8-pin DIL package: the TDA7052. This is a neat little chip that requires no external components apart from a volume control (P1). It did appear that this integrated amplifier had a tendency to oscillate when long leads were used to connect the loudspeaker. Air-cored coil L1 (6 turns of 1 mm CuL wire, with an inner diameter of 10 mm) offers reasonable protection against this, but it is better still to use such a coil in each of the loudspeaker leads. A variant of the amplifier is the TDA7052A. This has an internal DC controlled volume control, which is controlled with a voltage on pin 4 (not connected here). Although we haven't tested this, this pin can probably be left unconnected, which means that the A-type can be used without having to make any changes. Connector K1 in the circuit diagram corresponds to K1 in the 'FM remote control'. Apart from the RSSI signal strength output, use is also made of the demodulated LF signal (AF output). It becomes clear that the transmitter and receiver modules of this remote control were not really designed for audio when the distortion of the signal is measured when it has gone through the transmitter and receiver. This turned out to be about 3%. The speech quality is good, due to the heavy use of filters in this circuit. IC1b is used to make a 3rd order highpass filter with a turnover frequency just above 200 Hz and round IC1c is a similar lowpass filter with a turnover frequency of about 5.5 kHz. Both filters are Cheby-

shev types with a ripple of only 1 dB. Figure A shows two graphs: I is measured with the transmitter/receiver/amplifier combination, II is just for the amplifier. The filters also reduce any noise introduced by the transmitter or receiver. As a comparison we have also taken some measurements using a previous generation of the transmitter and receiver modules (the transmitter was a TX2-433-40-5V and the receiver an RX2-433-14-5V, both made by Radiometrix).

With optimum modulation (1.2 V at the transmitter module, 4.5 V supply voltage, compressor set for fixed gain) we measured a distortion of only 0.4%; this is quite a big difference! And with a similar measurement as in figure A-I the noise floor at low frequencies seemed to have moved

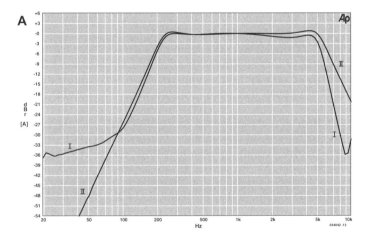

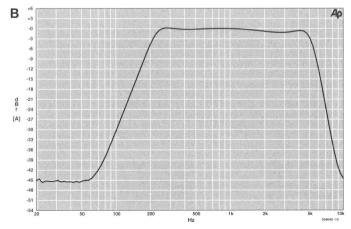

from –36 dB to –45 dB. The disadvantage is that these modules only have a carrier-detect output and require some extra circuitry to change this into a 'signal strength' output. The current consumption of this circuit when it fully drives an 8 Ω speaker is about 0.22 A.

The amplifier can then deliver 400 mW into 8 Ω. The 'FM Remote Control Receiver' is not capable of delivering such a large current, which is why jumper JP1 gives a choice of using either an internal or external power supply. The supply voltage for the filters and comparator can still be derived from the receiver via K1, but if use is made of an external 5 V source then it makes sense to use that to supply the whole circuit. Without an audio signal the current consumption is about 33 mA. When the signal strength is below the threshold the circuit goes into a mute state and the current consumption drops to just 4.5 mA.

T. Giesberts   (034042-1)

286
Unusual LED Blinker

This LED blinker manages with only a few components and is dimensioned to operate from an ac supply in the range of 4–16 V (6–24 V dc). As its current consumption is less than 1 mA, it's also suitable for long-term battery-powered operation. It thus offers several advantages in various applications, compared with using the well-known 555 timer IC as an astable multivibrator. Depending on the values of the timing components, the blinking rate ranges from 1 to 1.5 Hz. Although the duration of each blink is only a few milliseconds, a high brightness level is achieved by using a relatively high LED current. There are numerous potential applications for this circuit in model railway systems, in both stationary and moving equipment. An small, inexpensive thyristor serves as an oscillator. Voltage divider R1/R2 holds the voltage on the gate lead (G) to approximately 20 % of the supply voltage. Capacitor C2 charges via R3. This causes the volt-

age on the cathode (K) to drop until it is around 0.5 V to 1 V below the gate voltage (depending on the thyristor type), at which point the gate current is sufficient to trigger the thyristor. Capacitor C2 then discharges via the cathode-anode junction, R4 and the LED. The only purpose of R4 is to limit the LED current to a permissible value. After C2 has discharged, the cathode–anode junction is again cut off, since the resistance of R3 is so high that the sustaining current level (which is less than 5 mA for the BRX45–57 family) is not achieved.

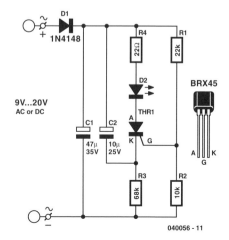

040056 - 11

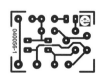 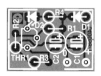

COMPONENTS LIST

Resistors:
R1 = 22kΩ
R2 = 10kΩ
R3 = 68kΩ
R4 = 22Ω

Capacitors:
C1 = 47µF 35V
C2 = 10µF 25V

Semiconductors:
D1 = 1N4148
D2 = low-current LED
THR1 = BRX45

Miscellaneous:
PCB, order code 040056-1 from The PCBShop

The next blink cannot occur until C2 has again charged. The blinking rate can be adjusted over a wide range by varying the values of R3 and/or C2. The listed thyristor is recommended for use in this circuit due to its high gate current sensitivity (<0.2 mA). Practically any model railway transformer or bell transformer can be used as a power source. A half-wave rectifier and small filter capacitor are adequate for rectifying the supply voltage. The components can be fitted to a small printed circuit board or a small piece of perforated board. The blinker can also be powered from a dc source (5–24 V). In that case, D1 provides reverse-polarity protection.

Robert Edlinger   (040056-1)

287
FM Remote Control Receiver

It is often required to switch electrical appliances from a distance (the garage light, when you're still in your car for example) without there being a direct line of sight between the transmitter and receiver. This rules out the use of infrared, so we have to use radio signals instead. An ideal solution for this application is provided by transmitter and receiver modules, which operate at a frequency of 433 MHz and are available ready-made. This circuit complements the small FM transmitter that uses such a module. Here we describe the associated receiver, which picks up the transmitted signals us-ing an R5-434-5-20 receiver module made by R.F. Solutions (this 20 kbps version is now also stocked by Farnell). This integrated receiver has been tuned to a frequency of 433.92 MHz, exactly the same

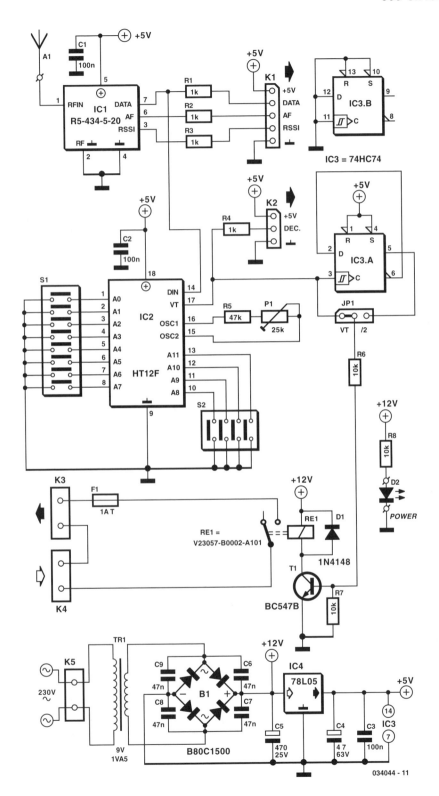

034044 - 11

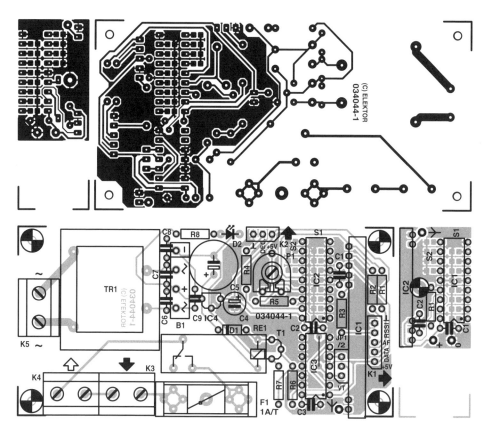

as for the transmitter. To prevent interference and unauthorised use the transmitter sends out a coded signal. This is processed in our receiver by decoder IC HT12F, made by Holtek. P1 and R5 are used to tune the oscillator frequency of the decoder to that of the encoder in the transmitter. In this way any possible variations due to tolerances or a different battery voltage can be compensated for by P1.

The data bits are set up using solder bridges on the PCB (S1 and S2). With jumper JP1 the output can be diverted through a divide-by-two circuit (half of IC3, a dual D-type flip-flop 74HC74), so that relay RE1 can remain energised without the presence of a signal. The make contact of the relay connects a contact of terminal block K4 to K3 via a fuse. The other contacts are connected directly. The relay and terminal blocks are placed far enough from mains voltages on the PCB to comply with

Class II safety specifications. It is also possible to connect terminal block K4 with K5 in order to safely switch mains voltages. In this case you should make sure that the 'live' mains voltage is connected via the relay and fuse. The circuit has a mains power supply on-board, making it ready for use as soon as the construction is complete. The relay and mains power indicator are driven directly from the smoothed transformer output, keeping the load on the small 5 V regulator (78L05) to a minimum. It should now only have to supply a few mAs. The use of a ready-made receiver module simplifies the construction of this circuit and also makes it more reliable. Apart from the RF input (RF IN) the module has a data and analogue output (AF). There is also a status output that gives an indication of the RF signal strength (RSSI). These last two signals are not used in this application. There is another article in this

COMPONENTS LIST

Resistors:
R1-R4 = 1kΩ
R5 = 47kΩ
R6,R7,R8 = 10kΩ
P1 = 25kΩ preset

Capacitors:
C1,C2,C3= 100nF ceramic
C4 = 4µF7 63V radial
C5 = 470µF 25V radial
C6-C9 = 47nF ceramic

Semiconductors:
D1 = 1N4148
D2 = LED, high-efficiency
T1 = BC547B
IC1 = R5-434-5-20 from R.F. Solutions
(Farnell # 352-4383)
IC2 = HT12F from Holtek (Maplin)
IC3 = 74HC74
IC4 = 78L05

Miscellaneous:
JP1 = 3-way pinheader with jumper
K1 = 5-way pinheader
K2 = 3-way pinheader
K3,K4,K5 = 2-way PCB terminal block, lead
pitch 7.5 mm
S1,S2 = solder links
B1 = B80C1500 (rectangular case) (80V
piv, 1.5A)
TR1 = mains transformer, 9V/1.5VA, e.g.,
Block type EI30-X6/4001/112 VV1109
F1 = fuse, 1AT (time lag), with PCB mount
holder
RE1 = V23057-B0002-A101 vertical card
relay 12 V/330 ?/ 8 A
PCB, order code 034044-1 (transmitter and
receiver)

issue (Amplifier with Squelch), which uses these two analogue signals with an extra circuit and a modified version of the transmitter. For this reason all outputs of the receiver module have been made available on a 5-way pin header (K1), along with the supply voltage. R1 to R3 protect the outputs from potential short circuits. The supply on this header is only meant to provide a few mAs! Just as for the transmitter, the aerial has to be mounted as close as possible to the 'RF IN' pin.

The construction of the aerial is described in the article for the transmitter. The output of the decoder (as well as the supply voltage) is brought out on a separate pin header (K2), making the logical signal available to circuits that need it. The transmitter and receiver PCBs have been combined on one board to keep the cost down. The small transmitter section can be cut easily from the board.

The remaining receiver PCB is well-organised and easily populated (don't forget the wire-link underneath IC3!). It is possible that jumper JP1 may be a tight fit between IC3 and receiver module IC1. All connectors have been placed along the edges of the PCB. If you make a connection to K1 you should take care that you don't make contact with the metal shielding of the module.

T. Giesberts (034044-1)

288
Flickering Light II

In the July/August 2004 issue we published a flickering light for use in models that uses a microcontroller to minimise the component count. Regardless of whether you want to effectively imitate a house fire, a campfire, or light from welding, the circuit described here fills the bill without using a microcontroller, although it does use a larger number of components (including some truly uncommon ones). The circuit is based on three oscillators, which are built using unijunction transistors (UJTs). Each

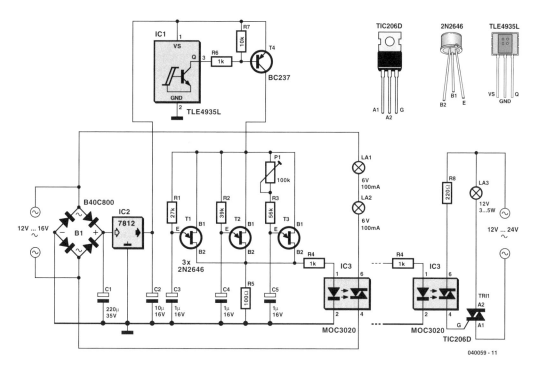

040059 - 11

oscillator has a different frequency. The output voltages are mixed, which produces the flickering effect. A uni-junction transistor consists of an n-type bar of silicon between two ohmic (non-barrier) base contacts (B1 and B2).

The effective resistance is controlled by the p-type emitter region. The designation 'transistor' is a somewhat unfortunate choice, since it cannot be used for linear amplification. UJTs are suitable for use as pulse generators, monostable multivibrators, trigger elements and pulse-width modulators. If a positive voltage is applied to the emitter (E), the capacitor charges via the resistor.

As soon as the voltage on the emitter reaches approximately half the supply voltage (for a 2N2645, the value lies in the range of 56–75 %), the UJT 'fires' and the capacitor discharges via base B1 and the resistor, generating a positive pulse. The UJT then returns to the non-conduct state, and the process just described repeats periodically.

The frequency can be approximately given by the formula:

$$f \approx \frac{1}{R \cdot C}$$

The frequency is independent of the value of the supply voltage (which must not exceed 35 V). The maximum emitter blocking voltage is 30 V, and the maximum permissible emitter current is 50 mA. The values of resistors R1, R2 and R3 can lie between 3 kΩ and 500 kΩ. If necessary, the frequency can be varied over a range of 100:1 by using a trimpot instead of a fixed resistor. The frequencies from the three pulse generators are mixed by connecting them to the IR diode of a triac optocoupler via R4. The optocoupler, a type MOC3020, K3030P or MCP3020, can handle a maximum load current of 100 mA. The triac triggers at irregular intervals and generates the desired flickering light in the two small lamps, L1 and L2, which are connected in series to the transformer secondary. The light effect can be noticeably improved by using a MOC3040, which contains a zero-voltage switch, since it generates irregular pauses of various lengths when suitable frequencies occur in the individual oscillators. The zero-voltage switch does not switch while the current is flowing, but only when the applied ac voltage passes through zero. An integrated drive circuit (zero crossing unit) allows full half-waves or full cycles to pass (pulse-burst control) Due to the flickering effect arising from its switching behaviour, it is not suitable for normal lighting, but here this just what we want.

This version of the optocoupler is also designed for a maximum current of 100 mA. For a small roof fire or the light of a welding torch in a workshop, two small incandescent lamps connected in series and rated at 6 V / 0.6 A (bicycle taillight bulbs) or a single 12 V lamp (rated at 100 mA) is adequate. If it is desired to simulate a large fire, a triac (TIC206D, rated at 400 V / 4 A, with a trigger current of 5 mA) can be connected to the output of the circuit and used to control a more powerful incandescent lamp. As continuous flickering looses its attraction for an interested observer after a while (since no house burns for ever, and welders also take breaks), it's a good idea

to vary the on and off times of the circuit. This is handled by a bipolar Hall switch (TLE4935L), which has such a small package that it can fitted between the sleepers of all model railway gauges, including Miniclub (Z Gauge), or even placed alongside the track if a strong permanent magnet is used. If a magnet is fixed somewhere on the base of a locomotive such that the south pole points toward the package of the Hall switch (the flattened front face with the type marking), the integrated npn transistor will switch on and pull the base of the external pnp transistor negative, causing the collector–emitter junction to conduct and provide the necessary 'juice' for the uni-junction transistors.

If another traction unit whose magnet has it s north pole pointing toward the Hall switch passes a while later, the switch will be cut off and the flickering light will go out. Of course, you can also do without this form of triggering and operate the device manually.

Robert Edlinger   (040059-1)

289
FM Remote Control
Transmitter

This extremely simple transmitter, consisting of only an encoder IC and a 433 MHz licence-exempt TX module, was designed to remotely switch simple appliances on and off. The associated receiver has a relay that can be switched permanently or momentarily and is suitable for various applications (including mains). For the transmission of a unique signal an encoder is indispensable.

For this we've used the HT12E made by Holtek, which we've used previously. Since this is an old favourite there is no need to go into detail and we'll just mention that the oscillator is set to about 3 kHz

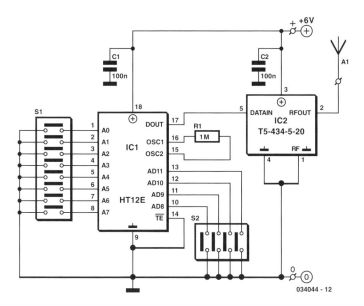

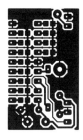

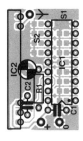

034044 - 12

by R1. The radio frequency part consists of a standard FM transmitter module from R.F. Solutions, part number T5-434-5-20, which makes the circuit easy to construct and improves its reliability. The transmitter module works at a frequency of 433.92 MHz and has a range of about 400 m according to the manufacturer. The transmitter module has five pins. Apart from 'data in' (pin 5) and the supply (pin 3), there is a common ground for data and supply (pin 4). Last, but not least, is the RF output (pin 2) with its associated ground (pin 1). The best results are achieved when a 15.5 cm length of stiff wire is used for the aerial. If you want to keep the circuit compact and

build it into a small handheld case, you should use a helix aerial. This is wound like a coil, which will be about 34 mm long and consists of 17 turns with an inner diameter of 5 mm. We've used 0.9 mm enamelled copper wire (ECW) for this, which keeps its shape reasonably well (you could use the smooth end of a 5 mm drill bit for winding the wire). The aerial has to be mounted as close as possible to pin 2 of the module. The transmitter module is also available from several other sources, although the part number may be slightly different. Examples are the QFMT5-434-5 from Warwick Wireless Limited and the QFMT5-433B5 from OKWelectronics (www.okwelectronics.com). A small PCB has been designed for the transmitter, with room for the transmitter module (IC2) to lie flat along the length of the board. For practical reasons the board is combined with that for the receiver PCB (see 'FM Remote Control Receiver') and needs to be cut from it. Solder bridges on S1 and S2 are used to set the address and data bits. The current consumption with a supply voltage of 4.5 V (three AA cells for example) is about 9 mA. With a supply voltage of 6 V the current consumption rises to 12.5 mA. These figures apply to the 5 V version of the transmitter module. This is specified

COMPONENTS LIST

Resistors:
R1 = 1MΩ

Capacitors:
C1,C2 = 100nF ceramic

Semiconductors:
IC1 = HT12E (Holtek)
IC2 = T5-434-5-50 R.F. Solutions (Farnell # 352-4371)

Miscellaneous:
S1,S2 = solder bridges
PCB: see 'FM Remote Control Receiver'

for use with a supply between 4.5 and 5.5 V, but has an absolute maximum rating of 10 V, so 6 V won't do it any harm. Furthermore, the 5 V version will still function with a supply of only 2 V, although the range is then much less. Since the circuit consumes very little current the power can also be provided by button cells. Elsewhere in this issue is a circuit that combines a battery holder for one or two lithium button cells with a miniature automatic switch on a small PCB (see 'Battery Saver Switch). That circuit is ideal for use as a power source for the transmitter.

T. Giesberts (034044-2)

290
Car Central-Locking System

For a few pounds you can buy a kit from any automotive accessory shop that will allow your car to be fitted with a central-locking door system. Such a kit essentially comprises a number of motors. There is also a control unit that enables the whole system to function. Here we show an example of such a unit. There are 5 wire motors and 2 wire motors. The 5 wire version is used in doors that have a key-lock. There are 2 connections for the motor itself and 3 connections for the sensor part (an 'open' and a 'close' contact). These sensors determine whether the door is to be unlocked or locked. If there is no key lock in the door, these sensors are superfluous and a 2 wire motor can be used. The polarity of the motor determines whether the locking mechanism goes up or down. By making a circuit that simply reverses the polarity of the motor, the door can be either locked or unlocked. The winding of the motor is connected between M1 and M2 in the schematic. When relay Re1 is energised,

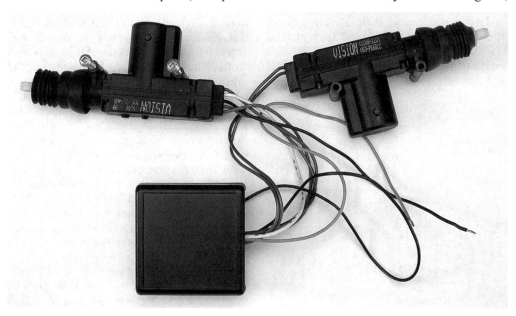

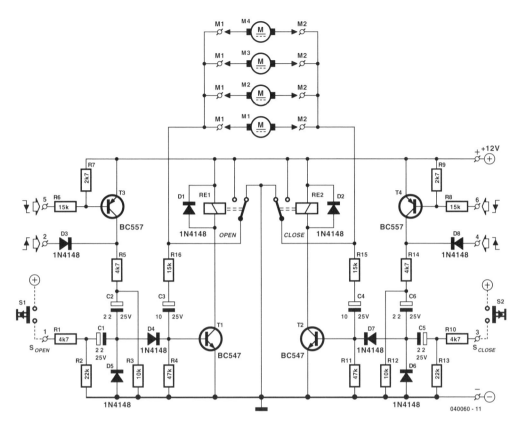

all motors will, for example, rotate anti-clockwise. By activating Re2 the motors will rotate clockwise. This depends on the actual polarity of the motor, of course. The sensors are connected to R1 and R10. Here you have to pay careful attention. If Re1 causes the door to unlock, then Re1 has must obviously be connected to the

'open' contact. In that case, Re2 is for locking the doors and R10 is then connected to the 'close' contact. The R/C-combinations R16/C3 and R15/C4 ensure that the relays are energised for a certain amount of time (obviously this can be changed if this time is too short or too long for your doors). This time has to be just long enough to lock

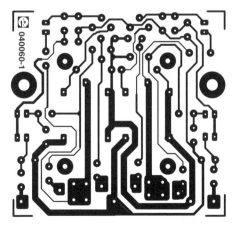

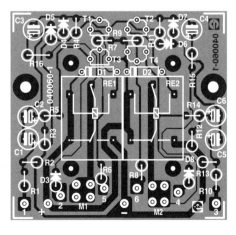

COMPONENTS LIST

Resistors:
R1,R5,R10,R14 = 4kΩ7
R2,R13 = 22kΩ
R3,R12 = 10kΩ
R4,R11 = 47kΩ
R6,R8,R15,R16 = 15kΩ
R7,R9 = 2kΩ7

Capacitors:
C1,C2,C5,C6 = 2µF2 25V radial
C3,C4 = 10µF 25V radial

Semiconductors:
D1-D8 = 1N4148
T1,T2 = BC547B
T3,T4 = BC557B

Miscellaneous:
RE1,RE2 = PCB mount relay, 12 V, 1 x
changeover, size 19x15.5x15.5 mm (e.g.
Conrad Electronics # 504289)
PCB, order code 040060-1 from The
PCBShop

or unlock the doors. The third wire of the sensors is the common and has to be connected to +12 V. The RC circuits at the inputs Sopen en Sclose ensure that the motors are driven only once when the door is locked or unlocked. In addition, there is a provision to allow the unit to be connected to a car alarm. There are two types of alarm available, with positive or negative control. In order to make the unit universally applicable, both types of alarm can be used. The circuit around T3 and T4 makes this possible. The diode inputs (D3 and D8) react to a rising edge, R6 and R8 react to a falling edge.

An RC time constant is used here as well to ensure that both relays are energised only once. Maybe this is stating the obvious: a motor unit has to be built into each door. All motor wires and sensor wires are connected in parallel to the electronics. The actual type of relay is not critical. The type indicated has the following properties: coil 12 V/400 Ω; max. switching current 12 A (AC), max. switching power 1200 VA. Finally, a car is a hostile environment for electronics. Ensure good connections, use automotive connectors and crimp these on the wires using the appropriate crimping tool. Solder connections in wires are best avoided. They have the tendency to break where the wire transitions into the solder connection when the wire is subject to vibration. Fasten the wires at regular intervals.

Christian Vossen (040060-1)

291
Compressor for Electret Microphone

The 'FM Remote Control Receiver' has a connector where an analogue output is made available. To make a simple intercom or P.A. system the associated transmitter needs a microphone pre-amplifier that outputs a signal at the correct level. And that is exactly the function of this circuit. Actually, this design is adapted from a circuit published last year ('AM Modulator for Intercom'). A few things have been changed so that it can work with the 5 V supply from the transmitter module. The OTA (IC1) used here is the single version (CA3080), which has slightly different characteristics from the dual CA3280. The quad opamp is the same rail-to-rail TS924IN, made by ST. The turnover frequency of the filter (3rd order 1 dB Chebyshev) has been increased slightly to improve the intelligibility of speech and is now about 5.5 kHz. The filter now amplifies the signal by a factor of 10. In practice

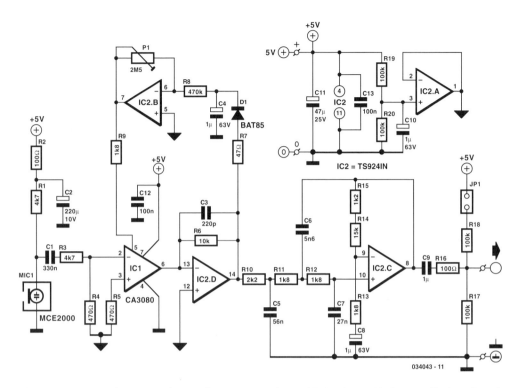

034043 - 11

it is possible that due to various tolerances and the fact that the opamp is not perfect, the filter characteristic shows some deviation from that required. In our prototype it was necessary to change R15 into 2k7 to straighten the response curve. The DC current variation at the output of the OTA and the resulting offset variation at the output of current/voltage converter IC2d is such that the gain of IC2d has to be substantially smaller than in the 'old' design. Otherwise the output could easily rise to the supply voltage at low signal levels. The value of R6 has therefore been made smaller by a factor of 10. This has reduced the gain of the circuit by 20 dB, which is compensated for in the filter. The amplitude of the signal from IC2d is fed back as a control current to the OTA by peak rectifier D1/C3 and inverting amplifier IC2b. R7 limits the loading on IC2d. P1 can be used to adjust the amplifier between a fixed gain and maximum compression. Figure A shows clearly what effect the circuit has. 0 dBr corresponds to 100 mV. The maximum gain, with P1 set to maximum compression, is

about 48 dB (250 Ω) for small signals. The minimum gain is about 20 dB (10 Ω). The OTA is then slightly overdriven and the distortion becomes several percent! With a fixed gain selected (P1 shorted) the gain is about 42 dB (125 ×). The middle curve was measured with P1 in its central position. The curve drawn for a fixed gain (the straight line) doesn't finish at the edge of the graph because the end of the line corresponds to the maximum possible output level, which is 25 dBr ($\approx$1.76 V or 5 / 2√2). Figure B shows the frequency response. The low turnover frequency is mainly determined by C8 (and to a lesser extent by C1) and is about 120 Hz. The current consumption is about 7 mA When the circuit is battery powered we recommend the use of three AA cells, because the circuit still works perfectly at 4.5 V. If you want to use a higher supply voltage (maximum 12 V for the de TS924IN and 30 V for the CA3080, but you should also think of the voltage across the electret microphone!) you have to keep in mind that the maximum current through R9 (which is I_{ABC}) is

395

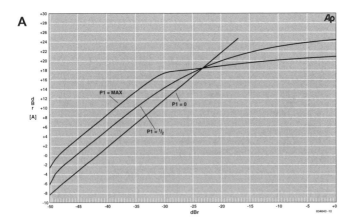

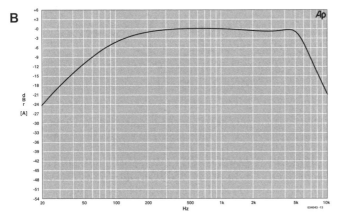

The value of 0.7 V corresponds to the potential between pin 5 and earth. For a larger safety margin R9 is calculated with the full supply voltage and a current of 2 mA:

$(5 – 0.7)$ V $/ 2$ mA $= 2k2$ (rounded upwards).

Of course the regulation will then be different (a little less gain). This circuit and the transmitter module can therefore be fed from the same 5 V supply. Because the transmitter requires a DC offset at its input, a resistor is connected to +5 V via a jumper, which biases the output to half the supply voltage. With the jumper open R17 functions as a load resistor when the output is not connected, because C9 still has to charge up even without a load. If you're designing a PCB for this compressor then it makes sense to include the

only 2 mA. When we consider a maximum chosen current of 1 mA and the maximum output voltage of IC2b (half the supply voltage, which is 2.5 V), then the value of R9 should be $(2.5 – 0.7)$ V $/ 1$ mA $= 1.8$ kΩ.

transmitter module as well. The current consumption then increases by about 10 mA.

T. Giesberts   (034043-1)

292
Stable Filament Supply

Valves are enjoying increasing popularity in audio systems. With the European 'E' series of valves, such as the ECC83 (12AX7) and EL84 (6BQ5), the filament voltage is 6.3 V. Depending on how the circuit is wired, the ECC 81–83 series of twin triodes can also be used with a filament voltage of 12.6 V. In earlier times, the filament voltage was usually taken directly from a separate transformer winding, which (in part) was responsible for the well known 'valve hum'. With regard to the sig-

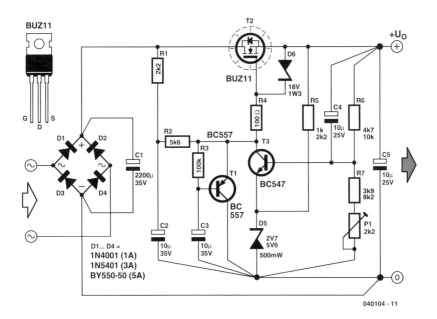

040104 - 11

nal path, current valve circuits have hardly experienced any fundamental changes. In high-quality valve equipment, though, it is relatively common to find a stabilised anode supply. Mains hum can have a measurable and audible effect on input stages whose filaments are heated by an ac voltage. The remedy described here is a stabilised and precisely regulated dc filament voltage. The slow rise of the filament voltage after switching on is also beneficial. The exact setting of the voltage level and the soft start have a positive effect on the useful life of the valves. Figure 1 shows a voltage regulator meeting these requirements that is built from discrete components. The two sets of component values are for a voltage of 6.3 V (upper) and 12.6 V (lower). Thanks to the fact that the supply works with a constant load, it can do without special protective circuits and the additional complexity of optimum regulation characteristics for dynamic loads. The circuit in Figure 1 consists of a power MOSFET configured as a series-pass regulator and a conventional control amplifier. Zener diode (D5) sets the reference potential. A constant voltage is thus present at the emitter of the BC547 control amplifier (T3). The current through D5 is set to ap-

proximately 4–5 mA by series resistor R5. The output voltage U_O (the controlled variable) acts on the base of the control amplifier (T3) via voltage divider R6/R7. If the output voltage drops, the collector current of T3 also decreases, and with it the voltage drop across load resistors R1 and R2. The voltage on the gate of the MOSFET thus increases. This closes the control loop. The values of the resistors forming the voltage divider are chosen for the usual tolerances of Zener diodes, but they must be adjusted if the diode is out of spec (which can happen). The load resis-

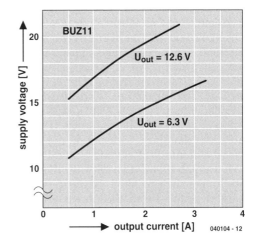

040104 - 12

tance of the control amplifier is divided between R1 and R2. The current through the load resistance and the collector current of T3 are practically the same, since the MOSFET draws almost no gate current. Filter capacitor C2 is connected to the junction of R1 and R2 to reduce residual hum. Electrolytic capacitor C4 and power supply filter capacitor C1 serve the same purpose. The hum voltage also depends on the magnitude of the load current. The voltage drop over the series-pass regulator is nearly the same for an output voltage of 6.3 V or 12.6 V. With a BUZ11 and a load of 1 A at 6.3 V, for instance, the average voltage across the source–drain channel is approximately 7 V. The power dissipation of 7 W requires a corresponding heat sink. The slow rise of the output voltage is due to the presence of timing network R3/C3 and

T1. When power is switched on, T1 holds the gate of the MOSFET at nearly ground level. As C3 charges, T1 conducts increasingly less current, so ultimately only the control transistor affects the gate voltage. The mains transformer must be selected according to the required load current. The required value of the input voltage can be read from the chart. The transformer should have a power rating at least 30 % greater than what is necessary based on the calculated load dissipation. Where possible, preference should be given to a filament voltage of 12.6 V. When twin triodes in the ECC81–83 series are used, for example, the power dissipation in the series pass transistor is lower with a voltage of 12.6 V.

Dr Alexander Voigt   (040104-1)

293
NiMH Charger for up to six Cells

It is impossible to imagine present day society without any batteries. Count the number of gadgets in your house that are powered from batteries, you will be stunned by the number of batteries you will find. The majority of these devices use penlight batteries and if you're a little environmentally conscious you will be using rechargeable batteries. A few years ago these batteries were invariably NiCd types. However, these batteries suffer from a relatively high rate of self-discharge and from the so-called memory-effect. It is now more common to use NiMH batteries. The advantage is that these batteries do not suffer from the memory-effect and generally also have a much higher capacity, so that they last longer before they have to be recharged again. From the above you can conclude that every household these days needs, or could use, a battery charger. A

good charger needs to keep an eye on several things to ensure that the batteries are charged properly. For one, the charger has to make sure that the voltage per cell is not too high. It also needs to check the charging curve to determine when the battery is fully charged. If the charging process is taking too long, this is an indication that something is wrong and the charger must stop charging. Sometimes it is also useful to monitor the temperature of the cells to ensure that they do not get too hot. The circuit presented here is intended for charging NiMH batteries. The MAX712 IC used here contains all the necessary functionality to make sure that this happens in a controlled manner. Figure 1 shows the schematic of the charger. The heart of the circuit is easily recognised: everything is arranged around IC1, a MAX712 from Maxim. This IC is available in a standard

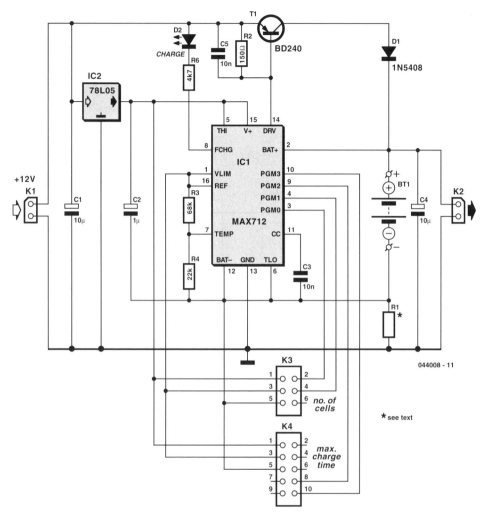

044008 - 11

★ see text

DIP package, which is convenient for the hobbyist because it can be directly fitted on standard though-hole prototyping board. IC1 uses T1 to regulate the current in the battery. R1 is used by IC1 to measure the current. While charging, IC1 attempts to maintain a constant voltage, equal to 250 mV, across R1. By adjusting the value of R1 the charging current can be set. The value of R1 can be calculated using the formula below:

$$R1 = 250 \text{ mV} / I_{charge}$$

For a charging current of 1 A, the value of R1 has to be 250 mV / 1 A = 0.25 Ω. The power dissipated by R1 equals U × I = 0.25 × 1 = 250 mW. A 0.5 watt resistor will therefore suffice for R1. Transistor T1 may need a small heatsink depending on the charging current and supply voltage. IC1 needs a small amount of user input regarding the maximum charging time and the number of cells in the battery to be charged. IC1 has four inputs, PGM0 to PGM3, for this purpose. These are not ordinary digital inputs (which recognise only 2 states) but special inputs that recognise 4 different states, namely V+, Vref, BATT– or not connected. To make this a little bit more user friendly, we've brought out the necessary connections to 2 connectors (K3

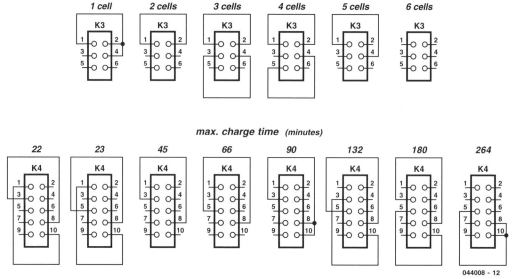

044008 - 12

and K4). A number of dongles have been made (Figure 2) that can be plugged into these connectors and set the number of cells and the maximum charging time. When determining the maximum charging time we have to take into account the charging current and the capacity of the cells that are connected. The charging time can be calculated with the formula:

$$T_{charge} = C_{cell} / I_{charge} \times 1.2$$

where Ccell is the capacity in Ah (e.g., 1200 mAh = 1.2 Ah). After the nominal charging time has been calculated, we can use the first dongle that has a value that is equal or greater than the calculated charg-ing time. For example, if we calculated a maximum charging time of 38 minutes, we have to select the dongle for 45 minutes. When IC1 is replaced by a MAX713, the charger becomes suitable for charging NiCd batteries (but not suitable for NiMH batteries any more!).

The only difference between these two ICs is the value of the detection point at which the cell(s) are considered to be completely charged. The ICs are otherwise identical with regard to pin-out, method of adjust-ment, etc. To make it easy to swap between the ICs, we recommend an IC-socket for IC1.

Paul Goossens   (044008-1)

294
LM4906 Boomer® Audio Power Amp

The well-known LM386 is an excellent choice for many designs requiring a small audio power amplifier (1 watt) in a single chip. However, the LM386 requires quite a few external parts including some electro-lytic capacitors, which unfortunately add

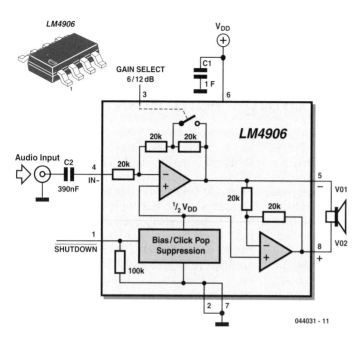

LM4906

044031 - 11

specified supply voltage. Four times the output power is possible as compared to a single-ended amplifier under the same conditions (particularly when considering the low supply voltage of 5 to 6 volts). When pushed for output power, the small SMD case has to be assisted in keeping a cool head. By adding copper foil, the thermal resistance of the application can be reduced from the free air value, resulting in higher PDMAX values without thermal shutdown protection circuitry being activated. Additional copper foil can be added to any of the leads connected to the LM4906. It is especially effective when connected to VDD, GND, and the output pins. A bridge configuration, such as the one used in LM4906, also creates a second advantage over single-ended amplifiers. Since the differential outputs, V01 and V02, are biased at half-supply, no net DC voltage exists across the load. This eliminates the need for an output coupling capacitor which is required in a single supply, single-ended amplifier configuration. Large input capacitors are both expensive and space hungry for portable designs. Clearly, a certain sized capacitor is needed to couple in low frequencies without severe attenuation. But in many cases the speakers used in portable systems, whether internal or external, have little ability to reproduce signals below 100 Hz to 150 Hz. Thus, using a large input capacitor may not increase actual system performance. Also, by minimizing the capacitor size based on necessary low frequency response, turn-on pops can be minimized. Further information from: www.national.com.

volume and cost to the circuit. National Semiconductor recently introduced its Boomer® audio integrated circuits which were designed specifically to provide high quality audio while requiring a minimum amount of external components (in surface mount packaging only). The LM4906 is capable of delivering 1 watt of continuous average power to an 8-ohm load with less than 1% distortion (THD+N) from a +5 V power supply. The chip happily works with an external PSRR (Power Supply Rejection Ratio) bypass capacitor of just 1 µF minimum. In addition, no output coupling capacitors or bootstrap capacitors are required which makes the LM4906 ideally suited for cellphone and other low voltage portable applications. The LM4906 features a low-power consumption shutdown mode (the part is enabled by pulling the SD pin high). Additionally, an internal thermal shutdown protection mechanism is provided. The LM4906 also has an internal selectable gain of either 6 dB or 12 dB. A bridge amplifier design has a few distinct advantages over the single-ended configuration, as it provides differential drive to the load, thus doubling output swing for a

Source: National Semiconductor   (044031-1)

295
High Voltage Amplifier

Audio analysers, such as the Audio Precision series have signal generators that for certain test purposes do not have a sufficiently high output voltage. That is why we have designed this 'booster' amplifier stage. It provides the same amount of voltage as a 300 W amplifier into 8 Ω. Applications for this amplifier include the testing of measuring filters or an automatic range-switching circuit. The amplifier can, with a ±75 V power supply, generate 50 V_{RMS}, with an input signal of 5 V_{RMS} from the signal generator. In other words, the voltage gain is set to 10 times. In many cases the full bandwidth of the generator is not available at maximum output voltage. For this reason the amplifier gain is a little more than what is actually required for this application. The graph shows the distortion (THD+N) as a function of output voltage. It is obvious that at 1 kHz (curve A), from about 10 V with 10 kΩ load, the limit of the Audio Precision has been reached. The

steps in the curve are caused by range switching in the analyser. At less then 10 V, the measurement is mostly noise. From 20 kHz (curve B) the distortion increases slowly, but with only 0.008% at 50 V remains small even at nearly full output! Both curves were measured with a bandwidth of 80 kHz. The amplifier is built around the old faithful NE5534 and a discrete buffer stage. The buffer stage has been designed as a symmetrical compound stage, which maximises the output voltage. The advantage of the compound stage is that it is possible to have voltage gain. This possibility has been taken advantage of, because the NE5534 can operate from a maximum of only ±22 V (and that is pushing it!). We make the assumption that the opamp can supply 15 Vpeak without distortion. There is, therefore, 4.7-times gain required in the compound-stage. This is set with the local negative feedback R15, R16 and R17. The values are as small as possi-

ble to enable the use of normal resistors. At first glance, the gain should in theory be 6 times, but the stages T3 and T4 influence the local feedback. When driven with DC, the dissipation in R16 (R17) will be a maximum of about 0.3 W. For T5 and T6, the MJE340 and MJE350 were selected. This pair has been practically unrivalled for many years. At a collector/emitter voltage of 150 V, the MJE350 can carry a collector current of 40 mA. The output stage is, with an idle current of 35 mA through T5 and T6, well into class-A

operation. Local and total feedback requires a maximum of 20 mA. Despite this
'no load' the output transistors are well
within their safe operation area. The minimum load is a few kΩ. The amplifier is not
short-circuit proof. If necessary, R20 can

be increased to 1 kΩ or a current limit circuit can be added. However, this is likely to
reduce the quality of the amplifier. The
drivers T3 and T4 are a couple of new transistors from Sanyo. These have a considerably better linearity (h_{FE}) and have a lower

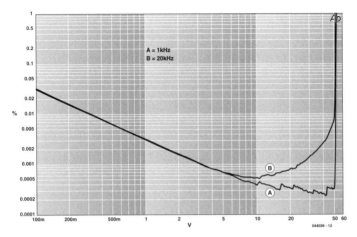

capacitance than the MJEs. The maximum collector/emitter voltage is 160 V. At this value (DC), up to 7 mA may be processed (or 20 mA for 1 s). The actual operating voltage for these will in practice be about 100 V. With a current setting of 7 mA, these transistors are operating well inside their safe operating area. These are details that have to be taken into account because of the high power supply voltage. T3 and T4 are, with the aid of four 1N4148 diodes, biased at a fixed current, so it is not necessary to adjust the quiescent current. The current through these diodes is set with two symmetrical current sources to a value of about 4.5 mA (T1 and T2). BF-series devices were selected for this application because they have even lower capacitance. Should the availability of the BF469 and BF470 prove to be a problem then it is possible to substitute the same devices as used for T3 and T4, i.e. 2SC2911 (NPN) and 2SA1209 (PNP) respectively. This stage is driven from the middle of D5 through D8, so that the opamp only needs to deliver the amount of current to compensate for the difference in current between T3 and T4. R6 protects the output of IC1 from any possible capacitive feedback from the output stage. The overall feedback loop is set with R4 and R3. This has been quite accurately adjusted for a gain of 10 times (A = 1 + R4/R3), so the actual value is 10.09 times. R5 and C1 are the compensation network for the entire amplifier by providing the opamp with local feedback. Note that if you change the gain

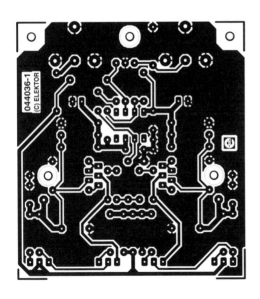

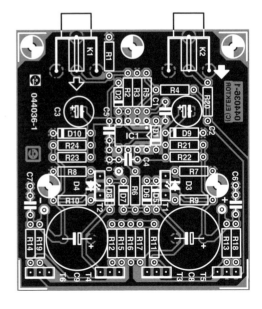

COMPONENTS LIST

Resistors:
R1,R16,R17 = 10kΩ
R2,R6 = 100Ω
R3 = 1kΩ10
R4 = 10kΩ0
R5 = 5kΩ6
R7,R8 = 47kΩ
R9,R10 = 270Ω
R11,R12 = 82Ω
R13,R14 = 220Ω
R15 = 1kΩ
R18,R19 = 27Ω
R20 = 47Ω
R21-R24 = 12kΩ

Capacitors:
C1 = 22pF
C2,C3 = 47µF 25V radial
C4-C7 = 100nF (C6/C7: 100V!)

C8,C9 = 470µF 100V radial

Semiconductors:
D1,D2,D5-D8 = 1N4148
D3,D4 = rectangular LED, red
D9,D10 = zener diode 22V 1.3W
T1 = BF470
T2 = BF469
T3 = 2SC2911
T4 = 2SA1209
T5 = MJE350
T6 = MJE340
IC1 = NE5534

Miscellaneous:
K1,K2 = cinch socket, PCB mount, T-709G
from Monacor/Monarch Heatsink, 2.5 K/W
(e.g. Fischer type SK100, 50mm height)
PCB, order code 044036-1 from The
PCBShop

of the amplifier, the compensation network has to change as well. An NE5534 has internal compensation when the gain is 3 or greater. So the ratio between R5 and R3 must be greater than 2. The bandwidth of the amplifier at 11 MHz is quite good (measured at 40 Vrms). R2, D1 and D2 provide input protection. R1 determines the input resistance, which is 10 kΩ. The value of R1 may be increased, but the result will be a higher output offset. The bias current of IC1 can easily be 0.5 µA and that explains the offset of 50 mV at the output. For most applications this value will not cause any problems. The power supply for the opamp is derived from the ±75 V power supply using two zener diodes. This is designed with two parallel resistors so that it is not necessary to use special power resistors. C6 through C9 decouple the output stage and C2 through C5 decouple the opamp. The total current draw of the prototype, after it had warmed up, was about 57 mA. The 'High voltage supply' elsewhere in this issue can be used as the power supply.

(044036-1)

296
Resistor-Equipped Transistors (RETs)

Developments in the electronics area (or is it an arena?) are never at a standstill. Whenever there is something substantially new to report it is almost always related to complex chips. However, Philips proves that new developments are possible even with what you thought were dead-standard components. For example, take the new range of transistors with the name 'RET' – meaning Resistor Equipped Transistor. The novelty with this new range of transistors is that they have a base resistor included. Some versions also have a resistor from base to emitter. These new transistors

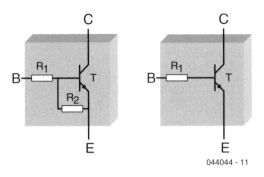

044044 - 11

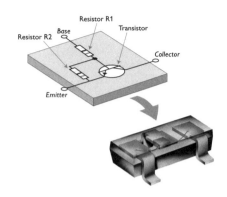

enable engineers to design even smaller devices because the space required on the PCB has been reduced as a consequence. An additional advantage is that the total component count is reduced.

The new transistors are available in both PNP and NPN versions. For the package, the designer can choose from a number of SMD packages as well as the familiar TO-92. The maximum power dissipation depends on the package and varies from 150 mW to 500 mW for the single transis-

tor chips. There are also chips with two transistors, where each chip is provided with either one or two resistors. With these you can choose between a maximum dissipation of 300 mW or 600 mW. A complete overview of the available RETs can be found on the Internet at: http://www.semiconductors.com/acrobat/-literature/9397/75012514.pdf.

source: Philips
Paul Goossens   (044044-1)

297
Dimmer with a MOSFET

This circuit shows that dimmers intended for use at mains voltage do not always have to contain a triac. Here, a MOSFET (BUZ41A, 500 V/4.5A) in a diode bridge is used to control the voltage across an incandescent bulb with pulsewidth modulation (PWM). A useful PWM controller can be found elsewhere in this issue. The power supply voltage for driving the gate is supplied by the voltage across the MOSFET. D6, R5 and C2 form a rectifier. R5 limits the current pulses through D6 to about 1.5 A (as a consequence it is no longer a pure peak rectifier). The voltage across C2 is regulated to a maximum value of 10 V by R3, R4, C1 and D1. An opto-

coupler and resistor (R2) are used for driving the gate. R1 is intended as protection for the LED in the optocoupler. R1 also functions as a normal current limiting device so that a 'hard' voltage can be applied safely. The optocoupler is an old acquaintance, the CNY65, which provides class-II isolation. This ensures the safety of the regulator. The transistor in the optocoupler is connected to the positive power supply so that T1 can be brought into conduction as quickly as possible. In order to reduce switching spikes as a consequence of parasitic inductance, the value of R2 has been selected to be not too low: 22 kΩ is a compromise between inductive voltages and

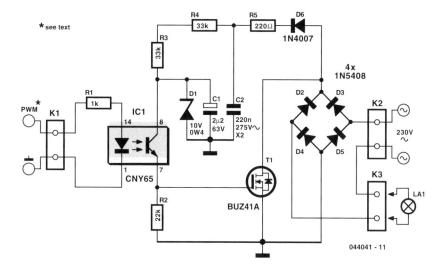

044041 - 11

switching loss when going into and out of conduction. An additional effect is that T1 will conduct a little longer than what may be expected from the PWM signal only. When the voltage across T1 reduces, the voltage across D1 remains equal to 10 V up to a duty cycle of 88 %. A higher duty cycle results in a lower voltage. At 94 % the voltage of 4.8 V proved to be just enough to cause T1 to conduct sufficiently. This value may be considered the maximum duty cycle. At this value the transistor is just about 100 % in conduction. At 230 V mains voltage, the voltage across the lamp is only 2.5 V lower, measured with a 100 W lamp. Just to be clear, note that this circuit cannot be used to control inductive loads. T1 is switched asynchronously with the mains frequency and this can cause DC current to flow. Electronic lamps, such as

the PL types, cannot be dimmed with this circuit either. These lamps use a rectifier and internally they actually operate off DC. A few remarks about the size of R3 and R4. This is a compromise between the lowest possible current consumption (when the lamp is off) and the highest possible duty cycle that is allowed. When the duty cycle is zero, the voltage across the resistors is at maximum, around 128 V with a mains voltage of 230 V. Because (depending on the actual resistor) the voltage rating of the resistor may be less than 300 V, two resistors are connected in series. The power that each resistor dissipates amounts to a maximum of 0.5 W. With an eye on the life expectancy, it would be wise to use two 1 W rated resistors here.

Ton Giesberts   (044041-1)

298
Modern SMD Packages

Parts in SMD packages are becoming more and more frequent in DIY circuits. Sometimes they are used because of better per-

formance, usually as a consequence of the shorter PCB traces between the various parts. The result of this is a lower self-in-

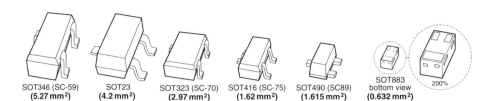

| SOT346 (SC-59) | SOT23 | SOT323 (SC-70) | SOT416 (SC-75) | SOT490 (SC89) | SOT883 bottom view |
| (5.27 mm²) | (4.2 mm²) | (2.97 mm²) | (1.62 mm²) | (1.615 mm²) | (0.632 mm²) |

ductance so the circuit exhibits better RF behaviour. Another reason components in an SMD package are selected is simply because the equivalent parts in a 'through-hole' package are no longer being manufactured. All those different packages have naturally resulted in several new standards for packages.

Here we show a number of common SMD packages with three terminals. In particular, note the difference in dimensions for the various packages. Arguably a SOT346 package is a lot easier to solder than a SOT490 package. It is obvious then, that a SOT883 package is going to be the most difficult to solder. Not only is the package extremely small, but in addition, the terminals are underneath. This means that it cannot be soldered with a soldering iron. To solder these, you would have to resort to using a hot plate, an iron or even better: an SMD oven.

Paul Goossens   (044045-1)

299
IR Transmitter with HT12E

IR (infra-red) transmitters containing the encoder-IC HT12E from Holtek have been published in Elektor Electronics on previous occasions. The interesting aspect of this design is that the entire IRtransmitter has been squeezed into a handy key ring. The operation of the encoder IC HT12E has already been thoroughly dealt with in earlier publications. It will suffice to mention here that an address can be programmed on inputs A1 through to A8. This address has to be set the same as in the receiver. On the circuit board, the address lines are connected to ground with a thin piece of track so that address zero is selected. By cutting the track with a sharp knife the corresponding input can be made logic high. This change of address is only necessary when more than one transmitter is active in the same house. The same story applies to the four data lines, but that is not relevant if you use the IR Multi-Position Switch published on page 414 in this book. The generated code is available on pin 17 of IC1, which is modulated with the aid of IC2 on a carrier of 40 kHz. Transistor T1 drives the infrared LED via R4, so that an IR code is transmitted. Two Lithium cells

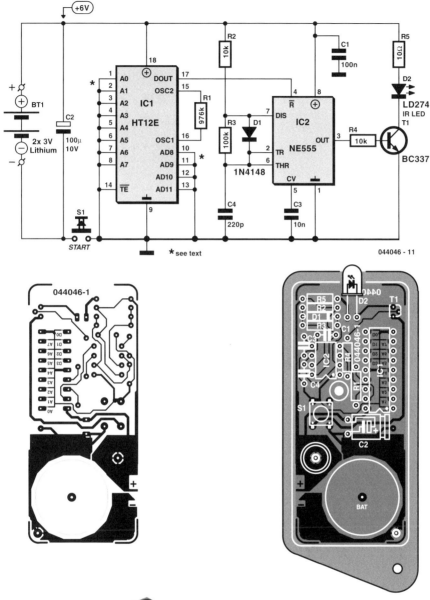

044046 - 11

of 3 V each power the circuit. The circuit is only powered when transmit pushbutton S2 is pressed. The life expectancy of the batteries is during normal use more than a year! The construction itself should be no problem. However, note the height of the components. For this reason an IC socket cannot be used. A paperclip soldered to the bottom acts as a battery holder, this is

COMPONENTS LIST

Resistors:
R1 = 976kΩ
R2,R4 = 10kΩ
R3 = 100kΩ
R5 = 10Ω

Capacitors:
C1 = 100nF
C2 = 100µF 10V radial
C3 = 10nF
C4 = 220pF

Semiconductors:
D1 = 1N4148
D2 = IRED (e.g., LD274)
IC1 = HT12E (Holtek)
IC2 = TLC555
T1 = BC337

Miscellaneous:
S1 = miniature pushbutton (HAK)
Enclosure: KM series Box UM14
PCB, available from The PCBShop

ground. For the top part of the battery holder we use an automotive connector and an M3 bolt. Because the height of T1 can be a problem when closing the enclosure, the device is best bent flat on the PCB before soldering.

Peter Verhoosel   (044046-1)

300
5 Volts from the Mains

Sometimes we would like to hide certain equipment to prevent cluttering up our living room or any interior in general. A number of devices lend themselves to be built into a wall adapter. Think of a remote-control extender, for example. All these de-vices need a power supply and we would prefer to use the mains so that no external connections are required for the power supply. The power supply in this article is intended for exactly these situations, namely converting the mains voltage to a

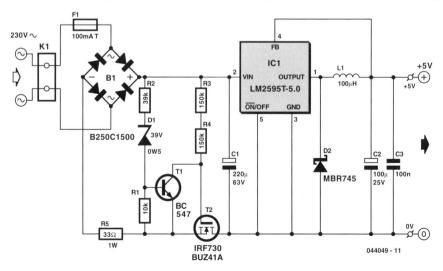

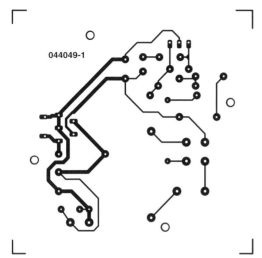

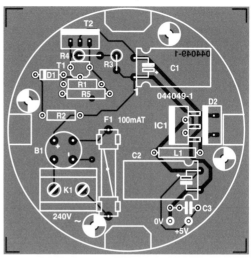

5 V power supply voltage. The accompanying PCB fits exactly in a round wall socket enclosure. A power supply is usually fitted with a transformer to reduce the voltage and also provide galvanic isolation between the device and the mains. In this power supply a transformer cannot be used

COMPONENTS LIST

Resistors:
R1 = 10kΩ
R2 = 39kΩ
R3,R4 = 150kΩ
R5 = 33Ω

Capacitors:
C1 = 220µF 63V radial
C2 = 120µF 25V radial
C3 = 100nF

Semiconductors:
B1 = B250C1500, round case
D1 = zener diode 39V, 0.5W
D2 = MBR745
IC1 = LM2595T-5.0
T1 = BC547B
T2 = IRF730 or BUZ41A

Miscellaneous:
F1 = fuse, 100 mAT (slow), with PCB mount holder and cap
K1 = 2-way PCB terminal block, lead pitch 7.5mm
L1 = 100µH choke
PCB, available via The PCBShop

because of lack of space. That is why we use a step-down regulator here. A problem with most stepdown regulators is that they cannot be supplied directly from the mains. Hence, in this schematic (Figure 1) we first create a rough power supply voltage of around 40 V using passive components and subsequently present it to step-down converter IC1. The converter can operate from a maximum input voltage of 45 V. The mains voltage is first fused by fuse F1 and then converted to full rectified sine wave by the bridge rectifier. FET T2 is used as a switch, which is Turned off when the voltage is greater than 40 V. The gate of T2 is driven via R3 and R4. As soon as the voltage exceeds the value of 40 V, transistor

T1 will conduct which causes the gate-drain voltage of T2 to be so small that T2 stops conducting. Because of this, electrolytic capacitor C1 cannot charge any further and the maximum voltage across C1 is therefore limited to about 40 V. This voltage is converted by IC1 and surrounding components to a 5 V power supply voltage. The maximum output current is 1 A. The PCB (Figure 2) that has been designed for the power supply fits, as mentioned before, in a round wall socket enclosure. Note that resistor R1 and the link have to be soldered first, then comes R5, in the air above R1 because of the tight space. Populating the remainder of the PCB should not present any difficulties. When mounting the PCB, do note that it is directly connected to the mains, so make sure no conductive parts can be touched when the circuit is in use. The mounting holes are separated less than 6 mm from the PCB traces so the board has to be secured with plastic screws in order to satisfy the safety requirements. Also, after fitting the board, a cover has to be mounted over the PCB so that it is impossible to touch the PCB when the wall socket is opened. As always, you cannot be too careful when dealing with mains voltages!

Paul Goossens   (044049-1)

301
On-line Conversions

From time to time you may come across a unit of measure that's completely unfamiliar. It also happens that the name sounds vaguely familiar, but you just can't remember the fine points. Although the Internet probably contains numerous sites that can help out, the most comprehensive one is probably www.onlineconversion.com.

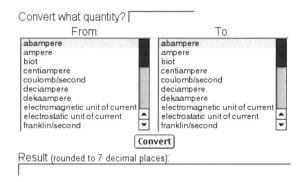

Here you can convert physics quantities such as, for example, lengths, velocity and forces. There are also conversion tools for currencies, clothing sizes and terminology used in cookbooks. Although the title of this article may imply otherwise, some parts can be downloaded and used off-line.

Luc Lemmens   (044052-1)

302
SATA

There is a new standard these days for connecting hard disks and DVD drives: SATA. You may already have noticed that new PCs don't contain the wide 40- or 80-pin ribbon cables any more, but the hard disk is connected with a slender 7-way cable. Even the corresponding power supply connector is different with a SATA drive and now has 15 connections instead of 4. This is clearly illustrated in the drawing: The large connector provides power and the small connector is the data connection. The power supply connector provides 12, 5 and 3.3 V and two ground connections. SATA means Serial ATA. ATA is the connection standard that at present has been the most common for hard disks, CD and DVD drives. At the time of writing the ATA100 bus with a data transfer speed of 100 Mbyte/s is the most widespread. A number of manufacturers use the slightly faster ATA133 bus. That is pretty much the end of the line; it has not been possible for the industry to increase the speed even further using simple means. In order to increase the speed of the interface between motherboard and HDD, a serial connection was selected. At first glance this appears to be an perplexing choice!

The parallel ATA bus operates with 16 bits simultaneously; the serial bus therefore has to be at least 16 times faster to obtain the same equivalent speed. That is why SATA operates at a clock speed of 1500 MHz and reaches a data transfer speed of 150 MByte/s; versions with 3 GHz and 6 GHz clock speed are planned. The industry made this choice because, in the end, it is cheaper to have a solution with one data channel at very high speed than trying to increase the speed of all 33 signals in the ATA connection. The data connection has a balanced outgoing and a balanced return connection, plus three ground connections. The signal amplitude is not 5 volts any more, but only 0.25 volts. The symmetrical connection and the small signal amplitude together result in a fast, low interference and energy efficient connection. In addition, the cable is allowed to be much longer: 1 m instead of 45 cm with ATA. Another big difference is that each device (HDD, DVD) has its own cable and connection to the motherboard. With the current ATA standard two devices can share the same cable. As a consequence various jumpers (master, slave) disappear and the interface can (in principle) always operate at maximum speed. Additionally, the contacts in the connectors are of different length, so that hotplugging (connecting or disconnection without switching the power off) becomes possible. The expectation is that this interface will satisfy the requirements for the next ten years.

Karel Walraven   (044053-1)

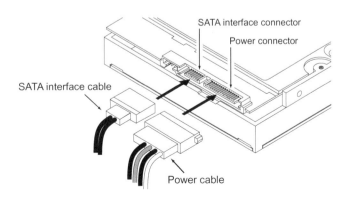

SATA interface connector

Power connector

SATA interface cable

Power cable

303
IR Multi-Position Switch

This multi-position switch is operated by the IR transmitter with HT12R described on page 408 in this book. The signal transmitted by the transmitter enters the circuit through infrared detector IC4. At the output the demodulated signal is available, that via R2 and T1 is routed to the input of IC1 (decoder type HT12D). Like the encoder IC, this IC has already been described in *Elektor Electronics* on multiple occasions and we will therefore not describe it again here. On the PCB the address lines of IC1 are by default connected with a thin trace to ground. The default address is therefore zero. By cutting the trace with a sharp knife the corresponding address input can be made logic high. This

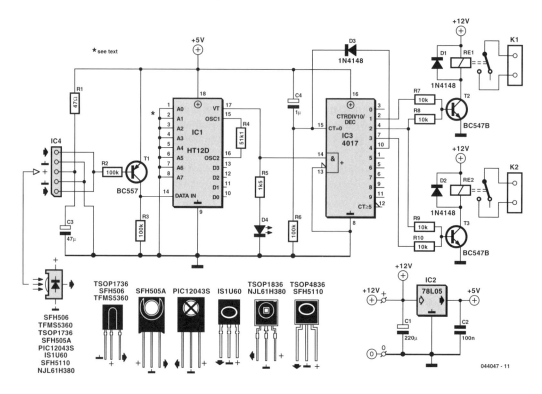

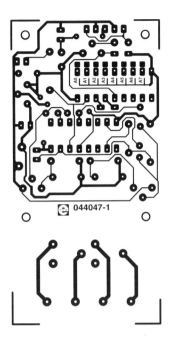

044047-1

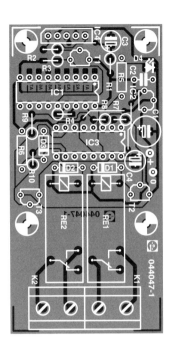

044047-1

COMPONENTS LIST

Resistors:
R1 = 47?
R2,R3,R6 = 100k?
R4 = 51k?1
R5 = 1k?5
R7-10 = 10k?

Capacitors:
C1 = 220µF 16V radial
C2 = 100nF
C3 = 47µF 10V radial
C4 = 1µF 10V radial

Semiconductors:
D1,D2,D3 = 1N4148
D4 = LED, red, high-efficiency
T1 = BC557
T2,T2 = BC547B
IC1 = HT12D (Holtek)
IC2 = 78L05
IC3 = 4017
IC4 = SFH506-40 or equivalent

Miscellaneous:
Re1,Re2 = 12 V relay, e.g., Siemens
V23057-B0002-A101
K1,K2 = 2-way PCB terminal block, lead
pitch 7.5mm
PCB, available from The PCBShop

change of address is only necessary if multiple transmitters are active in the same house.

When the programmed address code corresponds with the address code of the IR transmitter, pin 17 of IC1 will be high for as long as the transmitted signal is available. LED D4 will light up. This pulse is presented to the clock input of IC3, a decade counter. After each pulse the decade counter will make the next output high and the previous output will go low again. When the power supply is switched on, C4 and R6 ensure that the IC is reset; the first output is high and all the others are low. This is the reason that relays are connected only from the second output onwards. After the first pulse is transmitted, Re1 will be switched on via R7 and T1. The next pulse causes both Re1 and Re2 to be switched on. After the next pulse only Re2 will be on. The reset-input of the decade counter is connected via D3 to output Q4. This causes the switch to return to the rest position after the next pulse.

Peter Verhoosel   (044047-1)

304
Voltage Monitor

It is often necessary to monitor the power supply voltage in a piece of equipment. When the device takes its power from a battery, the input voltage can change and provide an indication of how much energy is left. Even a device powered from the mains can benefit from keeping an eye on the (various) power supply voltages in the circuit and take the necessary steps in the event of a fault condition. If a slightly intelligent reaction is required for this situation, it is obvious to do this with a microcontroller. This requires the microcontroller to periodically measure the voltages with one or more A/D converters and decide whether the values are correct or whether something is the matter.

This naturally costs processor time and makes the firmware a little more complicated. This can become a problem, particularly when other functions have to be carried out on a regular basis as well. Fortunately, chip manufacturer Maxim has a number of ICs in its line-up specifically for this purpose.

These ICs all have as task to monitor a number of voltages and when these do not conform to certain requirements the IC generates an interrupt, which indicates to the microcontroller that something has gone wrong. The MAX1153 and MAX1154 are provided with a 10-bit ADC, while the MAX1253 and MAX1254 are provided with a 12-bit ADC. All these

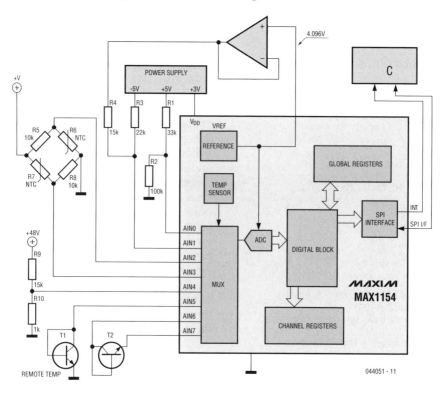

044051 - 11

ICs are capable of monitoring 8 external voltages and 2 internal voltages. These 8 external voltages can also be programmed for temperature sensors, where an external diode can be used as a simple and cheap sensor. For each channel, a maximum and minimum voltage or temperature can be specified. As soon as the input voltage or temperature falls outside this range the interrupt output goes low. If desired, a glitch suppressor can be individually configured for each channel. A recursive averaging filter is also available and it is even possible to specify how many successive samples have to be outside the range before the interrupt is generated. The sample-rate of the ADC is also adjustable, to a maximum of 90 ksamples/s. The SPI bus serves as interface to the controller and requires only four signals to communicate. For more information you can visit the Maxim website: www.maxim-ic.com.

Paul Goosens   (044051-1)

305
BTX

What did you say BTX was again? Well, it's just the successor of ATX! ATX is the collective term for the electrical and mechanical characteristics of the present generation of motherboards in your PC. There is now a new standard in its place, BTX (Balanced Technology eXtended Form Factor), with the main advantages of more compact dimensions and better heat removal. When grouping the components on the PCB, careful attention is paid to power dissipation and the possibility of efficiently removing the heat. This is necessary because the CPU meanwhile needs to dissipate 100 W and a respectable video card can easily dissipate 50 W. That is why there has to be serious consideration of how the heat can be removed from the case, where wide cables and other obstacles are no longer allowed to be in the way (also refer to SATA elsewhere in this issue). That the energy consumption is continually increasing can be deduced from the ever more powerful PSU units. While 10 years ago a 150 W power supply was quite adequate, these days 350 or 400 W is the norm. With the BTX standard, there is the assumption of a fan for the power supply (just as before), but also a separate second cooling circuit for the motherboard, drawing in cold air from outside the case and blowing out warm air via a separate path, completely independent of the fan in the power supply.

BTX power supplies will also be able to supply more power. There is a new standard for the power supply connector that increases the number of pins from the current 20 to 24 ('Main power connector'). The first 20 pins keep the same functionality as with ATX motherboards (only −5 V

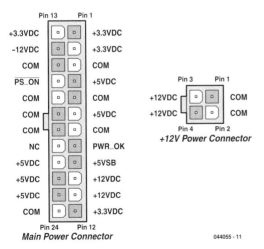

Main Power Connector

+12V Power Connector

044055 - 11

has been omitted!). The extra 4 pins are for more 3.3 V, 5 V and 12 V connections (refer figure). In addition, there has to be a separate 4-way connecter ('+12 V power connector') that's intended to power the CPU only.

Karel Walraven   (044055-1)

306
Battery Polarity Protection

Look, no diodes, no relays!

With battery-operated equipment there's always the risk of batteries being inserted the wrong way around. Although a diode or a bridge rectifier can be used to solve the problem, the down side is considerable energy loss through dissipation of the forward biased device(s). An energy-wise alternative is now available in the form of an integrated circuit containing two analogue change-over switches.

For sure, switching devices using CMOS technology have been around for a long time, but it was not until the requirements of mobile phones and portable test equipment forced chip designers to make substantial improvements to the venerable 4000 series of CMOS logic ICs. The 'retrofit' circuit described in this short article is based on the MAX4684 (from Maxim) and is capable of automatically swapping (or, if you like, correcting) the polarity of a battery set. The circuit can work from a voltage as low as 1.8 V, which may be supplied by two totally exhausted dry batteries, NiCd or NiMH cells in series. The internal architecture of the chip as pictured in Figure 1 shows two singlepole changeover switches realised using P-channel and N-channel MOSFETs. These devices are marked by a extremely low 'on' resistance, while acting very fast and being capable of carrying and switching high currents. Functionally, the two switches mimic a bridge rectifier fed with a direct voltage. The MAX4684 not only protects the equipment being powered, but also arranges for an incorrectly polarized battery voltage to be swapped very quickly. The operation of the IC with the correct or the wrong supply polarity is illustrated in Figure 2. For clarity's sake, the current paths are highlighted. With no battery (or battery pack) connected, the two switches are in the 'inactive' position, so that the COM pins are effectively connected to NC (normally closed). Figure 2a shows the switch positions when the right polarity is applied: control input IN1 (upper switch) is tied to the negative battery terminal and leaves the switch in

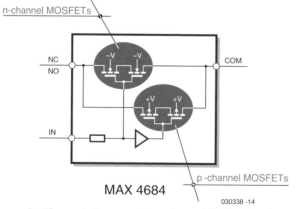

Figure 1: The switches consist of two MOSFETs (P- and N-channel).

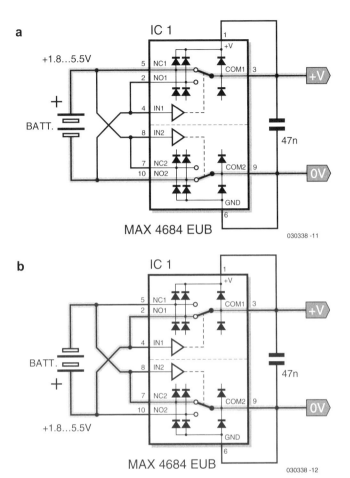

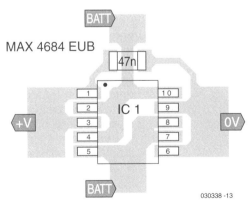

Figure 2: Current paths with correctly (a) and ... ...incorrectly (b) polarized battery.

Figure 3. Suggested PCB layout with generous copper areas.

its inactive position. The lower switch, however, toggles because IN2 is connected to the positive battery terminal. The switch now connects COM2 with NO (normally open), taking the battery negative terminal to the 0 V output.

The other drawing shows the effects of the battery terminals being swapped over. The upper switch changes state, while the lower switch remains in the inactive state. Internal protection diodes at all inputs and outputs of the chip guarantee that the MAX4684 will always start properly the instant the batteries are inserted into the equipment. Before the relevant switch changes over, both outputs COM1 and COM2 (and with them, +V and ground) are short-circuited. The MAX4684 has a maximum switching time of 60 ns. Within that period, the internal protection diodes provide the supply voltage for the IC itself, independent of the battery polarity.

The internal MOSFETs are driven and the switch toggles only when the relevant control input is connected to the positive terminal of a battery supplying at least 1.4 V. Only then does the correctly polarized battery voltage appear at the chip outputs. The internal switching MOSFETs are controlled using a certain 'dead time' during which all switch contacts are opened. This so-called break-before-make mode of at least 2 ns prevents a short-circuit on the battery in case the output capacitor is still charged, supplying the IC during a battery change. As opposed to conducting diodes with their virtually constant voltage drop, the voltage across the MOSFET switches is dependent on the

Component	Voltage drop	Dissipation P_{tot}
4 Si diodes	2 x 0.65 V = 1.30 V	130 mW
4 Schottky diodes	2 x 0.40 V = 0.80 V	80 mW
MAX4684	2 x 0.02 V = 0.04 V	4 mW
2 MAX4624	2 x 0.08 V = 0.16 V	16 mW
2 ADG819	2 x 0.05 V = 0.10 V	10 mW

Table 1: Comparison between diodes and analogue switches (at 100 mA and 2.4 Vcc)

current being passed. At just 0.3 Ω, the RDSon specification at V+ = 1.8 V is remarkably low. At 5 V, the switch resistance drops to 0.2 Ω.

The two switches and the protection diodes are capable of passing currents up to ±300 mA. The above features make he MAX4684 an excellent choice for lots of equipment running off AA or AAA cells, including portable audio equipment, mo-

bile test and measurement apparatus and cordless telephones.

Table 1 shows the maximum voltage drop across the two switches as compared with silicon and Schottky diodes in a bridge rectifier configuration. The datasheet of the MAX4684 may be found at www.maxim-ic.com. The IC is supplied in a miniature 10-pin μMAX case (3 × 5 mm). For space critical applications, the MAX4684 also comes in an even smaller 'UCSP' case which measures just 1.5 × 2 mm. Figure 3 shows a suggested PCB layout for the MAX case. Finally, it should be noted that the MAX4684 can not be used to rectify alternating voltages.

K.J. Thiesler

(030338-1)

307
IIR Tool

In the digital domain, analogue signals can be easily manipulated without requiring

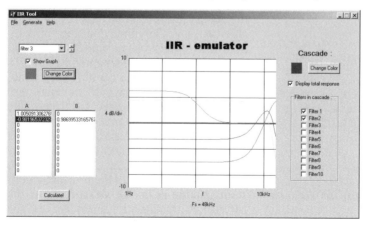

different hardware for each operation (as is necessary for analogue circuits). Other important advantages are that no noise is added during the operation (provided it is programmed correctly) and that mathematical algorithms are easier to implement. Unfortunately, the design of a digital filter is not that straightforward. There are several methods for implementing a digital filter. A relatively efficient filter implementation is the IIR (Infinite Impulse Response) type. The filter is mathematically represented as:

$$x[n] = a0 \cdot y[n] + a1 \cdot y[n-1] \ldots - b1 \cdot x[n-1] - b2 \cdot x[n-2] \ldots$$

Where x[n] is the output signal and y[n] is the input signal. The values of the coefficients ax and bx determine the transfer function and therefore the characteristics of the filter. Calculating the coefficients for a particular filter is often a stumbling block for designers. To make this task much easier, we have written a program that not only calculates the coefficients for simple filters, but can also determine the frequency characteristic of IIR filters where the user calculated the coefficients themselves.

This software is available from the Free Downloads section at www.elektro-electronics.co.uk under number 044050-1 (select month of publication). It does not need to be installed – double clicking the filename IIRTool.exe is sufficient to launch the program. This program can simulate up to 10 different IIR filters. As an additional bonus, it is possible to cascade multiple filters in order to examine the total frequency characteristic. Initially, after starting the program, all filters are set as all-pass and without delay. At the top left you can see that filter1 is selected. The coefficients that are shown on the screen belong to this filter. To enter your own coefficients for this filter you only need to click the desired coefficient after which you can type in the new value.

For example, try changing a1 (the second value at the top of the column near a) from 0 to 0.5. Immediately after this change you will note that the graph of the frequency characteristic has changed. At top left you can select another filter. Choose filter2. In the windows below it, the coefficients belonging to filter 2 appear. This time change a1 to –0.5. There now appears a second frequency characteristic. This looks like the reverse of filter1. The colour of each filter curve can be changed by clicking the button change color. This makes it easier to distinguish the separate graphs. In order to know what the total frequency characteristic will look like when a signal propagates through both these filters we can select filters 1 and 2 under the button cascade. Another curve will appear in the output window, which is the result of the two filters cascaded.

The program is not only able to simulate filters but can also generate 3 different types of simple filters. These generators can be found under the menu generate. Here you can choose from bass-, midor treble-filters. With the bass and treble filter you can choose the cut-off frequency and the desired gain or attenuation. The bass filter provides gain or attenuation of frequencies lower than the cut-off point. The treble filter does the same for frequencies higher than the cut-off point. A practical example: first select filter number 3, then choose generate followed by bass. In the window that appears, give the parameter frequency a value of 100 and gain a value of 5. Gain is the location where we enter the gain or attenuation (in dB). Now click the button OK. The program will now quickly calculate the required coefficients. As a consequence the coefficients of filter 4 are changed and the result is shown in the graph.

Generating a treble filter follows exactly the same method. Finally, the program can also design a mid filter. This filter provides gain/attenuation of signals around a specific frequency. When entering the parameters you will find, besides frequency and gain, a third parameter, namely Q. This represents the quality factor of the filter. The higher the Q factor the narrower will be the frequency range of the filter. A Q factor between 0.6 and 2 is typical for audio applications.

Paul Goossens
(044050-1)

308
I2C and SMBus

The I2C-bus has been dealt with before in Elektor Electronics. Most readers will know that this bus requires only two signals to communicate between, for example, a controller and one or more ICs in a circuit. We may also assume that it is well known that this bus was developed by Philips and is used mainly in televisions, video recorders, tuners, etc. Unfortunately, the I2C-bus lacks a number of features that manufacturers of computer motherboards require. They therefore developed their own bus, which has been derived from the I2C-bus. This bus goes by the name SMBus (System Management Bus). The main task for the protocol is to provide communications between the processor and various temperature sensors, battery management/charging-chips and even memory modules, etc. in a PC. Because the SMBus has been derived from the I2C-bus, there are obviously a number of similarities. It is in many cases even possible to let components with an I2C-bus communicate with those with an SMBus and vice versa.

The similarities

Both bus protocols make use of the same start and stop mechanism. A start-condition can be recognised by the falling edge on the data line while the clock line remains high. With a stop-condition, a rising edge appears on the data line while the

clock line is high. These two conditions are the only situations where the data line may change level when the clock line is high. This can be seen in Figure 1. Following the start condition there is the 7-bit address that has to be unique for each device connected to the bus.

Another similarity is that the bus has a single pull-up resistor for each signal and the connected chips drive these signal lines with an open-source output, as can be seen in Figure 2. This forms a so-called wired-OR configuration. As soon as one of the chips pulls a signal line to ground, the signal will have a low level, irrespective of what the other chips try to do.

The differences

Naturally there are also a number of differences between the two protocols. We start with the clock speed. With I2C-bus this can vary from 0 to 100 kHz in NORMAL mode and up to 400 kHz in FAST mode. With the SMBus, the clock speed has to be between 10 kHz and 100 kHz.

The minimum speed is the result of an additional provision of the SMBus. If the clock line is low for more than 35 ms a time-out situation occurs. All chips on the bus have to detect this time-out and ignore all communications until a valid start-condition occurs. This prevents one chip from disabling the entire communication of the

1

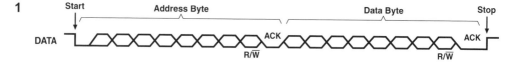

044054 - 11

2

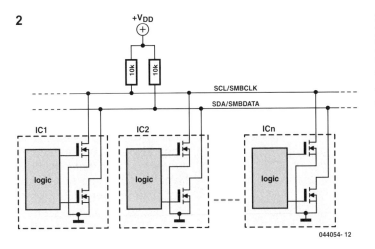

044054- 12

SMBus is also used in laptops and therefore have to be a little thrifty with power consumption. This means a higher pull-up resistor is required. A value of 10 kΩ is usually a good starting value so that both SMBus-ICs and I2C-ICs can communicate together on the same bus. There are also a number of subtle differences, such as maximum bus capacitance, rise- and fall-times, etc, but nothing that will prevent you from connecting an I2C chip to the SMBus or the other way around.

bus. Another difference is the specified levels for logic high and logic low. In the I2Cspecifications a low level is defined as 1.5 V, while with the SMBus a level less than 0.8 V is required. The high-level is different as well: with I2C this has to be a minimum of 3 V, while with the SMBus a level of 2.1V or higher is considered a high level.

This difference in the required voltage levels would indicate that both protocols cannot be used with each other. In practice this is not so much of a problem, since most chips will generate signals that range from about zero to nearly the power supply voltage. The outputs of an SMBus IC are specified to be able to sink at least 100 µA, while I2C requires a minimum sink current of 3 mA. All this is related to the fact that the

Finding the correct signals in a PC

To experiment (at your own risk!) with the SMBus (and I2C) it is necessary to find the signals Vcc, GND, SMBCLK and SMBDATA in the PC. The easiest place to find these signals is on the memory modules. Every memory module has besides the memory chips a small EEPROM, which contains the necessary information about the memory module for the BIOS. These EEPROMs all communicate via the SMBus. This is therefore the ideal location to tap into the required signals.

Paul Goossens   (044054-1)

309
High-Q Notch Filter

Without close-tolerance components

A notch for a narrow frequency band of a few per cent or even less normally requires

close-tolerance components. At least, that's what we thought until we came across a special opamp IC from Maxim. In filters with steep slopes, the component tolerances will interact in the complex fre-

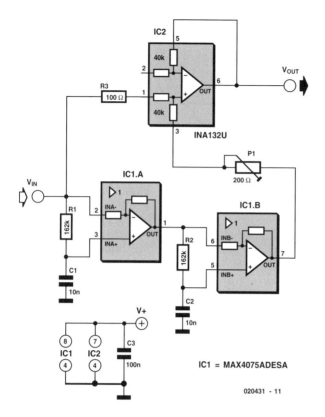

Figure 1: Special opamps incorporating laser-trimmed resistors

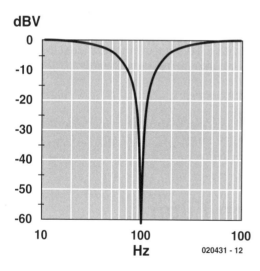

Figure 2: This deep notch is within reach using just 5%-tolerance resistors and 20%-tolerance capacitors

quency response. This effect rules out the use of standard tolerance components if any useful result is to be achieved. The circuit shown here relocates the issue of the value-sensitive resistors that determine the filter response from 'visible' resistors to ready available integrated circuits which also make the PCB layout for the filter much simpler. The operational amplifiers we've in mind contain laser-trimmed resistors that maintain their nominal value within 1‰ or less. For the same accuracy, the effort that goes into matching individual precision resistors would be far more costly and time consuming. The desired notch (rejection) frequency is easily calculated for both R-C sections shown in Figure 1.

Dividing the workload

The circuit separates the amplitude and frequency domains using two frequency-determining R-C networks and two level-determining feedback networks of summing amplifier IC2, which suppresses the frequency component to be eliminated from the input signal by simple phase shifting. IC1 contains two operational amplifiers complete with a feedback network. The MAX4075 is available in no fewer than 54 different gain specifications ranging from 0.25 V/V to 100 V/V, or +1.25 V/V to 101 V/V when non-inverting. The suffix AD indicates that we are employing the inverting version here $(G = -1)$. These ICs operate as all-pass filters producing a phase shift of exactly 180 degrees at the roll-off frequency f0. The integrated amplifier resistors can be trusted to introduce a gain variation of less than 0.1 %. They are responsible for the signal level (at the notch frequency) which is

Desired notch frequency, f$_0$, in Hz	R1 / R2 Nearest value (E12, ±5%) in kΩ	C1 / C2 Selected value (±20 %) in nF	Actual notch frequency f$_0$ (using ±0.1 % R + C)	Theoretical maximum tolerance on f0 in Hz (using ±5 % R and ±20 % C)
50	300	10	53	42 - 70
60	270	10	59	47 - 77
100	150	10	106	85 - 140
120	56	22	128	100 - 170
400	18	22	401	100 - 170

Table 1: Selection of standard SMD parts

added to the input signal by IC2 by a summing operation. However, they do not affect the notch frequency proper – that is the domain of the two external R-C sections which, in turn, do not affect the degree of signal suppression. In general, SMDs (surface mount devices) have smaller production tolerance than their leaded counterparts. Because the two ICs in this circuit are only available in an 8-pin SOIC enclosure anyway, it seems logical to employ SMDs in the rest of the circuit as well. Preset P1 allows the filter to be adjusted for maximum rejection of the unwanted frequency component.

R-C notch filter

Using standard-tolerance resistors for R1 and R2 (i.e., 1%, 0806 style) and 10%-toleance capacitors for C1 and C2 (X7R ceramic) an amount of rejection better than that shown in Figure 2 may be achieved.

The notch frequency proper may be defined more accurately by the use of selected R-C sections. Pin 3 of IC2 receives a signal that's been 90-degrees phase shifted twice at the notch frequency, while pin 1 is fed with the input signal. These two signals are added by way of the two on-chip resistors. IC2 is a differential precision operational amplifier containing precision resistor networks trimmed to an error not exceeding ±0.2 ‰. Here, it is configured as a modified summing amplifier with its inverting input, pin 2, left open.

For frequencies considerably lower than the resonance frequency $f_0 = 1 / (2 \pi R C)$ the capacitors present a high impedance, preventing the inverting voltage followers from phase-shifting the signal. At higher frequencies than f_0, each inverting voltage follower shifts its input signal by 180 degrees, producing a total shift of 360 degrees which (electrically) equals 0 degrees. The phases of each all-pass filter behave like a simple R-C pole, hence shift the signal at the resonance frequency by 90 degrees each. The three precision amplifier ICs can handle signals up to 100 kHz at remarkably low distortion. The supply voltage may be anything between 2.7 V and 5.5 V. Current consumption will be of the order of 250 µA.

Design by K.-J. Thiesler   (020431-1)

Contents A ... Z